MOUSTERIAN LITHIC TECHNOLOGY

STEVEN L. KUHN

MOUSTERIAN LITHIC TECHNOLOGY

An Ecological Perspective

PRINCETON UNIVERSITY PRESS
PRINCETON, N.J.

Published by Princeton University Press, 41 William Street,
Princeton, New Jersey 08540
In the United Kingdom: Princeton University Press,
Chichester, West Sussex

Library of Congress Cataloging-in-Publication Data

Kuhn, Steven L., 1956–
Mousterian lithic technology : an ecological perspective / Steven L. Kuhn
p.   cm.
Includes bibliographical references (p.   ) and index.
ISBN 0–691–03615–2 (cloth : acid-free paper)
1. Mousterian culture.   2. Mousterian culture—Italy—Lazio.
3. Stone implements—Italy—Lazio.   4. Lazio (Italy)—Antiquities.
I. Title.
GN772.22.M6K84   1995
930.1′2—dc20   94–32822

This book has been composed in Times Roman
Designed by Jan Lilly

Princeton University Press books are printed on
acid-free paper and meet the guidelines for permanence
and durability of the Committee on Production
Guidelines for Book Longevity of the
Council on Library Resources

Printed in the United States of America

10 9 8 7 6 5 4 3 2 1

CONTENTS

FIGURES

TABLES

PREFACE

This book is about the ways Mousterian hominids living in the region of Latium, Italy, between 110,000 and 35,000 years ago made and used stone tools. I seek to understand not only what flint artifacts look like and how they were produced, but why varying tactics for manufacturing and refurbishing tools may have been adopted at different times and places. Explanations for the actions of Mousterian tool makers are pursued within an explicitly ecological framework. I approach the organization of tool manufacture and use as the accommodation of toolmaking to the larger time and energy "budget" of foraging groups. Information gleaned from studies of faunas associated with the lithic materials, as well as knowledge about the nature and distribution of stone resources, help to define the factors that would have influenced and constrained the behavior of Middle Paleolithic populations. This research differs from other work on the ecology of Paleolithic hominids (e.g., Gamble 1984, 1986), which tend towards a broad geographical and climatic perspective on variability. Because of the nature of the Middle Paleolithic technological record, evidence for the direct influence of environmental factors on toolmaking is minimal, except insofar as the natural forms and distributions of stone constrained the options available to tool makers. Instead, hominid foraging and land use patterns represent the real "ecological context" of Middle Paleolithic technologies.

In adopting an ecological perspective to the study of Mousterian stone tool technology, I have taken one of a number of possible explanatory positions. I argue that lithic technology can be understood, at least in part, as a series of strategic choices made in response to the opportunities and limitations presented by the environment and a population's general mode of making a living. This is not to say that technology can only be understood this way, or even that technology is in its essence an ecological or economic phenomenon. The ways people make and use tools may also be influenced by the social context, active symbolic manipulation, and a multiplicity of other agents lying outside the realm of group-wide time and energy budgets. Having recognized that we are faced with a manifold subject, however, there is much to recommend studying it from a more limited perspective. The aim of this kind of research is not to produce a complete "explanation" for stone tools or toolmaking, but to use evidence derived from the study of stone tools to help in learning about specific dimensions of hominid existence. I am most interested in human evolutionary ecology, and in long-term changes in human behavior over the course of the Upper Pleistocene. The research described here is an explicit attempt to explore how toolmaking was linked to and influenced by hominid subsistence and land use. The choice of an analytical methodology follows from this general orientation, not from a belief that all aspects of lithic technology must reflect the ecological circumstances of toolmaking.

Because technology interfaces with so many domains of human existence, a focused approach to studying the remains of prehistoric technologies requires more than simply embracing a canon for the codification and interpretation of patterned variation in the material record. The strength of any interpretive framework depends in large part on the theoretical grounding, on explicit arguments about *why* certain archaeological facts are pertinent to the cultural or behavioral phenomena of interest. This study is based on a general theoretical model that links the economics of stone tool manufacture and maintenance to the ways mobile, tool-using populations get food and move about the landscape. Because they are structured in terms of universal, if abstract, currencies like time, energy, and risk, the ideas derived from the model provide a framework within which the behavior of pre-modern hominids—responsible for producing some if not all of the Mousterian assemblages discussed in this book—can eventually also be compared and contrasted with the behavior of anatomically modern populations.

If one is interested in hominid ecology and evolution, why even bother with technology? After all, one can investigate both foraging and its ecological context more directly by studying sediments, animal bones, pollen, and macrobotanical remains found in Pleistocene archaeological deposits. The answer to this glaringly rhetorical question is that human ecology and technology are inextricably intertwined. Although a variety of animals manipulate objects in foraging and social contexts, it is obvious that modern humans as a species are *dependent* on technology to a degree not approached by any other organism: the ecology of modern humans is technologically mediated in a very fundamental way. At the same time, the simplicity and monotony of early Pleistocene stone artifacts suggest that the earliest human ancestors could not have been so utterly dependent on technology for their survival as are modern humans. Thus, the long-term evolution of hominid technologies, as manifest in both their changing aspects and their developing *roles* within the hominid adaptive repertoire, is of necessity a central issue in human evolutionary studies.

These days, it is difficult to mention the words "Mousterian" and "evolution" in public without being drawn into a heated debate about the biological origins of modern humans, and the pros and cons of the two dominant positions on the subject. This is a discussion I openly attempt to avoid, as I believe that the genetic and archaeological or behavioral issues must be decoupled in order to conduct productive research. The work reported here explores how Middle Paleolithic stone tool technology may have been linked with foraging and land use in one small part of the world. Although the results bear on issues relating to human origins, the work is not intended to support a particular point of view about the Middle to Upper Paleolithic "transition" or the origins and spread of *Homo sapiens sapiens*.

Organization

The first chapter in this book situates the research within the broader issues and problems that have come to dominate Paleolithic studies in general, and studies of the Mousterian in particular. This introductory chapter also outlines some widespread properties of the Mousterian archaeological record and the anatomy of Neandertals that are of special relevance to this research. The subsequent section establishes the theoretical and methodological framework for the study. The heart of Chapter 2 is a strategic model of how mobile foraging populations keep themselves supplied with artifacts and raw materials under varying conditions, and how this can be monitored in the archaeological record. If it takes a long time to get around to the presentation and analysis of data, it is because I believe that a coherent theoretical perspective and an explicit model of the way the world works are vital to, but too often are missing from, paleoanthropological research.

Chapter 3 describes the sites and the study area (the coastal regions to the south of Rome, Italy) in some detail. Previous research on the area has been published mostly in Italian, less frequently in French, so that the data base may not be familiar to most English speakers. Chapters 4 and 5 contain the preponderance of the analyses of archaeological materials. Chapter 4 addresses variation in core reduction technology. The approach adopted here is of necessity based on formal and metrical analyses of flint cores, flakes, and tools, rather than on extensive refitting studies and replicative experiments. Chapter 5 is an analysis of lithic raw material economies, examining how patterns of stone artifact modification, renewal, and transport were influenced by both the natural distribution of raw materials and hominid land use. Data on faunal exploitation are used as a source of partial control over foraging and mobility. Taken together, the faunal and lithic data furnish a much richer picture of patterns of adaptive variation than could be obtained from either data set alone. The final chapter discusses the findings of the study as they compare with results from work on the Mousterian in other regions, and explores implications for several important general issues within Middle Paleolithic research.

ACKNOWLEDGMENTS

This work could have been completed only with the much-needed and greatly appreciated assistance, advice, support, and criticism of a large number of individuals. Lewis Binford, Lawrence Straus, Erik Trinkaus, and Don Henry, members of my dissertation committee, provided invaluable guidance and critical insight throughout the course of my Ph.D. research and beyond. Lew Binford's never-ending quest to "push the envelope" of research, to come to grips with what we do not know, has been a special inspiration to me.

Conducting research overseas has given me the opportunity to work with and come to know a number of estimable individuals. More than anyone else, Amilcare Bietti has made it possible for me to do research in Italy. He has been, and remains, both a highly respected colleague and a valued friend. Aldo Segre and Eugenia Segre-Naldini generously allowed me to study the many collections of the Istituto Italiano di Paleontologia Umana in Rome, also sharing with me their vast first-hand knowledge of both the geology and the history of archaeological research in Italy. Marcello Piperno and Graziella Bulgarelli were also of great assistance in arranging access to collections at the Museo Pigorini in Rome. While at the University of Pisa, Carlo Tozzi not only freely granted me access to the collections from Grotta di Sant'Agostino, but also patiently tolerated the presence of a barely intelligible American student occupying half the table space in his office for several months. Albertus Voorrips and Susan Loving generously allowed me to study material from their Agro Pontino survey, imparting many insights about the materials, and even put me up in Amsterdam. I have learned a great deal about the archaeology, geology, and physical anthropology of central Italy from a number of Italian colleagues, including Piero Cassoli, Giorgio Manzi, Fabio Parenti, and Silvana Vitagliano. I should also thank Graeme Barker, former director of the British School at Rome, for providing a safe haven during my first trips to Rome, as well as for sharing his considerable knowledge about the Italian archaeological scene. Finally, while it may be unusual to make posthumous acknowledgment of individuals one never had the chance to meet, I emphasize that the well-documented and well-curated archaeological record upon which this research is based exists only by virtue of the energy and skills of Alberto Carlo Blanc, Luigi Cardini, and Mariella Taschini.

Regarding the production of this book, I should first thank Paul Mellars, Olga Soffer, Catherine Perlès, and one anonymous reviewer for their pithy comments and cogent critiques of the first version of the manuscript. Dr. Perlès in particular went well beyond the normal call of duty for a reviewer, providing not only very extensive and well-directed comments, but also corrections of my egregious misspelling of French terms and titles. Liliane Meignen also read and critiqued several sections of the manuscript. The Loyola University Center for Instructional Design and the University of Chicago Libraries provided invaluable aid in composing and assembling artifact illustrations.

Funding for research in Italy has been provided at various times by the L.S.B. Leakey Foundation, the Institute for International Education (Fulbright Program), and the National Science Foundation (NSF/NATO Postdoctoral Research Fellowships Program). I gratefully acknowledge the generous support of these institutions.

Finally, the importance of Mary Stiner, my research partner and my wife, in all facets of my existence cannot be measured.

FIGURE CREDITS

Acknowledgment is due to several individuals and organizations for their generous permission to reprint the following illustrations.

Figure 4.2, nos. 1 and 2; Figure 4.4, no. 4; Figure 4.6, no. 8; Figure 4.14, no. 2; Figure 5.4, no. 5: **from** Bietti, A., S. Kuhn, A. Segre, and M. Stiner

(1990–91). Grotta Breuil: Introduction and stratigraphy. *Quaternaria Nova* 1:305–24, reprinted with the permission of the editors of *Quaternaria Nova.*

Figure 3.4, nos. 4, 5, and 22; Figure 3.5, nos. 1 and 2; Figure 4.1, nos. 2, 4, 9, and 13; Figure 4.6, no. 6; Figure 4.14, no. 6; Figure 4.15, no. 13; Figure 4.16, nos. 4 and 11: **from** Bietti, A., G. Manzi, P. Passarello, A. Segre, and M. Stiner (1988a). The 1986 excavation campaign at Grotta Breuil (Monte Circeo, LT). *Quaderni del Centro di Studio per l'Archeologia Etrusco-Italica* 16:372–88, reprinted with the permission of the editors of *Quaderni del Centro di Studio per l'Archeologia Etrusco-Italica* and of the authors.

Figure 4.1, nos. 3, 5, and 7; Figure 4.2, nos. 7 and 9; Figure 4.4, no. 9; Figure 4.6, no. 9; Figure 4.8, nos. 1 and 2: **from** Laj-Pannocchia, F. (1950). L'industria Pontiniana della Grotta di S. Agostino (Gaeta). *Rivista di Scienze Preistoriche* 5:67–86, reprinted with the permission of the editors of *Rivista di Scienze Preistoriche.*

Figure 3.4, no. 13; Figure 4.1, no. 8; Figure 4.2, no. 8; Figure 4.4, no. 10; Figure 4.14, nos. 7 and 8; Figure 4.15, nos. 11 and 12; Figure 5.4, no. 6: **from** Taschini, M. (1970). La Grotta Breuil al Monte Circeo, per una impostazione dello studio del Pontiniano. *Origini* 4:45–78, reprinted with the permission of the editors of *Origini.*

Figure 3.4, nos. 2, 3, 7, 14, 15, 17, 21, and 23; Figure 3.5, nos. 6, 9, 11, 16, 19–23, 27; Figure 4.1, nos. 1, 12, and 14; Figure 4.4, no. 6; Figure 4.6, nos. 1–5; Figure 4.8, no. 5; Figure 4.15, nos. 1–8; Figure 4.16, no. 10; Figure 5.4, nos. 1–4: **from** Taschini, M. (1979). L'industrie lithique de Grotta Guattari au Mont Circé (Latium): Définition culturelle, typologique et chronologique du Pontinien. *Quaternaria* 12:179–247, reprinted with the permission of the editors of *Quaternaria.*

Figure 3.4, nos. 1, 6, 9–12, 16, 18–21, and 24; Figure 3.5, nos. 3, 4, 7, 8, 10, 12–15, 17, 18, 25, and 28; Figure 4.1, nos. 6, 10, and 11; Figure 4.2, no. 10; Figure 4.4, nos. 1–3; Figure 4.6, no. 7; Figure 4.8, nos. 3 and 4; Figure 4.14, nos. 9–19; Figure 4.15, nos. 14–18; Figure 4.16, nos. 2, 3, and 5–9; Figure 5.4, nos. 7 and 8: **from** Tozzi, C. (1970). La Grotta di S. Agostino (Gaeta). *Rivista di Scienze Preistoriche* 25:3–87, reprinted with the permission of the editors of *Rivista di Scienze Preistoriche* and the author.

Figure 3.5, nos. 5, 24, 27; Figure 4.2, nos. 3–6; Figure 4.4, nos. 6–8; Figure 4.8, nos. 6–10; Figure 4.14, nos. 3–5; Figure 4.15, nos. 9 and 10; Figure 4.16, no. 1; Figure 6.3: original drawings by Mary C. Stiner.

Figures 3.6, 3.7, 3.9, 3.11, 3.12a, 3.12b, and 3.14 reprinted with the permission of Princeton University Press and Mary C. Stiner.

Figures 5.1 and 5.2 reprinted courtesy of Academic Press and *Journal of Archaeological Science.*

MOUSTERIAN LITHIC TECHNOLOGY

1

NEANDERTALS AND THE MOUSTERIAN

The research reported in this book is an inquiry into the ecological significance of variation in stone tool technologies during the Upper Pleistocene. Working with information from a series of Italian Middle Paleolithic sites, I explore the relationships between toolmaking and food-getting, and in particular the influence that patterns of land use might have had on the economics of stone tool manufacture and use. Information about three phenomena—flake production technology, artifact reduction or resharpening, and artifact transport—helps to define variation in Mousterian stone technologies. When findings from independent research on patterns of game procurement and exploitation are integrated into the "map" of technological variability, a number of interesting relationships emerge, showing how the manufacture and maintenance of stone tools were impacted by foraging and mobility patterns. These connections between technological and subsistence behavior provide some intriguing indications as to the evolutionary implications of long-term variation in human technologies.

The main data base for this research consists of Mousterian assemblages from four cave sites located on the Tyrrhenian coast of central Italy. The assemblages discussed here belong to what is known as the "Pontinian," a distinctive, regionally bounded "facies" of the Mousterian found only in the littoral margins of the regions of Latium and Tuscany. Although they are typically Mousterian in many aspects, the Pontinian assemblages also represent a unique case. The great majority of artifacts are manufactured using pebble flints, which are the only raw materials available in the central Italian coastal zone. The pebbles are quite small, which limited the sizes of tools that could be made from them. Depending on the locality, the pebbles may have been difficult to locate and collect as well. As such, the Pontinian provides an excellent context in which to study Middle Paleolithic technological behavior in an unusual, somewhat "stressed" raw material environment.

The Mousterian is familiar territory to many, even if the central Italian sites discussed in this book are not. Throughout the long history of research, the Middle Paleolithic has often served as the background for significant theoretical and methodological developments. Today, much work on the Mousterian is closely bound up with one hotly contested issue, the origins of anatomically modern humans. At the same time, studies of Paleolithic stone technologies are witnessing a veritable methodological revolution, though not necessarily in concert with the dominant research concerns. This study draws on or addresses many of these current issues, but the perspective adopted is unique. In short, while the landscape may be familiar to some, the way of navigating is somewhat unusual. Because the historical and methodological issues are so commanding, it is necessary to set the stage for the research described in subsequent chapters.

The Mousterian Muddle in the Middle Paleolithic

The origin of anatomically modern populations is one of the most compelling as well as one of the most publicized issues in paleoanthropology today, and the Mousterian is right in the middle of it. Questions about where, when, and how modern folk first appeared in various parts of the globe have spawned lively, frequently acrimonious debate. The appearance of the first studies of mitochondrial DNA, which suggested a single, southern African origin for all modern populations (Cann 1988; Cann et al. 1987; Stoneking and Cann 1989), along with new and surprising dates from Near

Eastern Mousterian sites (Bar-Yosef 1988; Bar-Yosef et al. 1986; Stringer 1988; Stringer and Andrews 1988; Valladas et al. 1987, 1988), set off a virtual firestorm of reaction and counter-reaction. Starting around 1987, the archaeological and physical anthropological literatures contained increasing numbers of highly polarized position statements, summaries of archaeological and/or fossil evidence asserted to either support or contradict the models based on mitochondrial data. Some researchers (e.g., Stringer 1988; Stringer and Andrews 1988) took the position, apparently supported by the DNA evidence, that our species came to be present throughout the globe because one relatively distinct population spread rapidly from a single African "homeland," swamping the genetic contributions of preexisting hominid populations: this is commonly known as the "Out of Africa" or "replacement" model. The contrary view, the so-called "multi-regional" or "continuity" model (e.g., Clark and Lindly 1989a, 1989b; Frayer et al. 1993; Wolpoff 1989), holds that the broad distribution of *Homo sapiens sapiens* came about as the result of a gradual, worldwide diffusion of genetic material across reproductively non-isolated regional groups. Although there are alternative viewpoints (e.g., Smith et al. 1989), and while not all researchers found themselves so polarized (e.g., Trinkaus 1989a), the extreme positions did drive much of the polemic.

The past several years seem to have witnessed a marked reduction in the volatility of the debate surrounding human origins. In part this reflects a general tendency to reevaluate and reformulate the original arguments. Simultaneously, researchers have begun to question the relevance of the facts that have been martialed in support of opposing models of human origins, including the mitochondrial data themselves (e.g., Spuhler 1988; Templeton 1993), and have embarked on the search for more pertinent and critical information. Nonetheless, the theme of replacement versus continuity is still prominent in both the archaeological and the human paleontological literature.

The topic of modern human origins confronts us with a fascinating set of questions. Nonetheless, the current state of affairs in Upper Pleistocene research is somewhat problematic. Several distinct disciplines with largely independent realms of inquiry and relevance have focused in on a single problem, the origins of *Homo sapiens sapiens* as a species. This problem is structured in largely biological terms, that is, as population replacement versus *in situ* evolutionary change across a broad front. While the three realms of scientific inquiry most commonly brought to bear on the problem of human origins—genetics, human paleontology/physical anthropology, and archaeology—do share a great deal of common ground, their empirical domains are largely distinct. Roughly speaking, archaeologists study the material evidence of human behavior, human paleontologists study the results of behavioral and genetic influences on the hominid skeleton, and the geneticists (in this context at least) try to reconstruct phylogeny and past demographic events from modern gene frequencies and distributions. Phylogeny, anatomy, and behavior *are* connected in important ways, but the links are much too poorly understood to allow one discipline to simply borrow the research priorities of another. Archaeologists are in no position to evaluate the genetic relatedness of populations, any more than geneticists can realistically endeavor to reconstruct the foraging patterns of archaic *Homo sapiens*.

There is no question that archaeological data can and do feature in debates over modern human origins, but such observations are employed primarily as fuel for one fire or another. Hypothetical links between behavior and phylogeny are so tenuous that the archaeological record can be used only in advocating a particular position, and never as a means of potentially falsifying it. It is certainly rational to equate evidence of rapid changes in prehistoric human behavior with population replacement, or to argue that gradual change indicates evolutionary continuity, but by that same token each of us can easily imagine a situation in which a precipitous behavioral change occurred without population replacement, and vice versa. Similarly, if one could identify an unambiguous archaeological "signature" for the very first populations of modern humans, it might be possible to follow their expansion across the globe. But I seriously doubt that advocates of an African origin of anatomically modern humans would accept the absence of an easily isolated archaeological signature as disproof of their position.

I would argue that the "mission" of archaeology within human evolution research is to document and explain long-term evolutionary changes in hominid behavior, irrespective of biological changes. Archaeologists are in a unique position to study trends in human subsistence, social life, and symbolic behavior during the Upper Pleistocene. How the behavioral changes archaeologists observe might relate to changes in human physiology is an interesting question, but one that will be answered once cultural and biological evolutionary histories of our species are themselves better understood. Filling archaeology's role in paleoanthropology requires three things. First, the behavioral and biological issues must be made separate. Second, archaeologists must cease to treat the Mousterian and other broadly defined units as internally homogeneous and stable units: they might have been so, but we cannot assume that this was the case. Evolution occurs as the result of selection on variation, so understanding the nature and extent of variation in human behavior across time and space is crucial for explaining either long-term change or long-term stability. Finally, archaeologists must reevaluate the phenomena they choose to study, and how they analyze them. Too often, research is framed in terms of "inherited" questions, using methods developed for entirely different purposes from those at hand (Kuhn 1991a). We need to think long and hard about how the things we study are related to evolutionary issues, and modify our approaches accordingly.

History has already at least partially taken care of the first step. For many years it was assumed that anatomically modern populations first appeared across Europe and Asia at the same time as Upper Paleolithic assemblages. The association of robust but essentially anatomically modern human remains with Levalloiso-Mousterian industries at the sites of Skhul (McCown and Keith 1939) and Qafzeh (Vandermeersch 1981) in the Near East has proven that the connection between hominids and assemblage types as they are conventionally defined is less secure than had been previously suspected. It remains true that the vast majority of fossils associated with Mousterian industries are attributable to archaic *Homo*, and that most fossils found with true Upper Paleolithic industries are fully modern. However, it can no longer be assumed that the anatomical transition occurred in perfect synchrony with the cultural transition as it is normally described. More than anything else, the invalidation of long-standing assumptions about the synchrony of conventional biological and cultural transitions should lead to the conclusion that conventional archaeological definitions and established chrono-stratigraphic units are not always extremely relevant to questions about what happened with human behavior and human biology during the Upper Pleistocene.

While it is more and more apparent that biological and cultural changes were not always coincident, we still have a long way to go in comprehending the scope of archaeological and behavioral variation across time and space, and (perhaps more important) in understanding this variation in evolutionarily relevant terms. Archaeological research on the period that witnessed the origins of anatomically modern humans often focuses on the presumed moment of change, the so-called "Middle to Upper Paleolithic transition." Variation before or after the transition is minimized as a matter of course, limiting our ability to investigate the causes and mechanisms of change, and leaving little option but to debate the timing and pace of the transition. Approaching human evolution in terms of transitions implies that one comparatively stable, consistent biological or cultural form metamorphosed into another stable, homogeneous form. Moreover, the attention paid to the "transition" sometimes leaves the impression that the only significant thing about the Mousterian was that it ended. Yet the Middle Paleolithic does not represent a transitory cultural stage, the material record of hominids "waiting to become modern." It is a tremendously widespread and persistent phenomenon, found across a remarkable array of environments over at least 200,000 years, perhaps much longer. We will never explain why the Mousterian disappeared, through whatever process, without understanding how it lasted so long.

In order to understand either stability or evolutionary change, we need to know about variability, the raw material upon which selection operates. The first step is to look closely at the "units" now used to partition the archaeological record, ranging from the Mousterian as a whole down through regional or temporal facies, extending even to the level of individual assemblages. Are these "modules" in fact single, uniform entities with regard to technology, subsistence, and other factors, or are they internally diverse and variable? At the same time, not all observations about variation are equally pertinent to evolutionary issues. It is necessary to understand how the scope of human cultural variation at any particular interval in the past might have been impacted by external (and internal) factors, leading to changes over time in what people did and how they responded to their biological and social environments. Therefore, it is vital to address the causes of variation, to investigate the selective pressures on human behavior and the nature of human responses to them; in other words, to study the processes of human adaptation.

Biologists and anthropologists alike have criticized what they term "Adaptationism" (e.g., Gould and Lewontin 1979; Leonard and Jones 1987). However, it is important to distinguish between the "ism" and the phenomenon itself. *Adaptation*—physiological or behavioral adjustment to prevailing conditions—occurs as one inevitable consequence of natural selection or, on a smaller scale, through phenotypic plasticity. *Adaptationism* is the tendency (chronic if not programmatic) to assume that all change and all variation came about in this way. Clearly, not all physical or behavioral change represents adaptation to any identifiable set of conditions. Having said that, I would maintain that, for many purposes, the most interesting and relevant changes are adaptive ones. Simply monitoring shifts in gene frequencies (or cultural traits) over time may be useful for tracing phylogeny, but the charting of history does not constitute an explanation. Moreover, a research program explicitly focusing on adaptive process can be potentially disconfirmed. In contrast, approaching everything as though it were selectively neutral leaves little ground for critical evaluation of one's ideas.

Critiques of adaptationism do highlight two very important points, namely that not all change over time can be assumed to represent adaption, and that similar developments are not necessarily adaptations to the same thing. Discussions of "Upper" or "Middle Paleolithic adaptations" have often taken the form of lists of features thought to distinguish each period (e.g., Clark and Lindly 1989a, 1989b; Mellars 1973, 1989; Straus 1983; White 1982). Frequently missing is any explicit argument about why a particular aspect of technology, game use, etc. was selectively advantageous at one point in time but not at another. Observing that Middle Paleolithic hominids often made blades or blade-like flakes (e.g., Conard 1990; Müller-Beck 1988), a component of stereotypical "Upper Paleolithic adaptations," shows that elongated tool blanks are not an unambiguous marker of the presence or actions of modern humans. However, since there was no systematic explanation offered for the evolution of blade technology in the first place, these facts alone provide no basis for asserting that Middle Paleolithic technological adaptations were like, or different from, those of Upper Paleolithic populations. Hominids may well have arrived independently at analogous technological solutions many times in the past, perhaps in response to quite different kinds of problems. The scientific challenge is to understand the factors that rendered a particular way of behaving advantageous in one context, but disadvantageous in another (see also Perlès 1992).

The Archaic and the Anomalous

Several distinctive features of both Neandertal physiology and the Mousterian archaeological record provide intriguing clues about how technology might have functioned within the larger domain of Middle Paleolithic human adaptations. This discussion is not intended to be a comprehensive review of the Middle and Upper Paleolithic evidence. A number of much more complete archaeological summaries of the Mousterian and later cultural manifestations have appeared over the past decade (e.g., Chase and Dibble 1987; Clark and Lindly 1989a, 1989b; Klein 1989b; Mellars 1973, 1989, 1991; Orquera 1984; Straus 1983; Stringer and

Gamble 1993; White 1982), and the proceedings of several symposia concerning both Neandertals and the origin of modern humans have been recently published (Dibble and Mellars 1992; Farizy 1990; Mellars 1990; Mellars and Stringer 1989; Otte 1986–89; Trinkaus 1989b). Instead, I concentrate on a few selected aspects of the osteological and material cultural evidence. I emphasize a limited array of observations for two reasons: they may express aspects of the behavior of Mousterian hominids that differed from what we know of later, anatomically modern populations; and they are of particular relevance to understanding how variation in Middle Paleolithic stone technologies might reflect on the ecology of Upper Pleistocene hominids.

Before proceeding with this discussion, it is necessary to emphasize the logical independence of contrasts between Neandertals and modern humans on one hand and views about the genetic relationship between the two hominid forms on the other. Frequently an author's position seems to spring directly from his or her views of the biological transition. Supporters of the continuity argument normally play down differences between archaic and modern *Homo sapiens*, while those in favor of the replacement model frequently emphasize them (Dennel 1992; Graves 1991). Neither position is a logical necessity. Demonstrating that archaic humans acted differently does not imply that they could not have been or were not ancestral to anatomically modern populations. If one holds that modern *Homo sapiens sapiens* evolved out of archaic humans *somewhere* in the world, then it is almost axiomatic that there was behavioral as well as biological change. It is simply unnecessary to assert that Neandertals and their contemporaries were just like modern humans in order to argue that they were our ancestors. Conversely, the existence of many broad similarities in the archaeological records of archaic and modern *Homo sapiens* does not necessarily imply genetic continuity. Neandertals and anatomically moderns were large-brained hominids with very similar (if not identical) physiological capabilities, and we should expect a great deal of overlap in how they used tools, procured game, etc. For example, the striking similarities in use wear found on Mousterian artifacts associated with both Neandertals and early *Homo sapiens sapiens* in the Near East (Shea 1989) probably stem from the facts that both forms of hominid lived in similar environments, and that neither habitually carved wood or cut meat without the aid of stone tools: it is a considerable stretch to argue that these similarities somehow relate to the genetic relationship between the two hominid types (e.g., Frayer et al. 1993).

Physiological evidence

Skeletal evidence cannot match the specificity of archaeological data in reconstructing patterns of behavior, but the human fossil record has important, if indirect, implications for the role of technology in the adaptations of archaic *Homo sapiens*. Habitual patterns of behavior affect functional anatomy either through direct (stress-related pathologies, remodeling) or selective processes. As such, the distinctive anatomical features of Neandertals, the primary (if not exclusive) authors of the Mousterian in Europe, offer hints about what technology did, and didn't, do for archaic *Homo sapiens*.

Arguably the most obvious general characteristic of the Neandertal anatomy is the high overall degree of skeletal robusticity and muscular hypertrophy, which must bespeak very high levels of habitual physical activity (Trinkaus 1983a, 1986). It was formerly thought that Neandertals and other archaic members of the genus *Homo* exhibited comparatively massive construction in their lower limbs (Lovejoy and Trinkaus 1980; Trinkaus 1983b), possibly as an adaptation to high levels of prolonged strain (Trinkaus 1986:205). It now appears that the greatest contrasts between the femorae of Neandertals and anatomically modern humans involve the cross-sectional shape of the diaphysis (Ruff et al. 1993). Because pilastering develops in response to habitual stresses, the shape differences may reflect habitual patterns of movement—not distinct types of locomotion, but a difference in the general way of moving about the landscape. The shape, cortical thickness, and sizes of muscle attachments on the humerus, scapula, and ulna also suggest that habitual ranges of movement and patterns of joint loading in the arm and shoulder differed somewhat from those of most anatomically modern populations (Churchill and Trinkaus 1990; Smith et al. 1989; Trinkaus 1983b). The general anatomy of the Neandertal hand suggests that they were capable of at least the same range of hand movements as modern humans, but may have regularly exerted more force in certain postures. For example, proportions of the manual phalanges indicate that Neandertals were able to exert as much force in a precision grip as modern humans, but possessed a much stronger power grip (Trinkaus 1983a, 1983b).

A shared characteristic of both the upper and lower limbs of Neandertals is the shortness of distal limb segments relative to proximal ones. This trait is present in both Near Eastern and European Neandertals and has been argued to represent an adaptation to cold environments (Trinkaus 1981, 1992). Because the earliest anatomically modern humans exhibit relatively long distal limb segments, regardless of the environment in which they are found, it is likely that the Neandertal pattern reflects relatively ineffective technologically aided thermal buffering (e.g., clothing, pyrotechnology) (Trinkaus 1981), particularly during crucial phases of growth.

Facial morphology and dental attrition also have important implications for behavioral adaptations and technology. Neandertal individuals often exhibit a very high degree of attrition to the anterior dentition, wear of a type often associated with the use of teeth for non-masticatory activities (Trinkaus 1983a; Smith 1976). Some researchers argue that the enlargement of the anterior dentition and the pronounced mid-facial prognathism in Neandertal populations represent an evolutionary response to the added mechanical and attritional stresses imposed by the use of teeth for non-masticatory purposes (e.g., Smith 1983; Spencer and Demes 1993). The actual developmental and selective influence of masticatory stress to the morphology of the mid-face region has been challenged, however (Rak 1986). Other factors, such as the morphology of the respiratory channels and growth and fusion sequences of different parts of the cranium, certainly did influence the morphology of this region.

The possible behavioral significance of individual features of Neandertal anatomy remains subject to debate. Overall, however, Neandertal functional anatomy does confer a strong impression of being adapted to directly absorbing a large amount of stress from the environment. Much of the robusticity of Neandertals was undoubtedly phylogenetic "baggage," a legacy from similarly hyper-robust archaic members of the genus *Homo* (i.e., *Homo erectus*). Yet phylogenetic inertia explains neither the maintenance of these traits for tens of thousands of years nor their eventual disappearance. Neandertal physiology as a whole seems to indicate that technology simply did not provide as universal an interface with the environment as it does for modern humans. Archaic *Homo sapiens* used their bodies differently, perhaps relying more extensively on the application of physical force and metabolic energy and less on technological support. This observation does not imply that Neandertals were unable to use tools effectively or efficiently, nor that they were incapable of producing clothing or fire sufficient to keep them from showing the physiological effects of exposure to cold. Archaic *Homo sapiens* didn't "lack" anything, in evolutionary terms. Rather, I would argue that they were physiologically adapted to a lifestyle that involved somewhat less habitual application of "technological buffering" than we expect to see among modern humans. In the long term, what really must be explained is how cultural adaptive patterns prevalent among anatomically modern populations may have relaxed selective pressure to maintain heavily built bodies, and even favored increasing gracility.

Two archaeological anomalies

Below, I discuss two especially anomalous aspects of the Middle Paleolithic archaeological record, one concerning technology and one relating to evidence of mobility patterns. In examining possible differences between Mousterian hominids and later humans, we cannot simply ask whether Middle Paleolithic groups did things differently from more recent populations. On one hand, if the archaeology of the Mousterian were just like the Aurignacian or the Magdalenian, no one would have ever identified it as something distinctive. On the other, modern human societies are astonishingly variable, yet no researcher today would argue that this reflects biological differences or fundamental inequities in the capacities of different populations. In looking for evidence of long-term evolutionary change, we really need to search for anomalies, aspects of the Mousterian that are inconsistent with our expectations about how modern people, *as a species*, are known to or ought to behave. Ideally, one would like to be able to conduct controlled experiments, to examine how different hominids acted in analogous contexts, how they responded to identical pressures and influences. However, it is often as difficult to reconstruct the context of prehistoric behavior as it is to reconstruct the behavior itself. Archaeologists can only work with what the record provides, and in comparing time periods or regions they are seldom in a position to control for all of the conditions that might have influenced human actions in the past. Alternatively, one can cast the net more widely, and look for generic tendencies or patterns that appear frequently and repeatedly among modern humans but that are either absent or highly divergent among earlier populations. The anomalies that one may perceive point to one of two possibilities: either Mousterian populations coped in unique and distinctive ways with the conditions they faced; or they faced conditions never encountered by anatomically modern populations. In either case, incongruities are central to understanding the place of the Mousterian, and the populations that created it, in human evolutionary history.

One surprising characteristic of Mousterian industries as a group is the scarcity or even absence of implements designed to aid in the procurement and processing of food resources. To the degree that functional inferences are justifiable, the Mousterian "toolkit" appears to have been made up primarily of implements used in processing and manufacture activities, an observation supported by use-wear studies (e.g., Anderson-Gerfaud 1990; Beyries 1987, 1988; Shea 1989). What has been referred to as "extractive technology" (Binford and Binford 1966), seems to be strongly under-represented in most Middle Paleolithic stone tool inventories. This is not to argue that no portion of the Mousterian toolkit was employed in the primary processing of food resources. Most Middle Paleolithic tool types—sidescrapers, denticulates, and notched pieces—are generalized in form, and they were probably employed in a vast number of contexts. Sharp edges of one sort or another are common in Middle Paleolithic assemblages, and these were undoubtedly used for butchering game or cutting up plant foods some of the time. But the kinds of specialized extractive artifacts characteristic of later time periods are notably scarce.

Technology related to the procurement of large game presents arguably the most conspicuous "gap" in the Mousterian extractive tool inventory. Because of the rather marked physical constraints on their design, weapons are comparatively easy to recognize. Most Middle Paleolithic assemblages do contain a certain number of pointed elements that could have served as weapon tips, namely unretouched Levallois points and retouched Mousterian points. Although some have assumed that these pointed artifacts resembled components of weapons (e.g., Binford and Binford 1966; Bordes 1968:44), only recently have the techniques of use-wear analysis been brought to bear on the problem. The results are highly ambiguous. Some researchers maintain that Levallois points in particular were in fact used as weapon tips (Shea 1989), while others question the existence of hafted projectiles in Mousterian assemblages (Anderson-Gerfaud 1990: 389; Holdaway 1989).

Irrespective of one's position on the question of whether or not Mousterian populations really did mount stone points on wooden shafts to make spears, two general facts are clear. First, the Mousterian pointed artifacts were apparently not *specialized* weapons. Even the strongest advocate for the use of Levallois points as spear heads observes that projectile-type damage patterns are present on at best a large plurality of specimens (Shea 1989). Secondly, there is a remarkable formal monotony within the class of potential Mousterian weapon components. From the Levant to northern France, to eastern Europe, one encounters basically the same two artifact forms: retouched Mousterian points and unretouched Levallois points. Both types are relatively large triangular pointed objects. They usually have thick bases, and while micro-wear researchers often report evidence of hafting wear (Beyries 1987, 1988; Shea 1989), thinning or other macroscopic hafting-related modifications are inconsistent at best. Both retouched and unretouched points vary somewhat in size and elongation, but this seems most directly attributable to raw material form and the use of different core reduction technologies rather than differences in function. This pattern presents a marked contrast to the plethora of probable bone and stone weapon tips found in even the earliest Upper Paleolithic assemblages (e.g., Knecht 1993; Peterkin 1993; Straus 1982a, 1990a, 1993). Even ambiguous or "transitional" late Middle or early Upper Paleolithic assemblages like the Castelperronian (Harrold 1983, 1989), the Szeletian (Allsworth-Jones 1986), the Aterian (Ferring 1975), and the Emiran (Copeland 1975; Volkman and Kaufman 1983) seem to have their own regionally distinctive pointed tool forms with apparent hafting modification.

Ambiguities about the existence and nature of potential hunting weapons do not cast doubt on the abilities of Middle Paleolithic hominids to procure game. Faunal studies leave little doubt that they could and did hunt successfully, with or without the aid of more sophisticated stone- or bone-tipped armaments (e.g., Binford 1984; Chase 1986a, 1989; Farizy and David 1992; Klein 1979, 1989a; Stiner 1990a, 1990b, 1994). What is remarkable is that the forms of potential weapons are so homogeneous across such a vast range. Looking at the Mousterian technological evidence from a purely formal, design-oriented perspective it would be difficult to find any evidence of any functional or strategic variation in potential weapon technology on either a local or a regional scale. According to the most basic ecological principles, the importance of, and the means for, obtaining animal protein should have varied dramatically across the extensive temporal and geographic span of the Middle Paleolithic. Much of the proliferation of point types during the Upper Paleolithic and later periods must reflect adjustment to changes in the array of species taken, their behavior and distribution, and the tactics used to exploit them (e.g., Gamble 1986; Peterkin 1993; Straus 1993). Yet given the fact that Middle Paleolithic and earlier industries are found in a wide variety of environments, including cold temperate and even subarctic contexts (Gamble 1984, 1986; Roebroeks et al. 1992), there is surprisingly little convincing evidence for investment in more complex projectile weapons, even at the northern limits of the distribution of the Mousterian. It may not be startling that there is comparatively little elaboration of projectile technology in Holocene Australia, where the largest terrestrial mammal weighs around 50 kilograms and where seeds and small game constitute an important part of the diet. In light of general expectations about both hunting behavior and the availability of vegetable foods, it *is* quite surprising to see so little investment and such limited variation in projectile technology across the entirety of temperate western Eurasia during the 200,000 or so years that Mousterian industries were made.

Obviously, stone is not the only raw material from which weapons can be constructed, but there is precious little evidence for tools of bone and wood during this period. Working of bone by Mousterian hominids appears to have been both infrequent and casual (Vincent 1988a, 1988b). Finds of two sharpened wooden staves in late Middle and early Upper Pleistocene deposits in Europe (Jacob-Friesen 1954; Oakley et al. 1977) suggest that at least untipped wooden spears were in use before the Mousterian (but see Gamble 1987 for an alternative functional interpretation). Moreover, use-wear analyses suggest that woodworking was a frequent activity in some European Middle Paleolithic sites (e.g., Anderson-Gerfaud 1990; Beyries 1987). However, presupposing a more elaborate but undocumentable wooden projectile technology only leaves us with another anomaly. The argument that for more than 200,000 years hominids relied almost exclusively on perishable materials to make different types of hunting tools, subsequently switching to more durable bone and stone less than 35,000 years ago, demands considerable explanation in itself.

If hunting is in fact limited by technology, the uniformity of "point" morphology in the Mousterian would suggest that restricted ranges of hunting techniques were used over a vast span of time and space. More important, however, it implies that Middle Paleolithic tool makers did not respond to the same forces which lead to variation in modern hunter-gatherer technologies. From a long-term evolutionary perspective, modern hunter-gatherers have shown the tendency to periodically increase their labor investment in technological aids facilitating the procurement of game, presumably in response to fluctuations in the food supply and/or human demands. We might think of this macro-strategy as "technologically aided intensification." From the perspective afforded by stone tool technology, there is nothing during the Middle Paleolithic, on even a small scale, analogous to the kind of "arms race" apparent during the Upper Paleolithic (Straus 1990a). In other words, while Mousterian populations certainly hunted, and in some contexts may have depended on meat for the majority of their caloric intake, there is very little evidence of a technological response to pressures to be more effective or efficient at procuring game.

The scarcity and monotony of food procurement technology in the Middle Paleolithic implies that Middle Paleolithic populations responded in a very different manner to inevitable variation in the abundance of food relative to the number of consumers. One could argue that Mousterian populations simply never found it necessary to increase the efficiency or effectiveness of game procurement, that they simply focused on other resources when game was harder to come by. However, as argued in greater detail below, corroborative evidence for this kind of resource switching is generally lacking. A more likely possibility is that hunting strategies varied not in terms of technological aids, but via patterns of co-operation and collective action (Stiner 1994), the kind of macro-strategy seen among non-human, non-tool-using predators. A third prospect is that, rather than investing in technology for increasing the effectiveness or efficiency with which resource patches (i.e., animals or groups of animals) were exploited, Mousterian hunters adjusted by expanding the *number* of patches encountered—that is, by adjusting coverage of the territory.

This last possible explanation for the scarcity and monotony of extractive technology in the Mousterian leads naturally to a second major Mousterian "anomaly," deriving from contrary indications of mobility and settlement permanence. At least since François Bordes (Bordes 1961b; Bordes and Sonneville-Bordes 1970), some researchers have maintained that Middle Paleolithic populations in western Europe were largely sedentary, or at least that they spent long periods of time living in particular sites. One observation thought to support this point of view pertains to the sheer density of many of the archaeological deposits (Gamble 1986:299). Depending on geological circumstances, sparse deposits, apparently representing very brief Mousterian occupations, have been isolated in some sites (e.g., Mora et al. 1988; Otte et al. 1988). However, many localities, most notably classic French cave sites such as Combe Grenal, Pech de l'Aze (Bordes 1972), and La Quina (Martin 1907–10), possess extraordinarily rich archaeological layers, containing truly enormous quantities of cultural debris deposited within thick, apparently homogeneous strata—unambiguous evidence that hominids spent a great deal of time in these places.

Another class of observation lending possible credibility to the idea of "sedentary Neandertals" pertains to the seasonality of occupations. In several cases, large ungulates from Mousterian sites appear to have been killed over a variety of seasons, leading to the inference that the site's occupants stayed there year-round (e.g., Bordes and Sonneville-Bordes 1970; Lieberman 1993). However, when the samples used in determining seasonality come from thick geological layers deposited over tens, hundreds, or even thousands of years, these observations are somewhat ambiguous. Given the protracted time spans involved, it would be impossible to differentiate permanent settlements from short-term occupations reused over the decades during a number of different seasons. Data on raw material transport are similarly equivocal. Middle Paleolithic sites in western Europe seldom contain raw materials originating more than 50 km away (see Roebroeks et al. 1988), whereas eastern and central European Middle Paleolithic sites more frequently contain materials derived from 100 km or more distance (Roebroeks et al. 1988; Simán 1991). It is not clear whether the contrasts between eastern and western Europe attest to differences in hominid behavior, or whether they reflect the distributions of identifiable raw material sources. Moreover, while the scale of raw material transport in the Mousterian does seem somewhat limited in comparison with some highly mobile populations like North American Paleo-indians (Ellis 1989; Frison 1991; Hoffman 1991; Tankersley 1991) and later Upper Paleolithic hunters on the loess plains of eastern and central Europe (e.g., Montet-White 1991; Schild 1987), the simple distances stone was displaced are not unambiguous indicators of mobility (Kuhn 1992b).

Other generalized patterns are more consistent with the idea that Mousterian hominids in Europe tended towards an extremely mobile way of life. One of the strongest arguments in favor of very short-term occupations is the extreme scarcity of recognizable features, structures, and facilities in Middle Paleolithic sites. Normally, the longer foragers stay in one place, or the more regularly they reoccupy a site, the more rigidly they structure the use of space and the more they invest in structures such as hearths, shelters, and storage features (Hitchcock 1987; Kelly 1992:56). In situations outside the Middle Paleolithic, archaeologists tend to equate constructed shelters and hearths and evidence for maintenance of sites (cleaning up) with increasing permanence of occupations and restricted mobility. While excavators have reported a variety of constructed hearths (Perlès 1976) and "pits" (Bordes 1972:131–35; Jelinek et al. 1989), as well as enigmatic stone walls (Freeman 1989) and other features (see Klein 1989b:313–17; Mellars 1989:360) from European Middle Paleolithic sites, these kinds of structures are rare and widely scattered (Rigaud 1989; Rigaud and Geneste 1988). Evidence for the improvement of life-space is markedly more abundant and more systematic in the earliest Upper Paleolithic (Gamble 1986: 250–72; Klein 1989b:315; Mellars 1989:358–59). There are probably more well-defined, *constructed* hearths and other features reported at a single Upper Paleolithic site like Abri Pataud (Movius 1966) or Le Flageolet (Rigaud 1989:150) than in *all* the Mousterian sites in southwest France. There is not even very much evidence that Mousterian hominids kept potential life-space cleared of debris within shelters, over and above moving large and intrusive objects to the margins of an occupied area (Mellars 1989; Simek 1987). Where appropriate spatial data are available, distributions of artifacts and bones certainly do appear spatially differentiated, but we cannot equate simple non-random distributions with the highly partitioned, regimented use of space seen in long-term human occupations. Even non-human animals that occupy constrained spaces such as dens and caves often move large, intrusive objects to the margins of sleeping or resting space (see review in Stiner 1994).

Evidence for dietary composition or diet breadth during the Mousterian is also difficult to reconcile with a sedentary lifestyle. Among later Pleistocene and Holocene foraging populations, decreased residential mobility is virtually always associated with an expansion of diet breadth, more specifically an increasing emphasis on smaller "package-size" resources. Causality aside, the appearance of more permanent settlements is almost inevitably accompanied by decreasing emphasis on large game and greater use of smaller terrestrial and aquatic animal and plant food resources (Binford 1968; Kelly 1992; Tchernov 1992:20). An analogous phenomenon occurs on a smaller scale among non-human predators. When tethered to breeding dens or other locations, wolves commonly switch from large ungulate prey to mice or other small animals (Peterson et al. 1984). This is because, while in many environments large game provide very high energy returns relative to investment in pursuit, capture, and processing, large game are also mobile and comparatively dispersed: all other things being equal, their predators must move around a good deal to keep themselves fed. Under conditions of restricted mobility, the costs of finding dispersed prey and transporting it back to a den (or base camp) become prohibitive. Thus, when mobility is restricted there is often greater relative payoff to exploiting smaller game and gathered resources, which require more time to collect and process, but which are available in greater abundance and (more important) nearer at hand (see Gamble 1986:103–15).

The Mousterian provides little or no evidence for the kinds of dietary emphases characteristic of sedentary or territorially restricted forager populations. On a global basis, Mousterian deposits do not attest to *intensive* use of resources other than large game. The remains of fish and marine mollusks have been recovered in a few Middle Paleolithic contexts (Le Gall 1992; Stiner 1990b, 1993a), but they always occur in relatively small quantities and are by no means ubiquitous, even in coastal sites. The bones of small terrestrial vertebrates (rodents, lagomorphs, etc.) are also sometimes found in Middle Paleolithic shelter sites (e.g., de Lumley 1972), but their origins are often problematic. In the few cases where taphonomic analyses of small-mammal faunas have been carried out, the bones have been shown to represent the feeding activities of non-human predators (i.e., Stiner 1993a). Finally, burnt seeds have been found in Near Eastern Mousterian sites, suggesting that vegetable resources were consumed (e.g., Bar-Yosef et al. 1992:509), but the kinds of artifacts needed to process and effectively exploit such foods are conspicuously scarce in Middle Paleolithic "toolkits" of Eurasia. It is safe to say that informal grinding or pounding tools are more common in the Oldowan and early Acheulean (e.g., Leakey 1971; Willoughby 1987), and probably even in modern chimpanzee technology (Boesch and Boesch 1983; Hannah and McGrew 1987; McGrew 1991), than in the Eurasian Mousterian. Equipment for processing vegetable foods might of course have been made of wood or other perishable materials. As in the case of weaponry, however, this only leaves one to wonder why Mousterian populations would have been so devoted to perishable technology for tens of thousands of years when stone was normally plentiful.

I do not wish to argue that Middle Paleolithic populations were exclusively large mammal predators. There are many reasons to think that their diets were much more generalized than this. The point is that while Mousterian foragers may well have utilized a wide variety of resources, the archaeological evidence suggests that they were acquired in a casual, perhaps opportunistic manner. There are neither substantial accumulations of food debris nor the kinds of processing technologies that are indicative of *intensive* exploitation of small-animal and vegetable foods. It is possible, even likely, that small resource "packages" were always processed in the field, outside the shelter sites where preservation is best. Still, in-field consumption is not consistent with intensive processing and heavy dependence on these kinds of foods.

The Mousterian was surely not uniform over its vast spatial and chronological range. We can anticipate that some Middle Paleolithic populations were less mobile than others, and that resource utilization patterns varied across space and time. However, the characteristics described above are widespread, even ubiquitous, within the European Middle Paleolithic. At present, there seems to be no great regional or temporal variation in the degree of investment in features, or evidence pertaining to the use of small-animal or vegetable resources within the European Mousterian at least. Moreover, contrary indications can be and often are obtained from the same data. A site or a stratum may contain a dense deposit of debris made almost exclusively from local raw materials, suggestive of restricted movement, while at the same time lacking any indications for the investment in features and facilities normally associated with long-term occupations. Existing models of human mobility patterns and their consequences may be somewhat simplistic, but it is still necessary to reconcile the kinds of conflicting observations presented above. Either European Mousterian populations moved very little—but didn't need to adjust resource exploitation patterns and didn't bother to improve their life-space—or they changed residential locations frequently, but repeatedly abandoned artifacts and bones in a limited number of locales. Arguing that Mousterian groups simply had different ways of treating living space or filling dietary needs than modern humans only begs the question of why these differences might have existed. The responses to decreased mobility cited above can be observed on a cross-cultural (or even cross-specific) basis: they are not attributable to cultural differences *per se*.

In my view, the apparently contradictory evidence is best reconciled with reference to a pattern of high mobility but frequent reoccupation of the same places. The common Eurasian Mousterian patterns could be attributable to a widely shared mobility strategy involving very frequent displacement among a series of "target locations" such as resource exploitation areas and natural shelters, similar to the movement patterns of other territorial, social animals. This alternative only presupposes a kind of land use not widely documented among modern human foragers, and does not require positing major exceptions to more general ecological and behavioral trends. Locally restricted ranges of artifact transport could reflect small territories, as has been argued for the Near Eastern Mousterian and South African Middle Stone Age (Ambrose and Lorenz 1990; Henry 1992; Marks 1988a; Marks and Freidel 1977). On the other hand, the distances artifacts appear to have been moved are strongly influenced by artifact use lives and the amount of time it actually took to get from one point on the landscape to another (Kuhn 1992b). If residential moves were very frequent but short-range, the stone tools people carried with them and used on a daily basis would have been more likely to wear out, break, or be lost before they traveled very far, regardless of the actual sizes of hominids' territories.

However one interprets them, the incongruous aspects of the Mousterian imply that we may be dealing with a set of cultural adaptations, a range of behavioral responses to the exigencies of survival, that differed in important ways from what archaeological and ethnographic experience with modern human foragers leads us to expect. However, demonstrating differences between archaic and modern forms of *Homo sapiens* does not by itself explain the replacement of one form by the other (regardless of the mechanism involved). Archaic *Homo sapiens* was an extremely successful organism, persisting for 200,000–300,000 years and colonizing most of Africa, Asia, and Europe. To argue that Neandertals became extinct or were subject to selective pressures that eventually produced modern humans because they were bad hunters, slow talkers, or poor housekeepers essentially assumes that evolution is teleological and progressive. No one would maintain that moles dig to catch insects because they don't have wings like bats, and we should not argue that Neandertals acted as they did because they somehow lacked the capabilities to behave in another manner.

More generally, we must be cautious in attributing the distinctive features of the Middle Paleolithic archaeological record to differences in the capacities of archaic and/or early anatomically modern *Homo sapiens*. The period between roughly 250,000 and 35,000 years ago also witnessed environmental and demographic conditions that may have been without parallel in the more recent past. The challenge is to understand how these different factors—ecology, demography, and hominid capacities—may have contributed to both the persistence and the ultimate disappearance of the behavior patterns that make up Mousterian "culture," and the hominid populations responsible for them.

The anomalous features of the Middle Paleolithic record have a number of more specific implications for how one might approach the evolution of human technology and subsistence adaptations during the Upper Pleistocene, implications that guide this research. First, it is clear that artifact *designs*, the shapes of stone tools, will not be especially informative about the relationships between technology and subsistence adaptations in the Mousterian. The majority of stone tools probably played little or no part in subsistence, and those artifacts that might have been used in resource procurement are both limited in variety and relatively invariate across time and space. Fully appreciating how lithic technology might have been linked to foraging will require looking beyond tool forms. I argue that it will be more profitable to focus on the economics of stone tool production, to examine how the manufacture and maintenance of tools might have been organized around the more fundamental and immediate concerns of making a living. Secondly, the ambiguous and often contrary indications about mobility suggest that land use patterns may be crucial to understanding long-term trends in human adaptations. Whether human mobility strategies were a driving force, or just symptomatic of changes in other domains, remains to be seen. However, changes in the use of both landscapes and places on them would appear to be pivotal developments during the Upper Pleistocene.

Technological Research and Paleoanthropology

It is no accident that prehistorians refer to the earliest phases of human cultural evolution as the Paleolithic, the Old Stone Age. Old stones, lithic artifacts, provide the most abundant and most widespread testimony to the behavior of Pleistocene hominids. Beginning with the first scholarly treatments of evidence for pre-modern humans in the 19th century, the analysis of chipped stone artifacts has been both axis and foundation of Paleolithic research. Paleolithic studies are now much more than the study of paleo-lithics, however, and analyses of stone tools are no longer the only, or even the most important, source of information about the behavior and lifeways of humans and human ancestors. Faunal remains are being analyzed in a variety new and highly informative ways, and are providing a wealth of often unexpected information about the lifeways of Pleistocene hominids (e.g., Binford 1984, 1988; Chase 1986a, 1989; Farizy and David 1992; Klein 1977, 1979, 1989a; Stiner 1990a, 1991b, 1993a, 1993b, 1994). While not yet widely applied, new categories of data pertaining to phenomena such as intrasite spatial distributions or "site structure" have the potential to provide a truly novel perspective on the behavior of Mousterian hominids (e.g., Binford 1978b; Rigaud and Geneste 1988; Simek 1987, 1988). To the extent that questions about evolution and adaptation have come to the forefront of Paleolithic research, the centrality and even the relevance of lithic studies can no longer be assumed. We need to ask how stone tools can be integrated into this changing research program.

New perspectives on old stones

Up through the early 1970s, Paleolithic research in Europe, North Africa, and the Near East was dominated by typological studies focusing on variation in artifact morphology. For the Mousterian or Middle Paleolithic, of course, François Bordes's well-known artifact taxonomy (Bordes 1953, 1961a) had emerged as the single dominant system by the late 1950s. Over the past decade or so, a variety of new methods and data sets has come to play an increasingly important role in research on Paleolithic stone tool technologies. The Bordes typology has not been entirely forsaken: it remains the preeminent descriptive language for Middle Paleolithic retouched tools, and the Mousterian "facies" that issue from its application continue to structure much comparative research. However, Bordes's typological method increasingly takes a back seat to other analytic approaches. The research presented in this book draws on three of the emerging themes in Paleolithic technological research: analyses of *chaînes opératoires*, studies of stone tool "use histories" (especially the so-called "reduction argument"), and lithic raw material exploitation.

The study of *chaînes opératoires* (operatory chains) is arguably the most important, and certainly the most widely diffused, of the new trends in Paleolithic technological research. In principle, a *chaîne opératoire* includes the entire sequence of technological acts beginning with the collection of the raw material, running through its transformation into tools, and ending with the eventual abandonment of a reworked and exhausted implement. In practice, the greatest energies have been concentrated on the middle part of the process, the methods for producing flakes and tool blanks from cores: thus, the most commonly used English equivalent for the term *chaîne opératoire* is "core reduction sequence." Having their origin in the pioneering experimental work of individuals like J. Tixier, D. Crabtree, and F. Bordes himself, analytical approaches to flake and tool-blank production systems have developed somewhat independently among French- and English-speaking researchers. European and American researchers tend to employ rather different methodologies and interpretive frameworks, but they agree on the importance of understanding how prehistoric humans went about the task of taking apart a piece of stone.

Without question, the best-known arguments about stone tool "use histories" in the Mousterian come from a series of papers published by H. Dibble (Dibble 1987a, 1987b, 1988). Though the notion that tool forms change with progressive use and renewal is not a new one (Cooper 1954; Frison 1968), Dibble has applied this idea in a most comprehensive manner to Middle Paleolithic assemblages. He proposes that much of the variation in artifact morphology described by Bordes's typology actually represents successive stages in the resharpening of a very few basic artifact forms. Many of the patterns of inter-assemblage variation once viewed as reflecting either ethnic identity or functional variation can consequently be reinterpreted as representing different intensities of raw material exploitation. The so-called "reduction argument" has been quickly and widely adopted by Paleolithic researchers in the United States, although its acceptance in Europe has been less universal.

It is important to differentiate between the general notion that there is a relationship between tool form and resharpening and the specific manner in which Dibble and his students have applied the concept to Mousterian assemblages. Few lithic researchers would dispute the notion that the forms of tools recovered from archaeological sites reflect the results of maintenance and use as well as, or even more so than, the design conventions of prehistoric humans. On the other hand, Dibble's transformational model for explaining inter- and intra-assemblage variability in the Mousterian can be challenged in its specifics (Bietti and Kuhn 1990–91; Kuhn 1992b). To this point, the "reduction argument" has been cited most often as an alternative explanation for formal variation in the forms of finished tools, as described by established typologies. Surely, however, there is more about the Paleolithic record to investigate than the typological categories developed by past generations of archaeologists. In the long run, the most significant contribution of the "reduction argument" has been to focus attention on the "life histories" of tools. Viewing artifacts as the end product of a long and dynamic process of manufacture, use, and renewal—rather than as fossilized conceptual blueprints—has helped to divert research in a very productive direction, one which is much more in keeping with other developments in Paleolithic studies.

The third important recent trend in research on Paleolithic technologies, the investigation of raw material exploitation, has developed more or less harmoniously in a number of different research groups. Researchers have concentrated their efforts towards two principal aims. One involves using patterns of raw material exploitation to reconstruct the scales and shapes of human territories (e.g., Demars 1982; Geneste 1985, 1986; Montet-White 1991; Wengler 1990, 1991; Wilson 1988). In other cases (Geneste 1989; Geneste and Rigaud 1989; Schild 1987; Simán 1991) greater attention has been paid to possible strategies of raw material conservation or cost minimization. The emergence of an adaptive/evolutionary paradigm in Paleolithic research during the 1970s placed new emphasis on phenomena like territoriality and mobility, making it of obvious interest to know where people got the stones they used to make tools. The representation of technology as a suite of organized responses to both internal and external exigencies—rather than as a static set of culturally prescribed behaviors—also provided a rich context within which to analyze the differential procurement, transport, and conservation of workable stone.

While studies of *chaînes opératoires*, artifact reduction, and raw material exploitation have provided a wealth of intriguing and useful data, the recent methodological "radiation" in technological studies has been accompanied by a certain theoretical fragmentation. For better or worse, typologically oriented prehistorians knew just what they were up to, namely "mapping" the boundaries of prehistoric ethnic units in time and space. In contrast, the various individuals involved in the development of new methods have different, often conflicting, ideas about the problems that should be addressed through their observations. The research presented in this book includes analyses of core technology, stone tool reduction, and raw material transport, not just because they are "in vogue," but because they are useful for addressing a particular range of issues.

Integrating technological studies into archaeology's mission within paleoanthropology, namely to document and explain evolutionary change in human behavior, inevitably leads beyond the confines of technology itself. In order to bring technological data to bear on major changes in human adaptations, it is essential to explore how toolmaking was related to hominid subsistence and foraging ecology. To date, archaeologists have had little success in identifying the links between technology and subsistence behavior during the Mousterian. At Combe Grenal, Bordes observed a correlation between the frequency of denticulate artifacts and the importance of equids in associated faunas (e.g., Bordes and Sonneville-Bordes 1970). However, it is difficult to imagine what this or corollary relationships (e.g., Chase 1986b) might represent, since there is no readily apparent functional connection between the exploitation of horses and the use of tools with sawtooth edges. Similarly, L. R. and S. R. Binford suggested a variety of subsistence-related explanations for possible functional variation in tool forms (Binford 1973; Binford and Binford 1966) but did not seek independent confirmation of the postulated functions.

In large part, the limited success of past attempts to draw connections between technology and subsistence is attributable to the reliance on the Bordes typology as the exclusive measure of technological variation. The forms of some tool types do hint at some kind of functional specificity, but there is no *a priori* reason that this should be linked with game species or any other tangible dimension of subsistence. On the other hand, if it is so difficult to link tool types to subsistence, why should we have an easier time with core technology, stone tool reduction, and raw material exploitation? After all, none of these spheres would seem to intersect in a direct and obvious manner with food procurement, for example. Indeed, the links must be pursued at a more fundamental economic level. One way of integrating studies of technology into a larger program of research on hominid cultural evolution and adaptations is to focus on technological responses to the exigencies of mobile foraging lifestyles. Mobility and land use patterns can serve as the essential economic bridge between technology and the larger sphere of subsistence adaptations. The ways foragers use the landscape are intimately connected with how they make their living. At the same time, the places people go to procure resources, and the scale and frequency of their movements, have direct consequences for the management of raw materials and the cost of transporting the technological support necessary for survival. In order to keep themselves supplied with necessary technological aids, human tool users must adopt strategies that reconcile the requirements of toolmaking with the demands of food-getting.

This is where I would argue that the emerging themes in lithic studies have their greatest potential. The ways people manufacture tool blanks, how much they rework their tools, and their strategies of raw material exploitation can all factor into responses to foraging patterns and mobility. Core technology has direct consequences for both the rates at which sometimes scant raw materials are used up, as well as for the functional properties of tool blanks. Resharpening or reduction of artifacts is simply a means of extending their useful lives. This can occur in response to raw material shortage, due either to general environmental scarcity or to mobility-related limitations in what people can keep on hand. Finally, patterns of raw material exploitation and artifact transport vary as a direct consequence of where people go, how they move about the landscape, and their ability to plan around future contingencies. A more detailed and explicit framework for exploring the links between foraging and mobility and patterns of stone tool procurement, manufacture, and use is presented in the next chapter.

2

TECHNOLOGY, FORAGING, AND LAND USE: A STRATEGIC APPROACH

> I am told that being ever prepared, *upterrlainarluta*, is a common caution from Yup'ik elders to young people, whether they are preparing for fishing or a trip to the city. Implicit is the understanding that one must be wise in knowing what to prepare for and equally wise in being prepared for the unknowable.
>
> (James H. Barker, 1993, *Always Getting Ready: "Upterrlainarluta"*)

In this chapter, I outline one model of how hunter-gatherer technologies are bound up with the concerns of making a living, and how those relationships can be operationalized for the study of technological changes over the course of human evolution. In keeping with the character of the Mousterian technological record, the primary focus of this discussion is on the management of raw material consumption, strategic responses to the problems of keeping mobile populations supplied with tools and toolmaking potential. As the above quotation implies, the issue of *planning*—anticipatory organization in the manufacture and treatment of artifacts—is integral to the discussion.

In constructing a model of how stone toolmaking and human subsistence interface, I make extensive use of recent ethnographic and archaeological comparative cases. Wobst (1978), referring to the "tyranny" of the ethnographic record, cautions archaeologists against relying exclusively on ethnographically documented societies as a source of models for interpreting prehistoric behavior. The groups described by ethnographers and other observers over the past two or three centuries encompass only a small part of the total gamut of social organization and subsistence that must have appeared over the course of human prehistory. How then can one justify using such a narrow and limited data set to learn about the Paleolithic, and about hominids that may well have been quite different from ourselves?

Biased or not, I would argue that we must use what we know about modern human foragers to help us learn about the Paleolithic. Neandertals were not Yup'ik, nor were they Pitjantjatjara. However, no archaeologist with whom I am acquainted has had to make and use stone tools in a mobile context, as Paleolithic humans did. If we base our models of the past on the "common sense" we receive from our sedentary, industrial, machine-age society, we are subjecting our research to an even greater tyranny. Nonetheless, the modern archaeological and ethnographic records must be employed with care. If we take ethnographically or archaeologically documented systems as direct analogs (see discussion in Tooby and DeVore 1987) for Paleolithic foragers, then it will indeed be impossible to transcend our limited scope of knowledge of the present. And, as the recent controversies over the history of San foragers demonstrate (e.g., Wilmsen and Denbow 1990, with commentary), there exist no fossilized Paleolithic hunter-gatherer societies that provide a direct window on the ancient past. On the other hand, mobile tool-using populations have had to overcome many of the same obstacles, problems that stem directly from mobility and the use of tools. Regardless of how they make their living and regardless of how long they have been at it, people that move about the landscape must cope with variable access to raw materials, with the need to carry and maintain transported toolkits, and with reconciling the schedules of food procurement with the making and mending of the necessary technological aids. The broad variation found within the archaeological and ethnographic records of recent hunter-gatherers provides a context in which it is possible to identify some of the basic problems that confront mobile tool users, and to gain some appreciation of the range of possible human responses to them. In other words, observations about modern humans provide archaeologists with clues about the basic rules of the game. However, it is incumbent on us to learn the stratagems employed by Pleistocene foragers.

Technological Strategies and Human Adaptations

How one defines technology, and how one identifies the key domains of technological behavior, depend largely on one's research goals. For this reason, it would be very difficult to extract from any large group of archaeologists a consensus view of what technology is and how it should best be studied. The majority of archaeologists working today would agree that technologies are not objects, that they are more than the sum of their material products. Moreover, most would probably say that technologies are best seen as information, systems of knowledge, sets of interrelated procedures or ways of doing things. Beyond that, however, we can anticipate little concurrence.

From the viewpoint of evolutionary ecology, technologies are best treated as *problem-solving strategies*. Tool makers expend time and effort to make artifacts that in turn provide some energetic, temporal, or other advantage at some later date. A wide variety of strategic obstacles must be overcome in order to get full return on the investments made in producing artifacts. Many of these challenges originate in the domains of foraging and land use. Understanding the constraints imposed on tool makers by subsistence concerns is vital to integrating the study of technology into a program of evolutionary research.

Analyzing technology, or any other behavioral phenomenon, from this sort of strategic perspective is simply a method of learning about the factors that might have influenced the actions of organisms, and about the evolutionary solutions that were rendered selectively advantageous (Foley 1987:61). While terms like "strategy" and "problem solving" refer to mental phenomena in common usage, they are employed in the present discussion as abstract analytical concepts, much as they are in evolutionary biology or game theory (e.g., Lewontin 1961; Maynard-Smith 1982). It is not assumed that prehistoric hominids (or other organisms) were fully rational beings behaving in terms of a set of economic rules, though they may well have been. Rather, Paleolithic tool makers are treated for heuristic purposes *as if they were* pursuing economic strategies, in an effort to identify the limiting currencies or resources in prehistoric adaptations. The question of whether the so-called "strategies" actually existed in the minds of prehistoric humans is another issue entirely (see discussion in Chapter 6). For the time being, questions about the rationality of Mousterian hominids are moot.

The subsistence sphere poses two main classes of problem for organisms that use technology. On one part is the issue of *design*, of making tools with properties appropriate to the intended use. On the other part are the problems of *supply*, of making sure technological aids are available when and where needs for them arise. The importance of keeping supplied with tools and raw materials varies as a direct function of the degree of dependence on artificial support in daily life. For an organism that—like modern humans—relies on technology in virtually every domain of its existence, keeping up the supply of tools and raw materials is a complex and crucial matter, as being caught without them can be disastrous. In contrast, if tools are incidental or peripheral to the main business of making a living, as among most other tool-using animals, then supply is a much less pivotal problem. Issues of supply are central to understanding relationships between technology and land use.

As argued above, principles of artifact design are of limited relevance to the study of Mousterian technologies. Studies of variation in artifact design have been most productive when applied to "extractive" tools, weapons in particular (e.g., Bleed 1986; Churchill 1993; Myers 1989; Torrence 1983). The procurement of resources is often associated with high-impact, "low-tolerance" applications, where subtle variations in tool forms have major functional consequences (Kuhn 1993). As we have seen, most Mousterian assemblages contain relatively small numbers of potential "extractive" tools or weapons: besides, the most likely weapons (Mousterian and Levallois points) show remarkably little formal variation over vast areas, suggesting that there is not a lot to be learned from studying their design parameters. The vast majority of Mousterian artifact forms—"sidescrapers," "notches," "denticulates," and the like—were most probably used to work other materials rather than to procure resources. The shape and angle of the working edge aside, the morphologies of artifacts used to work other materials have only limited consequences for their functionality or effectiveness (e.g., White et al. 1977).

The kinds of non-extractive tools that dominate most Middle Paleolithic assemblages were probably used to work materials that are hard, resilient, or tough; that is, materials not easily modified with hands or teeth. Such high-attrition manufacture and processing activities tend to produce a great deal of wear and tear on implements. When large numbers of tools are made, used, and used up, the rate at which raw material is consumed is a potential problem, particularly in environments where workable stone is hard to come by. In addition, if people are mobile, as Mousterian populations undoubtedly were to some degree, keeping an adequate supply of disposable tools and/or raw material on hand can also entail significant transport costs. Thus, the predominance of tools likely to have been used in manufacture and maintenance activities suggests that the greatest strategic problems for Mousterian tool makers involved the "delivery" of a ready supply of artifacts and fresh edges.

Over the long term, there is probably a fairly constant demand for general-purpose tools and sharp edges. The strategic element of technology comes into play in bridging the inevitable gaps between the scheduling and distribution of subsistence opportunities on one hand, and needs for tools on the other. It is reasonable to assume that where people go, how long they spend there, and the proportion of their time devoted to subsistence pursuits are all determined primarily by the distribution of crucial resources like food, water, and fuel. Social relationships and reproductive considerations may also play a major role in structuring settlement (e.g., Kelly 1992:48; Kent 1991; Wiessner 1982), but these are beyond the scope of the present discussion. This is not to say that hunter-gatherers don't make trips to collect flint or wood for making tools: it is simply that such concerns are secondary to securing the basic requirements for survival. Still, if raw materials were ubiquitous, and if people always had sufficient time at their disposal to make whatever tools they might need, then keeping themselves supplied with tools and raw materials would be a simple and straightforward business. In most normal contexts, however, this is simply not the case. Instead, mobile foragers inevitably encounter situations where they can secure plenty of food but no raw materials, or instances when toolkits need to be replaced or repaired but when time must be spent in other activities.

The likelihood that most, if not all, Mousterian artifacts were employed in manufacture, maintenance, and processing activities, rather than directly in the procurement of food, also has implications for the timing and distribution of needs for tools. Among mobile hunter-gatherers, manufacture and maintenance activities are most often organized *around* subsistence activities (Torrence 1983:13), carried out in the "down time" between food procurement and travel (e.g., Silberbauer 1981:243; Yellen 1977). The scheduling of free time for toolmaking and repair complements the pattern of labor investment in food-getting (Kuhn 1989). Among modern hunter-gatherers, most primary resource procurement takes place away from residential locations. The making and mending of things consequently tends to gravitate towards residential bases and other localities where people tend to have more free time on their hands. Because of the uneven distribution of different kinds of activities, factors such as the frequency of residential mobility, the duration of occupations, and the locations of living sites relative to sources of raw material must have played important roles in determining how Middle Paleolithic populations coped with maintaining a ready supply of "manufacture and maintenance" tools. Conversely, the technological strategies that prehistoric populations used to cope with discontinuities in the distributions of food and raw materials, or the scheduling conflicts between foraging and toolmaking, provide a host of clues about subsistence, foraging, and mobility. These are the links which have the potential to draw the technological data more closely into the general discourse on hominid evolution and adaptation during the Pleistocene.

Parenthetically, the actual time and energy devoted to stone tool manufacture itself are not expected to be a particularly crucial influence on Mousterian lithic technology. The activity of producing most flaked stone tools requires relatively little time and effort, particularly compared with the activities in which those same stone tools might be used. The manufacture of a fluted, bifacially flaked Clovis point takes an experienced knapper about 60 to 90 minutes (Callahan 1979:23), and the unifacially retouched flake tools that dominate Middle Paleolithic assemblages would require considerably less time to produce, even taking into account the use of elaborate Levallois core technology. In contrast, the manufacture of a simple wooden digging stick with stone tools, just the kind of activity in which Mousterian technologists might have used their scrapers and denticulates, requires around 90 minutes (Hayden 1979:111), just about the same as the Clovis point. Activities such as the production of complex composite artifacts, preparing hides, or sewing clothing require tens or even hundreds of hours of labor (e.g., Lee 1979; Osgood 1940). On the other hand, while the manufacture of stone tools may require comparatively little time or energy, procurement of raw materials has the potential to be a very time-intensive undertaking in some contexts.

Planning and Technological Provisioning

The problems involved in maintaining a ready supply of tools and raw materials are closely bound up with the issue of *planning* (Binford 1979, 1989). What is called planning refers to strategies for insuring the availability of tools in situations where it would not otherwise be possible to have them. Only if technology is peripheral to survival can tools be made and used purely as a matter of convenience, with no advance planning at all. It is quite unlikely that Mousterian hominids, or any other member of the species *Homo sapiens*, relied exclusively on tools made on the spot. At the same time, even fully technologically dependent modern humans do not always find it either useful or practical to fill their technological requirements far in advance. Moreover, modern foragers cope with anticipated demands for tools in a variety of different ways, depending on the conditions and contingencies they encounter.

As a preface to addressing variation in planned technologies, it is instructive to consider the implications of not planning at all. Expedient (Binford 1977) or "reactive" tool manufacture—the practice of making do with whatever materials come to hand, of meeting needs as they arise—is certainly a viable alternative in many situations. Making tools on the spot is easy and cheap, provided there is something to make them out of and time to do the work. Expedient or reactive toolmaking involves no transport costs, and since, by definition, necessary raw materials must be available on the spot, the cost of procuring them is minimal. We might expect tool makers to take advantage of locally plentiful materials to fill immediate needs whenever they can. If usable stone is ubiquitous, there is relatively little advantage to making and maintaining *most* implements long in advance of use (Bamforth 1986; Brink et al. 1986:163; Marks 1988b).

The down side of relying exclusively on expedient production is that it inevitably forces people to sometimes do without technological assistance. There are few real-world situations in which raw materials are so widely and evenly distributed that people can find what they require wherever they go, so tool users will invariably be caught short from time to time if they do not look ahead. Furthermore, many tasks can be effectively carried out only if the appropriate tools are already at hand. An obvious example is the procurement of mobile resources like large game. If hunters wait until prey is sighted to make weapons, they stand a good chance of losing the opportunity to make a kill (Larralde

1990:220). While the locations and timing of demands for material aids are not always predictable, people know that they will eventually want to have tools in situations where there is neither time nor raw material to make them (Binford 1977, 1979; Bamforth 1985; Kuhn 1990a:67; Nelson 1991). As a result, at least a limited component of all modern human technologies is made well in advance of use.

Although all human technologies, and likely those of our hominid forebears, incorporated a planned component, there is more than one means of coping with future needs, making sure tools and/or raw materials are available when and where they are needed. The concept of *technological provisioning* is useful for summarizing variation in planning strategies. The term "provisioning" crosscuts the more familiar terms "curation" and "expediency" (Binford 1973, 1977, 1979). It refers to the depth of planning in artifact production, transport, and maintenance, and the strategies by which potential needs are met. Different strategies of provisioning are associated with contrasting costs and benefits, and each imposes different problems for the procurement, modification, and transport of finished implements and raw materials.

Since artifacts are used by individuals, one way to make sure tools are available in advance of needs is for people to always have at least a limited toolkit on hand: this strategy is termed *provisioning individuals*. Durable implements can be manufactured, transported, and maintained in anticipation of a variety of exigencies. Specific anticipated needs can provisioned with specialized tools, while more general needs can be met with either multipurpose tools or toolmaking potential such as cores and raw materials. In referring to provisioning of individuals, I also intend to refer to situations in which co-operating groups of individuals may "own" items of technology and share in the cost of carrying and maintaining them (e.g., Marshall 1976:366–67; Tanaka 1980:44). The important distinction is the focus on supplying the tool user or users.

Activities requiring tools do not occur randomly in space. Many practices are strongly associated with particular places or types of locations on the landscape. As already argued, manufacture and processing tasks tend to gravitate towards locations like residential sites, where little direct resource procurement occurs. Resource procurement and processing activities may be quite closely linked with particular locations where food or other resources abound. As a consequence, another viable strategy for coping with anticipated requirements is to supply the places where tools are likely to be needed with the appropriate raw materials or implements: this strategy is termed *provisioning of places*.

Provisioning of individuals with some form of what Binford calls "personal gear" (Binford 1977, 1979) appears to be universal among mobile societies. The composition of individual toolkits varies widely according to mobility patterns, modes of transport, and the kinds of tasks in which people are likely to be engaged: it is instructive to compare descriptions provided by Binford (1980) of Nunamiut Eskimo logistical gear, Grey (1841) of the materials carried by women in Western Australia, Gould (1969:76) of Aborigine male hunting gear, Lee (1979) of a !Kung man's hunting equipment, and Hoffman (1986) of tools and weapons carried by Punan hunters. Nonetheless, it is safe to say that virtually all modern humans carry something wherever they go as a hedge against inevitable requirements for tools and weapons.

If individuals are kept supplied with tools, then it is fairly certain that some kind of technological support will be available when a need arises. However, the effectiveness of this strategy is limited by what people can practically haul around with them on a continuous basis. Portability is arguably the most important consideration in the design and assembly of mobile toolkits (Ebert 1979; Kuhn 1994; Nelson 1991; Shott 1986), and transported gear usually consists of a limited variety of general-purpose tools (Gould 1969:76–77; Hitchcock 1982: 370–80; Lee 1979:179). Because the number and variety of implements that can be transported over the long term is constrained, it is impossible to ensure that the most appropriate or effective tool for the job will always be available. Similarly, the cost of continuous transport prohibits carrying many backups and spares, so attrition and breakage can easily become a problem with toolkits carried by mobile individuals.

The strategy of provisioning places involves a somewhat different set of tactical problems. Keeping a supply of raw materials or finished implements on hand where they will be needed greatly increases the potential efficiency of conducting a broad spectrum of tasks. Depending on the duration of a stay, both the variety of potential tool uses and the number of potential tool users may be quite large, and the necessary quantities of raw material or tools correspondingly great. On the other hand, the cost of continuous transport is not a factor, so the volume of material and goods which can be brought to and kept at a place is effectively unlimited. The limiting factor is the spatial regularity or predictability of needs for tools. While it is safe to assume that all people will need certain kinds of tools at some time, all places are not used with equal frequency or regularity. It is a waste of time and energy to supply a location with gear that is never used, and efficient provisioning of places requires precise knowledge about where different activities might occur in the future. Particular tasks may be quite specific to certain locations, whereas other locations may be used in quite diverse ways. The strategy of provisioning places consequently requires some prior knowledge of both the timing and the probable location of future needs. Its utility and practicality depend on residential stability, on the duration of use of habitation sites, or (in the case of extra-residential caches) on the regularity with which resources are obtained at particular spots on the landscape.

Tactical correlates of alternative provisioning strategies

Future contingencies may be met either with implements or with the *potential* for making implements (raw materials in one form or another). The "choice" between finished implements and raw materials involves a series of trade-offs, hinging on the cost of transport and the predictability of the range of needs which will be encountered. Finished implements have relatively low transport costs, but are limited in their potential versatility. Minimally modified raw materials have the potential to fill a vast number of different functions, but are energetically costly to carry around. Moreover, time-stressed resource procurement activities—hunting, for example—require that a finished implement be available as soon as the need for it arises. Such exigencies can be met only with finished tools. On the other hand, most non-extractive activities are not expected to operate under such severe time constraints.

Individuals are mobile by definition, and must carry their toolkits with them more or less constantly. The costs resulting from continuous transport would seem to be the overriding concern in the design and assembly of toolkits used in provisioning individuals (Nelson 1991). In such a context, we might expect people to try and *optimize* the potential utility of artifacts carried with respect to the cost of carrying them. Because retouched tools or tool blanks deliver higher utility per unit weight than do cores, which contain much potentially wasted material, individuals should provision themselves with tools more often than with toolmaking potential (Kuhn 1994). The items that make up mobile toolkits should be light, versatile, and general-purpose (Gould 1969; Lee 1979; Nelson 1991). Various researchers emphasize portability or size (Ebert 1979), flexibility or versatility (e.g., Gould 1969:75; Kelly 1988), and durability (Goodyear 1989; Shott 1989a, 1989b), but clearly all three variables are potentially important.

The strategy of provisioning individuals also has implications for the treatment of artifacts. It is simply impractical to carry large numbers of spares or backups. Access to raw materials may also be quite unpredictable for people moving around in an environment where stone is not ubiquitous. Consequently, "mobile gear" should be subject to extensive maintenance and renewal (Kelly 1988; Kelly and Todd 1988:237; Goodyear 1989:3–4; Kuhn 1989; Shott 1989a, 1989b). In addition, unanticipated needs often have to be met by using whatever gear happens to be on hand (Bamforth 1985). "Situational" variability in tool use—improvisation to cope with unforeseen exigencies (Binford 1979: 226, 1989)—will result in modification or reuse of tools and cores in a variety of ways not consistent with their original design (Goodyear 1982; Gramly 1982; Nelson 1991).

The provisioning of places involves a very different set of constraints. Because many individuals may need to make and use tools at a particular location, and because the variety of activities in which those tools are used can be great, provisioning of places should favor *maximizing* the utility and potential versatility of the technological materials available there. Material goods can be amassed and cached at a place, so the energy cost of carrying them around presents less of a problem than it does in the context of provisioning mobile individuals. Because unmodified raw materials can potentially serve a wider range of functions than finished implements, it is expected that places should be provisioned more often with *toolmaking potential* (cores and raw materials). As the number of artifacts that can be kept at a place is very large, the use of light, general-purpose artifacts is not particularly advantageous. Similarly, tactics for extending the utility of tools—resharpening and situational reworking—would be less important in a situation where many functionally redundant implements or large quantities of raw material can be kept on hand. Comparatively frequent discard and replacement of worn or broken tools, and relatively little reuse or resharpening, can thus be expected to characterize the strategy of provisioning places.

Provisioning Strategies, Foraging, and Land Use

In order to provide a framework for studying the role of technology within Middle Paleolithic adaptive systems, it is necessary to examine expected links between the organization of subsistence and the planning of tool manufacture and use. For most human technologies, the use of non-planned (expedient) artifact production is not extremely informative. The practicality of truly expedient manufacture and use depends entirely on the likelihood that tool makers find themselves with a ready source of raw materials, a need for tools, and time on their hands—in other words, means, motive, and opportunity. If such lucky circumstances can be anticipated, people can plan to be expedient, can anticipate not having to worry about where their next tool will come from. Even if raw material is not ubiquitous, people can still take advantage of the happy coincidence of these three factors. Where, when, and how often this happens is more a function of the overall distribution of raw materials and the spatial coincidence between places where activities are conducted and localities where stone can be procured than it is symptomatic of any particular characteristic of human adaptations.

In discussing the relationship between foraging organization, mobility, and technological provisioning, it is important at the outset to distinguish between extractive tools and tools used for manufacture, processing, and maintenance. Resources are almost inevitably procured in the field, away from residential locations, and the technological support needed for such activities must be transported frequently and for prolonged periods. Needs for most extractive implements should consequently be met by provisioning of individuals more often than not. Moreover, many extractive tasks, particularly those involving the procurement of animal foods, also require comparatively elaborate tools, and considerable time and energy is expended in the design and manufacture of implements used in their execution (Bleed 1986; Torrence 1983, 1989). Such high investment in manufacture implies that tools will be "curated," transported, and maintained over long periods of time (Shott 1989b), again conforming most closely to the strategy of individual-level provisioning.

Strategies used to keep people and/or places supplied with general-purpose manufacture and maintenance technology are of far more immediate relevance to the Middle Paleolithic. As stated above, the scheduling of manufacture, maintenance, and processing activities is spatially and temporally complementary to extractive tasks or food procurement and processing. In a context of high residential mobility, manufacture and maintenance are most closely linked to residential locations, where most resting, recreation, and free time take place. The major exception would be places where bulky raw materials are procured and processed prior to transport back to base camps: examples include stone quarries (e.g., Binford and O'Connell 1984; Metcalfe and Barlow 1992) and sources of other desirable resources or materials (e.g., Akerman 1974; Hayden 1979). Binford (1980) refers to these specialized extractive sites as "locations." In highly differentiated, "logistically organized" systems, residential camps and extractive locations will also be the major focus of manufacture and maintenance activities. In addition, portable items may be manufactured or repaired at logistical nodes such as camps or observation stands, where people experience long intervals of waiting or "down time" (e.g., Binford 1978a; Rensink 1987).

Because manufacture and maintenance activities gravitate towards residential locations, the importance of strategies of provisioning places versus individuals with the tools needed in such tasks should vary directly with scales and frequencies of residential mobility (Binford 1980; Kelly 1983, 1992). The more frequently people move their residential locus, the more they must depend on strategies of provisioning individuals. Groups that often move from one base to another are in a poor position to predict where most needs for tools will come about. In contexts of very high mobility, a great many needs must be met with mobile toolkits, and a greater proportion of the toolkit must be carried from place to place. If, on the other hand, places are occupied for extended periods or are reoccupied often and regularly, the locations of many activities, including "gearing up" for logistical forays, can be predicted more securely. Moreover, foraging trips out from a stable residential base provide the opportunity to collect and bring back raw materials (so-called "embedded procurement" [Binford and Stone 1985]). Thus, a land use system (or portion thereof) involving repeated and/or extended occupations favors the strategy of provisioning places.

Broad patterns of variation in the New World archaeological record illustrate the general connection between provisioning strategies and patterns of land use. A number of researchers have noted that in several areas of North and Middle America apparently "wasteful" casual, unsystematic core reduction technologies became increasingly common as residential mobility decreased. In semi-sedentary contexts, virtually all domestic implements may be manufactured in this manner (Johnson 1986; Koldehoff 1987; Parry and Kelly 1987). Associated patterns of tool use are often characterized as casual or expedient (in an empirical rather than a processual sense), in that implements appear to have been made and discarded in a relatively short period of time, without extensive modification and renewal (Johnson 1986; Parry and Kelly 1987). Parry and Kelly argue that since activities were concentrated in long-term residential locations, people simply moved raw materials to these places, exploiting them casually as needed. In a sense, this can be seen as creating opportunities for expedient tool use in places with a predictable occupational history. It is also a paradigmatic example of provisioning places.

Paleoindian and Archaic period toolkits from western North America, thought to represent land use systems involving high levels of residential mobility, exhibit very different characteristics. Paleoindian retouched tool assemblages—non-extractive tools (scrapers, gravers, etc.) included—may be made up almost entirely of transported artifacts made on "exotic" raw materials (Goodyear 1989; Kelly and Todd 1988; Gramly 1982). Very frequently, Paleoindian retouched tools show signs of resharpening, reworking, and recycling (Goodyear 1982; Gramly 1982). This suggests that there was considerable emphasis on extending the utility of implements, as would be expected in a high-mobility context where much of the toolkit was transported by individuals (Kelly and Todd 1988). Although retouched tools predominate, transported bifacial cores, along with flakes and tools produced from them, are also frequent elements in the Paleoindian and Archaic archaeological record (Binford 1979:262; Goodyear 1989; Kelly 1988; Kelly and Todd 1988). Bifacial core reduction produces a relatively large number of usable blanks or edges per unit of raw material (Goodyear 1989; Morrow 1987). Because many potential tools can be made from a few small pieces of raw material, this approach to flake production would be ideal for provisioning individuals with toolkits, where maximizing utility per unit weight is of overriding importance. The small blade/bladelet cores of exotic raw materials often found in late Upper Paleolithic and Mesolithic sites in Europe (Gamble 1986; Schild 1987; Straus and Vierra 1989) may well represent a similar strategy of provisioning individuals with a limited degree of toolmaking potential.

It is important to emphasize that neither mobility patterns nor technological provisioning strategies are mutually exclusive. It is a gross oversimplification to classify either modern foraging groups or early humans as simply being either mobile or sedentary. Mobility patterns may vary over the course of a year and from place to place within the territory occupied by a single hunter-gatherer group (see also Bamforth 1991), and all peoples practiced a mixture of technological strategies. For example, northern hunters like the Netsilik (Rasmussen 1931), Nunamiut (Binford 1978a:255–344), and Iglulik Eskimo (Mathiassen 1928:53, 91) often split up into small, mobile groups during the warm season. Like more residentially mobile populations, they had to rely on whatever gear they could carry with them during these periods, caching large quantities of cold-season gear at frequently reoccupied winter villages or logistical nodes (e.g., Osgood 1958:43; Rasmussen 1931:167; Stefansson 1914:67). Conversely, highly mobile foraging groups tended to rely on what they can carry around with them, but did cache large, bulky items such as milling stones or mortars in the vicinity of seasonally predictable resource concentrations (e.g., Gould 1980:71; Hitchcock 1982:201; Silberbauer 1981:237), sometimes even storing general-purpose gear at locations that were frequently reoccupied (e.g., Smyth 1878:123).

Holen's study of protohistoric Pawnee sites is perhaps the best single archaeological example of variation within a cultural system in strategies for keeping people supplied with tools. The Pawnee inhabiting the eastern margins of the North American Great Plains were relatively settled village agriculturalists, but they also made regular forays into the grasslands to hunt bison. These prolonged trips involved many of the contingencies and restrictions faced by mobile foragers, despite the fact that the people were sedentary most of the year. Bison hunting expeditions were marked by a temporary switch to technological strategies resembling those normally associated with residentially mobile populations, such as the use of large, transported biface cores (Holen 1991).

Observations about variability within cultural systems are important because archaeologists study individual sites or groups of sites, but are seldom in a position to define an entire system. Evidence for different technological provisioning strategies should vary among individual occupations or segments of land use systems, as well as at the level of the entire technological and subsistence systems. Short-duration occupations and places used ephemerally should be less frequently provisioned, and should therefore yield relatively large numbers of tools carried by mobile individuals, regardless of the overall organization of the system. Conversely, places occupied for comparatively long periods are more likely to be provisioned with raw materials than ephemeral occupations, whether the people themselves would otherwise be characterized as residentially mobile or semi-sedentary. Artifact-poor but point-rich Upper Paleolithic assemblages in the mountains of Cantabrian Spain (Clark and Straus 1986) and Slovakia (Hahn 1977) are a case in point. These assemblages probably represent specialized, short-term hunting camps that contain little more than the debris from transported assemblages of weapons for game procurement: more substantial occupations dating to the same periods contain much more of the "fallout" from tool manufacture and processing activities.

In order to translate the relationship between provisioning strategies and land use in archaeologically relevant terms, it is necessary to consider how tools or raw materials reach sites. Invariably, mobile hunters and gatherers carry at least a minimal transported toolkit. Things brought to a place as parts of the mobile toolkits used to provision individuals may subsequently be used, broken and abandoned, lost, or cached at the site (Bamforth 1986; Stevenson 1985), along with a limited amount of debris from their repair or maintenance. If a stable residential base is subsequently established, the place may be provisioned with raw material from the surrounding area, or with objects brought for that purpose. Since the largest quantity of archaeological fallout will come from materials modified, used, and discarded at a site, then the amount of debris resulting from the provisioning of places with raw material will quickly swamp the few traces of mobile toolkits that may have entered the record.

Evidence for the movement of artifacts and raw materials can be important for distinguishing alternative strategies of technological provisioning. Conversely, however, the significance of such evidence varies according to the context in which artifacts were moved across the landscape. Hunter-gatherers move around as a matter of course, carrying at least some artifacts with them. The distance over which curated "personal gear" is moved before being abandoned depends primarily on the sizes of human territories and the durability of the artifacts in question. In contrast, if we assume for the sake of argument that most raw material procurement is "embedded" in forays aimed at the procurement of other vital resources (Binford 1979; Binford and Stone 1985; cf. Gould 1985; Gould and Saggers 1985), then the "catchment" area for stone transported in the context of provisioning places will reflect the scale of territorial coverage from a particular camp and, by implication, the duration of the occupation. In situations of high residential mobility, where most hunting or foraging is done within a day's round trip of the base camp, the scale of procurement of materials for use at a particular place will be quite limited. If raw material sources are widely dispersed, foragers may be unlikely to encounter even one in the course of a day. Very high levels of residential mobility have been proposed to account for the infrequent use of "local" raw materials at early Plains Paleoindian sites (Goodyear 1989; Kelly and Todd 1988). If, on the other hand, residential mobility is lower, and people engage in prolonged logistical forays from stable residential bases in order to obtain resources, the territory from which things can be "inexpensively" brought back to camp is greatly enlarged. Groups that regularly travel long distances in search of game will more frequently encounter opportunities to bring materials from relatively distant sources back to camp. Long-range logistical mobility may well explain the remarkable distances (sometimes exceeding 200 km) over which cores of "chocolate flint" were transported on the Polish Plains during the late Upper Paleolithic (see Schild 1987).

The predictability, as well as the duration, of occupations can have a pronounced effect on how people keep themselves supplied. The extent to which long-term occupations are planned, as opposed to chance events, will strongly influence the size of the catchment for materials used to provision places. Tool makers can plan to bring needed materials to a locus of long-term occupation only if the location is known in advance. If the decision to remain in one place is made only after arriving there—in reaction to an unforeseen overabundance of resources, for example—then groups are unlikely to arrive already "geared up" for a prolonged stay, and they will have to draw more on whatever technological materials they find in the immediate area. Conversely, if occupations are short but quite predictable, then it may be advantageous to collect and cache materials there in anticipation of future visits.

Parenthetically, this relationship between length of stay, foraging radii, and provisioning of places will also influence what archaeologists consider to be "local" raw material. There is little consensus regarding the significance of the distances stone tools found in archaeological contexts were moved. Instead, raw materials are usually termed local if there is evidence that they were worked in place, and that tools were seldom retouched and resharpened prior to discard. In other words, anything treated in a casual fashion, including raw materials used to provision places, may appear local by these criteria, regardless of where it comes from.

General expectations about relationships between land use and settlement patterns, provisioning strategies, and the scale of transport of raw materials are summarized in the graphs below. It is assumed that raw materials are not ubiquitous, and that higher-quality raw materials have a relatively limited, discrete distribution, while lower-quality materials are more common. This adequately characterizes many raw material environments, including the study area, coastal west-central Italy. In places where high-quality materials are more evenly distributed—the Dordogne valley in France and the Negev desert of Israel come to mind—the spatial scales of the various relationships would differ, but their basic structure should stay the same.

Figure 2.1 graphically displays the expected relationship between the complementary strategies of provisioning places and provisioning individuals. As the frequency of residential mobility increases, the duration and/or predictability of occupations decreases. In order to cope with potential needs for tools, tool users would have to carry an increasingly large proportion of their toolkit with them.

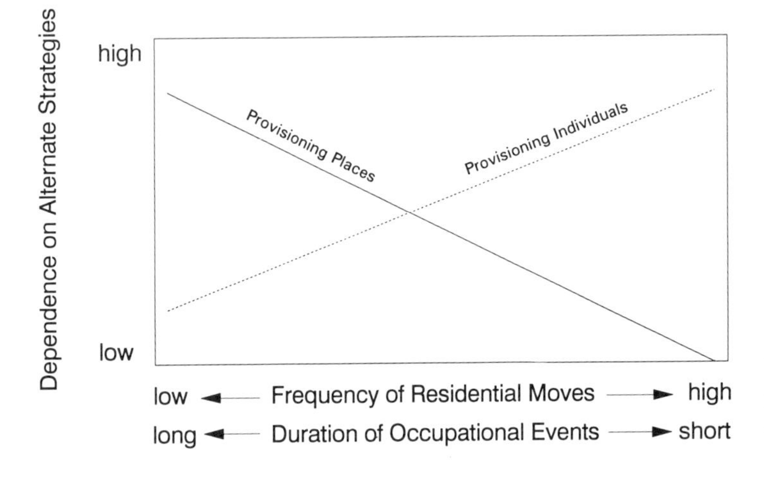

Figure 2.1. The expected relationship between mobility and provisioning strategies.

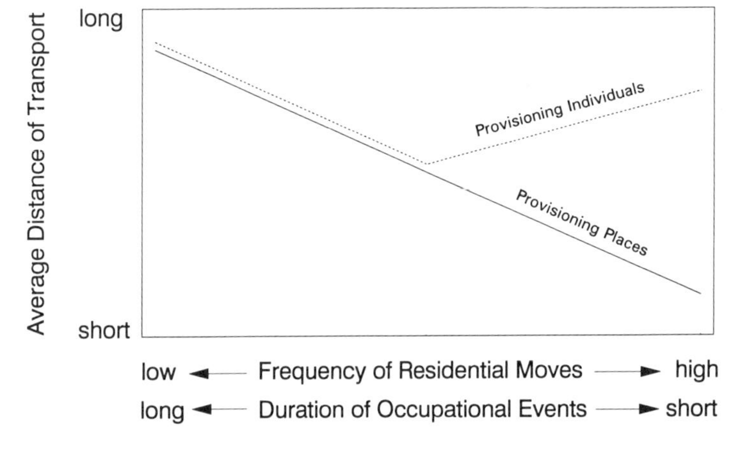

Figure 2.2. The expected relationship between residential mobility and transport of artifacts.

Moving in the opposite direction on the *x* axis, greater residential stability favors supplying places against potential needs for tools. It is worth noting that extremely frequent residential moves or very short occupations would render the scheme of provisioning places completely impractical. In contrast, because people continue to move around even when their living sites are fixed, some provisioning of individuals is expected to occur no matter how stable residential patterns become.

Expectations about the scale of raw material transport in different provisioning strategies are summarized in Figure 2.2. Decreasing residential mobility is generally accompanied by either an increased range of logistical mobility or expansion of the resource base, or both (Binford 1980; Henry 1983; Kelly 1983, 1992; Tchernov 1992). The distances materials are moved *to* places should increase as a function of both the scale of logistical mobility and the duration of use of residential locations. Longer foraging trips afford the opportunity to exploit more distant sources of raw material, and more stable residential patterns make it more profitable to collect technological goods at frequently visited places. The average transport distance for artifacts used to provision individuals (broken line) *may* increase with the frequency of residential mobility, depending on the size of territories and on patterns of seasonal movement. In principle, however, the number of residential moves occurring during the lifetime of a piece of transported personal gear should have the strongest influence on the distance it "travels" from the original point of raw material procurement or manufacture. In addition, such artifacts are unlikely to show less transport than the materials used to provision residential

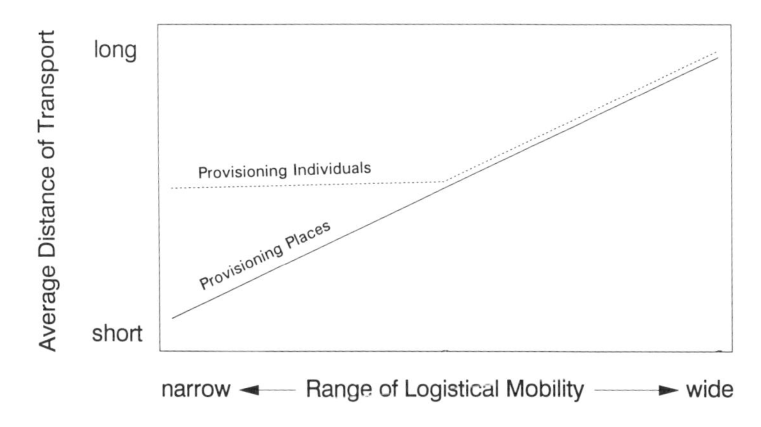

Figure 2.3. The expected relationship between logistical mobility and transport of artifacts.

occupations, since they would normally be manufactured out of goods stockpiled at residential sites.

Figure 2.3 shows an essentially complementary relationship to the previous graph, with the x axis representing the scale of logistical mobility, or the distances people regularly travel out from residential localities to collect resources. Relationships with the two y variables, distance of transport of personal gear (broken line) and scale of provisioning places (solid line), are somewhat different in this case. The catchment area from which places are provisioned may increase as a direct function of the scale of the logistical forays in which raw material procurement is embedded. Relatively large foraging and/or logistical territories ensure that materials can be brought back to camp from more distant sources, and from a greater variety of sources. The transport and disposal of gear used and regularly carried around by individuals are largely independent of this kind of mobility, although once again "personal gear" is unlikely to exhibit less extensive transport than the tools and raw materials brought to places.

It cannot be overly stressed that the importance of one provisioning strategy versus another will not be directly reflected in the absolute quantities of different kinds of debris in archaeological assemblages. Items recognizable as having been part of curated personal gear accumulate at a different rate than things produced, used, and discarded in place, whether the latter are made of local materials or of materials brought in. Because of the need for portability, the strategy of provisioning mobile individuals is expected to produce toolkits that are limited in both size and variety, and individual artifacts should be heavily maintained and reworked. Comparatively little archaeological debris will be generated as a result of this strategy. Both provisioning of places and the expedient use of local raw materials should result in larger quantities of manufacture debris and lightly used tools, which will numerically overwhelm evidence for provisioning of individuals as the duration of site use increases. In other words, one cannot monitor reliance on alternative strategies of technological provisioning in an isolated case simply by counting heavily maintained versus lightly used tools, or local versus "exotic" materials. Only broad inter-assemblage comparisons can provide estimates of the relative importance of different approaches to ensuring the "delivery" of technological support.

Some archaeological cases

Since so few living observers have actually witnessed mobile peoples dependent on a stone technology, ideas about technological provisioning can only be compared with archaeological case studies. Archaeological studies of Holocene and late Pleistocene hunter-gatherers in both the New and Old Worlds contain a number of examples of variation in technological provisioning strategies associated with broad variation in land use and subsistence patterns.

A particularly good illustration of differences in scales and patterns of raw material procurement and provisioning relating to the duration of site occupation comes from comparing patterns of raw material exploitation at two Holocene sites in arid regions, Puntutjarpa cave in Australia (Gould 1977) and Nightfire Island in eastern California (Sampson 1985). Both sites contain long occupational sequences, and in each case the raw materials preferred for toolmaking are found some distance from the site. However, the two localities were used by prehistoric foragers in very different ways, with important repercussions for the manufacture and treatment of stone tools.

Toolkits and raw material use patterns change little during the long occupational sequence at Puntutjarpa (Gould 1977:176). Overall patterns suggest a way of life similar to that of modern Australian desert aborigines, characterized by frequent residential mobility (*op. cit.*:179–80), and the shelter would probably have served as a short-term residential base. Quarries for white chert, preferred for making many types of tools, are located at a distance of between 22 and 32 km from the shelter. Materials classed as "exotic" probably come from locations found no closer than 40 km from the site, while quartz, quartzite, and agate are found quite close at hand.

Between 71.4% and 81.4% of large cores in the various levels at Puntutjarpa are made on *local* quartz, quartzite, and agate, while only between 3.6% and 12% are made on white chert and "exotic" materials; yet the latter were preferred for most tool manufacture. Samples of debris and unretouched flakes are overwhelmingly composed of local materials as well. Slightly more than 20% of micro-cores are made on white chert or exotic stone. In marked contrast, 52.9% of backed blades and 86.5% of hafted adzes are of white chert or "exotic" stone (Gould 1977:110–17). In concordance with expectations for comparatively high residential mobility, it would appear that Puntutjarpa was infrequently provisioned with anything but the most easily accessible raw materials for making tools on the spot. In contrast, *retouched tools* used in manufacture and processing (adzes) were quite often transported at least moderate distances and were heavily renewed and resharpened in the process, suggesting that they arrived at the site as components of transported toolkits. Since adzes are used in activities involving very high attrition (Bronstein 1977), it is not expected that they would ever travel very far before wearing out. The higher frequencies of white chert and "exotic" materials among the so-called micro-cores may indicate that people did carry at least some "toolmaking potential" as part of personal gear, a practice observed in western Australia (Grey 1841).

Patterns of raw material exploitation and tool use at Nightfire Island differ markedly. Although the site was used in a variety of ways, it functioned as a winter village or fishing camp through much of its occupational history (Sampson 1985:510). The data clearly indicate that the Nightfire Island site was frequently provisioned with obsidian raw material while it was being used as a stable residential location. In those levels corresponding to use as a long-term fishing camp, 71% of the cores are made of obsidian (*op. cit.*:272), derived from sources located between 37 and 193 kilometers from the site (Hughes 1985:256–57). The cores are often heavily reduced by bipolar technique, but it is clear from the debris that they were introduced as partially worked nodules (Sampson 1985:269–84). In other words, the cores and debris at Nightfire Island are derived from sources as far away from the site as most of the retouched tools at Puntutjarpa—or farther. It is interesting to note that bifacial thinning flakes (from transported bifacial cores or tools?) also decline in frequency after the site became a more stable residential base (*op. cit.*:293). Few data are presented on chipped stone manufacture and processing technology, which may well represent a dearth of formal tools in these categories, the product of expedient tool manufacture and use using obsidian cores stockpiled at the site.

Another interesting case of differential provisioning related to variation in the nature and duration of occupations involves two pleni-contemporaneous Magdalenian sites in the Bruder valley of Germany. Petersfels is a large, well-known Magdalenian habitation site associated with an extensive kill and butchery area (Albrecht 1979). Gnirshöhle, a small cave located nearby, contains more limited archaeological deposits and lacks evidence of an animal-processing area (Albrecht and Berke 1988). These two sites illustrate how debris from transported toolkits can constitute a substantial portion of assemblages in the context of very brief occupations, while material brought in to provision more prolonged stays quickly overwhelms the few "curated" artifacts present. Albrecht and Berke argue that a comparatively large group of hunters repeatedly occupied Petersfels, probably during the fall, using the site as the focus of intensive reindeer hunting and processing activities, perhaps associated with winter storage. The smaller Gnirshöhle cave was apparently the site of temporary, transitory occupations by a smaller group, perhaps during a different season. Evidence for technological provisioning corresponds well with presumed contrasts in the duration of occupations. Petersfels yielded hundreds of thousands of artifacts, almost exclusively (>95%) of local raw materials (Albrecht and Berke 1988:167–68). Apparently, hunters kept the camp supplied with flint from the surrounding river valley, engaging in extensive manufacture activities and producing vast amounts of debris. In contrast, the Gnirshöhle assemblage contains much less manufacture debris and furnishes a clearer picture of transported toolkits. The artifacts were apparently brought in as ready-made tools or at least blade blanks, and a large proportion of the raw material comes from outside the Bruder valley, some from as far away as the Paris Basin (*op. cit.*:170).

Methodological Implications

The studies of Middle Paleolithic stone tool assemblages presented in this book center on analyses of flake production technology, stone tool resharpening and reduction, and artifact transport, three of the most prominent themes in current research on Paleolithic technologies. As interesting and attractive as they are, these approaches to lithic studies are simply means to an end. In studying the ways cores were exploited, the progressive use and renewal of implements, or the origins of lithic raw materials, we are almost certainly on firmer inferential ground than typological studies founded on largely unquestioned assumptions linking tool forms and prehistoric design conventions. Ultimately, however, we want to do more than just catalog what people did and how they did it. The relevance of the methods and the data they produce must be justified in terms of the theoretical considerations outlined in the preceding sections.

Because the location and timing of requirements for technological aids, as well as the conditions needed for manufacture (free time and raw material), are determined in large part by the organization of foraging and land use, technological provisioning strategies serve as a kind of link between subsistence, land use, and technology. The aspects of lithic technology most sensitive to variation in provisioning strategies, and thus most pertinent to the relationship between technology, foraging, and land use, are those tactics that affect the immediate and potential *utility* of technological materials. For this reason, exploring links between technology and subsistence adaptations leads naturally to a consideration of lithic raw material economies, strategies for managing the consumption of raw materials. This is where core reduction technology, stone tool resharpening, and transport have the most to offer. All three phenomena can be integral to the management of raw material use, and to coping with the tactical demands of maintaining a supply of usable implements. Specific methodological considerations and expectations for links with alternative provisioning strategies are discussed below.

Core technology

European and American researchers have come to view methods of core reduction or flake production as central to a variety of issues in Paleolithic studies. For some, the subtle details of *chaînes opératoires* are most useful for establishing the "technological traditions" of tool makers (Bar-Yosef 1991; Bar-Yosef et al. 1992; Crew 1975; Van Peer 1991), or even the social relations among tool makers within a group (e.g., Pigeot et al. 1991; Ploux 1991). Other researchers are more interested in the economic or functional significance of different methods for producing flakes and tool blanks (e.g., Delagnes 1991; Henry 1989; Marks 1988a; Perlès 1991).

Both points of view have their merits. People manipulate stone in order to fill immediate and anticipated needs for tools. Insofar as their techniques are not constrained by the simple mechanics of stone fracture, the choice of methods must be determined by both the direct and the inherited (i.e., culturally transmitted) experience of tool makers. The fact that modern tool makers have succeeded over a generation or two in replicating virtually every stone-working technique devised over two million years of human prehistory demonstrates that we cannot place too much weight on cultural "inertia" as a force limiting technological options, at least on an evolutionary time scale. Hominids, particularly large-brained members of the genus *Homo*, were probably capable of developing new techniques over the course of a few hundred years, given sufficient pressures to do so. On the other hand, while every tool maker is to some extent an experimentalist, each individual also relies to a large extent on received knowledge about the "best" or most appropriate ways of doing things, and it cannot be assumed that members of prehistoric groups had unlimited arrays of technological options from which to choose. Ultimately, the factors deemed most important—shared technological traditions or functional/strategic considerations—depend on the problems of greatest interest. In this study I am concerned with integrating the study of Mousterian technology into a more general ecological and evolutionary perspective on human behavior during the Upper Pleistocene. For this reason, the economic and functional properties of core technologies are of primary concern.

The manner in which cores are shaped and the ways in which flakes are detached from them afford tool makers control over two factors. First, the method of core reduction determines the functional characteristics of blanks. The shapes of flakes, the lengths and forms of their edges, and the quantity and location of cortex are all intimately connected with how cores are prepared. Second, alternative methods of core reduction may vary in terms of their gross productivity—the number of usable flakes or tool blanks that can be extracted from a given mass of raw materials. There is an obvious trade-off between the number of flakes produced and their sizes: a core will yield either many small flakes or fewer large ones. Balancing the quantities, shapes, and sizes of blanks should be especially crucial where raw materials are small or difficult to obtain.

There is no single cost-benefit equation that guides all human behavior, no universal measure of absolute efficiency or effectiveness, and changes in technology over the course of human evolution did not occur because some techniques are essentially better, more efficient, or less costly than others (see also Perlès 1992:243). Although one can argue that efficiency is a "goal" of all living systems, the more important and interesting question remains efficiency to what end—in the optimization of which currencies? An elaborate method of core preparation that yields a few large, regularly shaped flakes per core may not be the most effective means of maximizing the number of sharp edges obtained per unit of stone. On the other hand, it would be uniquely suited to the production of tools needed for prolonged, heavy-duty use, where large size provided a distinct advantage. In the long run, some of the most interesting aspects of studying core technology will come from what the adoption of one method and the rejection of another can tell us about the functional or economic currencies that guided the choices of prehistoric tool makers.

Two complementary aspects of core technology—economy of production and the sizes and shapes of flake/blanks—are particularly relevant to the investigation of technological provisioning. One way of managing the consumption of raw material, whether in response to general scarcity or to high residential mobility, is to adopt techniques that maximize the number of flakes produced per unit of raw material. However, cores contain a large proportion of wasted material, and they are not generally expected to play a large role in transported toolkits, except when they also serve as heavy-duty tools (Kuhn 1994). Instead, provisioning of individuals should involve primarily retouched tools and flakes, which provide the best ratio of utility to unit weight. Because each flake tool contains an unusable portion, a hypothetical "slug" which cannot be further used or resharpened, it is not very efficient to transport very small artifacts. In theory, the optimal ratios of potential utility to unit weight are obtained with implements 1.5 to 3 times the minimum usable size (Kuhn 1994). Normally, this means that people would tend to produce and carry fairly small items in the context of provisioning individuals with mobile gear. However, where artifact sizes are limited by raw materials, as in the case of the Pontinian Mousterian, we might expect to see efforts to obtain larger-than-normal blanks, approaching the optimal size ranges, in the context of producing artifacts used to provision individuals.

In contrast, if artifacts are produced and used on the spot, with less possibility of prolonged use, neither durability nor utility/mass ratios of tools are overriding considerations. If the frequency of residential mobility is comparatively low, or if a particular occupational event is sufficiently prolonged to allow people to collect and stockpile tools and/or raw materials, there is less payoff to producing artifacts of the optimal sizes. "Expedient" or informal core reduction technology is frequently associated with sedentary communities (Parry and Kelly 1987), apparently because raw materials can be amassed and stored at sites in sufficient quantity to override any other economic factors. However, completely haphazard core technology is unlikely to occur in places like the study area, where raw materials are both small and relatively scarce. In such contexts, we might instead expect the focus of economizing behavior to shift from tools to cores as the duration of occupations increased. A greater emphasis on provisioning places might therefore favor core technologies that maximize the number of flakes (or edges) per core, at the expense of flake size.

Stone tool reduction and resharpening

The archaeological record is made up of discarded or abandoned artifacts. In many cases, these artifacts were modified, broken, repaired, and renewed before reaching their final resting places. A partial record of these events is recorded in the forms of the tools we dig up. Evidence of artifact "use histories" reflects past decisions about the immediate and future utility of tools. As such, the use histories of artifacts are linked indirectly to both planning and technological provisioning. Evidence for the reduction of artifacts reflects a simple behavioral phenomenon, however, not an organizational one. The fact that a scraper is intensively used and renewed does not imply that it was "curated" in any meaningful sense. Variation in the extent of artifact reduction may also reflect factors such as activity duration or even rates of "scavenging" and reuse of previously abandoned material. Inferences about technological strategies must be based on an understanding of the contexts in which artifacts were used and modified, and observations about stone tool reduction are most useful when studied in terms of co-variation with other variables. Relationships between extent of reduction and evidence of transport and/or settlement duration are especially pertinent to arguments about planning or technological provisioning.

Renewing the edge on a scraper or resharpening a hafted biface is but one possible tactic for producing a fresh working edge. One can also make new tools instead of reworking old ones. Neither resharpening nor replacing artifacts is inherently more "advanced" or more "economical." The advantages and disadvantages of different tactics for producing new edges can be exploited to serve different ends in different contexts. One strong point of resharpening things is that it minimizes the transport costs and raw material expenditures associated with making usable edges available when needs arise. The disadvantage of repeatedly reworking implements stems from the fact that the functional properties and effectiveness of tools change every time they are modified. With every resharpening the tool becomes smaller and the morphology of the edge changes; just as important, the potential for future renewal diminishes. Thus, heavy resharpening extends the *immediate* utility of tools at the expense of their functionality and future utility. The strategy of making new artifacts every time a fresh edge is needed involves a complementary set of advantages and drawbacks. Producing fresh tools allows one to maintain optimal functional properties in both whole tools and edges. It may also take more time than resharpening (Perlés 1992:239), although it is debatable whether or not the time differential would be significant with most common Middle Paleolithic artifacts. More important, constantly producing new tools as needs for them arise consumes larger quantities of raw materials. The strategy of replacement (as opposed to renewal) consequently requires that raw material be constantly available and not especially "costly."

Because reworking and reusing old tools consumes raw material at a relatively slow pace, we would expect artifacts to be more frequently and extensively resharpened where workable stone was comparatively hard to come by, other things being equal. However, extensive resharpening should also be characteristic of the strategy of provisioning of mobile individuals, since it is difficult to carry around large quantities of unmodified stone. In contrast, the strategy of provisioning regularly used places with raw material creates an artificial local abundance, making it practical to make new tools as needs arise and negating the benefits of extensively resharpening them. If workable stone is *universally* either difficult or easy to obtain, evidence of different provisioning strategies may well nullify any effects of differential access to raw materials, so that the extent to which artifacts were consumed depends more on the duration of occupations or regional patterns of land use than on immediate "cost" of flint. Conversely, pronounced inter-site variation in raw material availability may nullify the influence of alternative provisioning strategies. Determining whether the intensity of artifact reduction in a particular case is due to the natural scarcity of stone or to the strategies tool makers used to keep themselves supplied with tools is an analytical challenge. Sorting out these various effects requires most of all understanding the relationship between reduction intensity, raw material availability, artifact transport, patterns of mobility, and the nature of occupations (Kuhn 1991b).

Artifact transport

Evidence for the movement of tools and raw materials has been used both to reconstruct the territories used by Mousterian hominids (Geneste 1986; Wengler 1990; Wilson 1988) and as the basis for arguments about planning or anticipatory organization (Binford 1989; Kuhn 1992b; Roebroeks et al. 1988). Because of the difficulty in determining the precise origins of individual specimens of flint within the study area, it is very difficult to draw any reliable conclusions about the sizes and shapes of prehistoric territories based on raw materials. To the extent that they can be reconstructed, however, artifact transport patterns can be brought to bear on questions about planning and provisioning strategies.

Despite the difficulties in ascertaining where exactly raw materials came from, coastal Latium is a particularly interesting context in which to investigate artifact transport and raw material procurement. Locally available stone is limited to very small, and not especially abundant, flint pebbles, and there would have been both functional and strategic advantages to carrying in tools and cores from more distant sources of larger raw materials. Larger tools would probably have been more effective, and would certainly have been longer-lasting, than artifacts made from the diminutive local pebbles. Likewise, cores made from larger raw materials would have provided greater flexibility in the number, form, and size of blanks produced from them. Of course, transporting large artifacts from distant sources would have definite costs as well.

Like core technology or stone tool resharpening, transport is a problem-solving strategy. It is a tactic for making implements available in places where it would not otherwise be practical or convenient to produce them. Although evidence that artifacts were moved across the landscape does imply that prehistoric populations may have anticipated their needs to some extent, understanding the scale and nature of planning requires a more sophisticated approach. Observations about artifact transport are necessarily *post hoc*: they reflect what happened to an artifact, but not necessarily what the tool maker intended. Moreover, things can be carried around the landscape for a multitude of reasons. The distinction between provisioning places and provisioning individuals—the two major strategies of coping with future needs—is a somewhat arbitrary one, in the sense that a given implement or class of implements may be used in multiple strategic contexts. From an archaeological perspective, however, the conceptual distinction is useful as a means of understanding the different pathways by which artifacts can be moved around and eventually deposited at particular locations.

Taken in isolation, the observation that an object was moved from one point on the landscape to another shows only that tool users failed to abandon artifacts during the time it took to make the trip. We cannot assume that the actual location where the tool was used and/or abandoned was planned in advance. Chimpanzees sometimes carry wooden probes a kilometer or more as they move from one termite nest to another (Goodall 1964). It is easy to imagine how a tool-using hominid foraging among more widely scattered resource patches might transport artifacts much greater distances than that, all the while planning no farther ahead than the next foraging event. The fact that artifacts were moved a few kilometers, cited as evidence of a kind of planning among *Homo habilis* (e.g., Potts 1988; Toth 1985), could easily represent a chimpanzee-like level of anticipation (Wynn and McGrew 1989) in the context of exploiting highly dispersed resources such as scavengeable carcasses or vulnerable game. Similarly, the fact that unmodified stones were repeatedly abandoned in the same places does not necessitate the existence of a sophisticated strategic organization as Potts (1988) has argued. Over time, manuports could pile up in "magnet" locations simply as a by-product of other activities that brought hominids to those locations (Schick 1987; Sept 1992:188).

There is ample reason to believe that Mousterian hominids regularly anticipated needs over time periods well in excess of a single foraging event or day's foraging. Middle Paleolithic assemblages frequently contain artifacts and raw materials originating 50 km or more from the find spot, well outside any rational estimate of a day's travel (e.g., Roebroeks et al. 1988; Simán 1991; Wengler 1990). Even evidence that artifacts were transported farther than might be expected for a single day's travel, however, cannot prove the existence of highly forward-looking, planned activities. By themselves, such observations demonstrate only that tool users failed to abandon the implements they carried from one day to the next. This is not to say that Middle Paleolithic hominids never thought farther ahead than the next morning. These same patterns *could* reflect more forward-looking strategies, but other information is needed to determine the nature of planned behavior. Distinguishing the different strategies for filling future needs and sorting out their effects in the archaeological record require that we focus on the nature of the items transported and how they were treated in transit, not just how far they were carried. In other words, we need to examine the relationships between artifact form, reduction or resharpening, and transport. If the exotic artifacts present in an assemblage consist exclusively of highly portable items (i.e., tools and small cores) that were extensively reduced prior to arriving at a particular location, it is reasonable to infer that their movement occurred as a by-product of the normal treatment of "curated" gear. If bulkier items like raw stone, large cores, or unworked tool blanks were moved around, or if artifacts were not very extensively used and renewed (e.g., Henry 1989; Morrow and Jefferies 1989), then the evidence may reflect planned provisioning of places.

The significance of the scale of artifact movement in turn depends on the kinds of provisioning strategies represented. The distances artifacts were *carried to places* to provision occupations should vary more or less directly with scales of logistical mobility, the extent to which long-term occupations are anticipated, and the absolute sizes of territories. If people plan well in advance for a long stay, they may be able to cope with local shortages by bringing extra, high-quality raw materials from other parts of their territories. In contrast, the displacement of gear habitually carried around by mobile individuals has very little to do with "planning depth" (*sensu* Binford 1989). Artifacts used to provision individuals end up in archaeological sites as a result of being *carried along* and incidentally lost or abandoned there. The apparent distance this kind of "personal gear" was transported varies as a function of artifact turnover rates or use lives, the ranging patterns of tool users, and the distances between differentiable sources of raw material, rather than the scope of advanced planning. One can easily imagine how items of personal gear might travel great distances with their owner(s), yet end up being replaced and abandoned at the camp where they were originally produced.

In the subject assemblages, it is at best possible to assess the relative abundance of artifacts that are not likely to be of local origin. Based on the model outlined in this chapter, *frequencies* of exotic artifacts in an assemblage should depend in large part on the nature and duration of site occupations. Artifacts transported in the context of provisioning individuals with tools will be most abundant in assemblages resulting from a series of relatively ephemeral, short-term occupations, where they are less likely to be swamped by locally produced debris. Assemblages dominated by artifacts used to provision individuals should exhibit other characteristics reminiscent of mobile toolkits, including evidence of extensive maintenance (resharpening). More substantial, prolonged occupations should be accompanied by increased evidence for the provisioning of places. If the stay were anticipated well in advance, or if people regularly traveled long distances on logistical forays from the campsite in question, then exotic materials may be quite common and might include cores, unmodified flakes, and similar materials. However, if a prolonged stay was not anticipated, or if the normal range of logistical mobility did not encompass more distant sources of raw materials, the site would be provisioned from materials available closer at hand, and any transported gear would quickly be overwhelmed by debris generated from working local stone.

Summary

Any species that relies extensively on technology for its survival faces chronic problems in maintaining a ready supply of tools and raw materials. Tool makers will take advantage of opportunities to exploit raw materials that happen to be available near at hand. However, neither raw materials nor time for toolmaking are universally plentiful, so modern humans must employ a variety of strategies for ensuring that they have the technological support they require. I have identified two alternative responses to the problem of "delivering" technology. One is to keep individuals supplied with a portable toolkit; the other is to provision with tools and raw materials the places where technological activities are performed. The suitability of the alternative strategies of provisioning depends in large part on patterns of mobility and the duration of occupations. If people change residential locations frequently and rarely stay in one place for more than a few days, they must depend to a large degree on the limited array of things they are able to carry with them. More stable, long-term occupations and more predictable reuse of specific locations enable tool makers to amass artifacts and materials at particular places in anticipation of the needs that will inevitably arise. All foraging groups practice a mixture of these two technological provisioning strategies, depending both on overall frequencies of residential movement and on the nature of occupations at different times of the year and in different parts of their territories.

Alternative means of maintaining a ready supply of tools and raw materials demand disparate technological adjustments on the part of tool makers and tool users. Provisioning mobile individuals places a premium on portability, on maximizing the utility of artifacts per unit weight. The best utility/weight ratios are achieved by transporting retouched tools: to the extent that cores play a role in provisioning individuals, they should be exploited using highly efficient or productive techniques. Since opportunities to replace them may be sporadic and unpredictable, transported tools will generally be heavily reworked, and may be designed for durability as well.

The practice of amassing artifacts or artifact-making potential at frequently used places relaxes the constraints associated with continuous transport and the provisioning of mobile individuals. The diversity of activities and tool users may be high in prolonged occupations, and it is more practical to move cores and raw materials to places than it is to anticipate all future needs with appropriate finished implements. If places are provisioned in the course of normal foraging activities, substantial quantities of technological materials can be accumulated with relatively little expenditure of time and energy. This relaxes the need to repeatedly resharpen or rework tools, which has detrimental consequences for their functionality. Provisioning of places may consequently serve to counteract the effects of local raw material scarcity, facilitating apparently casual patterns of tool use even where workable stone is otherwise comparatively rare.

Archaeologists do not study "systems" or "strategies," but rather accumulations of objects at places on the landscape. Because it is so extensively conserved, "personal gear" used to provision mobile individuals is not expected to predominate in most archaeological assemblages. However, such artifacts should be most abundant in collections representing short-term occupations, and may even predominate in the case of extremely brief ephemeral occupations. Holding the overall availability of raw materials constant, assemblages containing a large proportion of "fallout" from mobile toolkits should exhibit high frequencies of retouched tools and comparatively little unmodified material, and artifacts should be extensively modified and resharpened. They may also contain comparatively large quantities of exotic raw materials, primarily in the form of extensively modified retouched tools. In contrast, artifacts and debris generated through provisioning of places should predominate in assemblages representing more prolonged occupational events or regular, predictable reuse of places. Because provisioning of places can help to counteract any immediate scarcity of workable stone, unmodified flakes and debris and minimally modified retouched pieces should be relatively common. If there is any response to the cost of procuring workable stone, it may be seen in how flakes and blanks were produced, since core technology probably has fewer negative effects on tool function than does repeated reworking or resharpening. In the context of provisioning of places, the frequency of exotic materials will depend on how far in advance people planned to come to a particular location, and how far out from it they regularly went in pursuit of food or other resources. If people stay a long time at a place and make extended trips out from it, they are more likely to encounter opportunities to bring back materials from distant sources. If the visit is not extended, or if there was no advance warning of an extended occupation, technological materials may only be collected from within an abbreviated foraging radius.

The model and expectations outlined in this chapter are highly abstract and general. Although they seem to be consistent with some aspects of the recent archaeological record, it remains to be seen whether or not they can also provide insight into the technological behavior of Mousterian hominids. The following chapter sets the environmental stage for a critical application of the model of provisioning strategies to investigating possible relationships between technology, foraging, and land use in the Mousterian of west-central Italy.

3

ARCHAEOLOGY AND ENVIRONMENT IN WEST-CENTRAL ITALY: BACKGROUND TO RESEARCH

The principal data base for this study consists of Paleolithic stone tool assemblages from a series of cave and open-air sites situated along the Tyrrhenian coast of Italy, not far from the city of Rome (Figure 3.1). Collections from four Middle Paleolithic cave sites furnished the largest part of the data base. Two of the cave sites, Grotta Guattari (Taschini 1979) and Grotta Breuil (Bietti et al. 1988a, 1990–91; Taschini 1970), are located on Monte Circeo, an isolated limestone promontory situated about 100 km southeast of Rome (Figure 3.2). The other two caves, Grotta dei Moscerini (Vitagliano 1984) and Grotta di Sant'Agostino (Laj-Pannocchia 1950; Tozzi 1970), are located about 45–50 km south of Monte Circeo, a short distance north of the town of Gaeta (Fig. 3.1). Materials from open-air sites consist of a sample of artifacts collected during a systematic survey of the coastal plain surrounding Monte Circeo (Voorrips et al. 1991), and less systematic studies of open-air localities at Tor Caldara and Nettuno, near Anzio, and Canale Mussolini/Canale delle Acque Alte, northwest of Monte Circeo (Fig. 3.1). Upper Paleolithic and Epipaleolithic assemblages from Le Grottacce, the Cavernette Falische, Grotta Jolanda, Riparo Salvini (Fig. 3.1), and Grotta del Fossellone (Fig. 3.2) also enter into the discussion at various points.

Physiography and Geology

The study area is geologically and topographically diverse. Low-lying coastal plains, principally the Agro Pontino or Pontine Plain and the Fondi Basin, are bounded on the east by a series of calcareous mountain ranges (the Monti Lepini and Monti Ausoni) belonging to the southern Apennines. To the north lie more recent volcanic mountains such as the Colli Albani, with their characteristic caldera lakes. The so-called Latial volcanoes, most active during the Middle and Upper Pleistocene (Parotto and Praturlon 1975), are responsible for the thick deposits of tuff that cover the hills north of the Pontine Plain. The Pontine Plain is actually a large graben or sunken fault block, formed by NW–SE trending faults running parallel to the mountains (Kamermans 1991), and the entire region was tectonically active during the earlier Pleistocene (Segre 1953, 1983).

The Mesozoic limestones and dolomites that form the coastal mountain ranges contain numerous caves and rockshelters (Sevink et al. 1984:7–10). Caves and shelters are most abundant, or at least most explored, in the precipitous cliffs that run along the present-day coastline, where they have been exposed and enlarged by wave action. More than 30 caves and shelters are cut into the roughly 9-km seaward face of Monte Circeo (Figure 3.3) (Blanc 1938). A survey of the coastline between Sperlonga and Gaeta conducted in the late 1940s revealed the existence of 113 additional karstic cavities, the vast majority at elevations less than 20 m above modern sea level (Blanc 1955a, 1955b; Blanc and Segre 1947). Only a small fraction of these caves and rockshelters preserve significant *in situ* Pleistocene deposits, however, as most were emptied of sediments by marine erosion during the Holocene. Nonetheless, the many karstic features of the Latium coast constitute a significant paleoanthropological resource.

The topography and geological history of the Agro Pontino are relatively well known, having

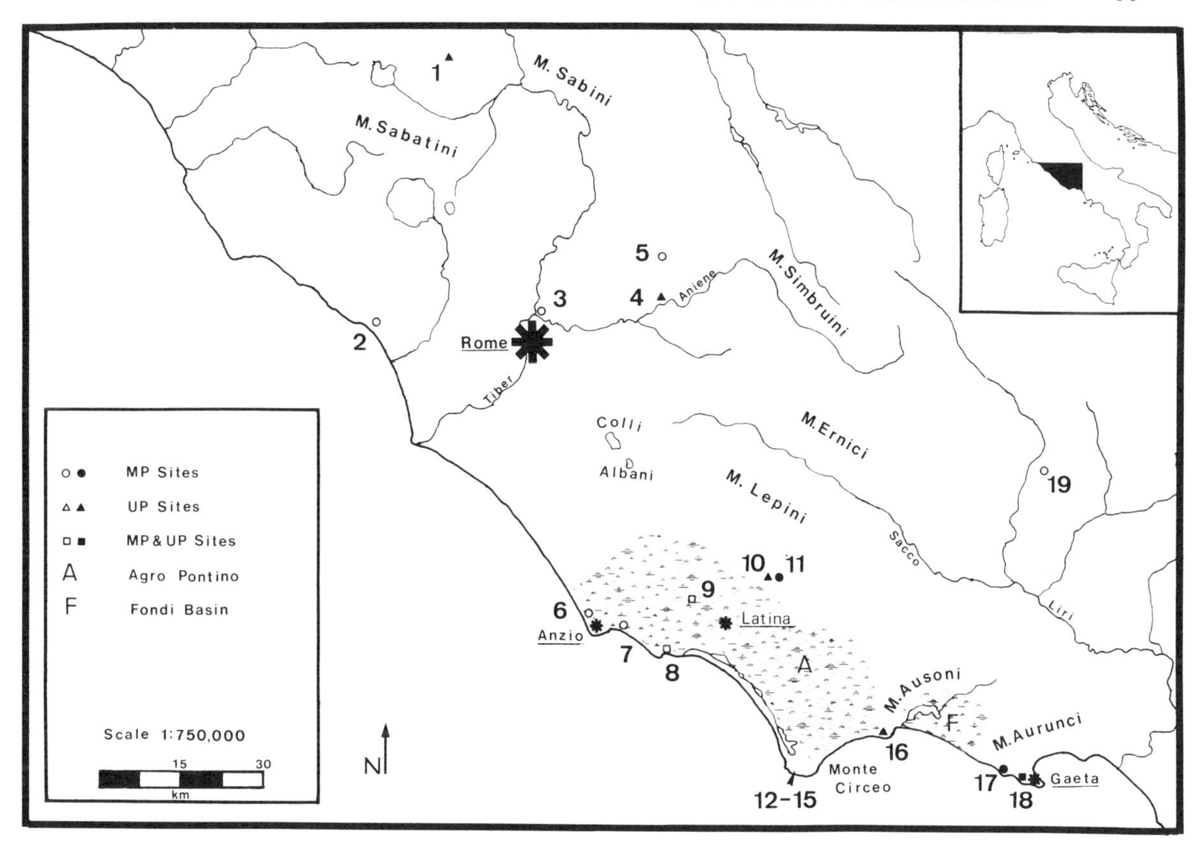

Figure 3.1. Map of study area showing sites discussed in text. 1=Cavernette Falische; 2=Torre in Pietra; 3=Saccopastore; 4=Grotta Polesini; 5=Monte Gennaro/Monte Pellechia; 6=Tor Caldara; 7=Nettuno; 8=Le Grottacce; 9=Canale Mussolini; 10=Grotta Jolanda; 11=Grotta della Cava; 12–15=Monte Circeo Caves (Grotta Breuil, Grotta Barbara, Grotta del Fossellone, Grotta Guattari: see Fig. 3.2); 16=Riparo Salvini; 17=Grotta dei Moscerini; 18=Grotta di Sant'Agostino; 19=Sora. Hollow symbols denote open-air sites; solid symbols denote caves and rockshelters.

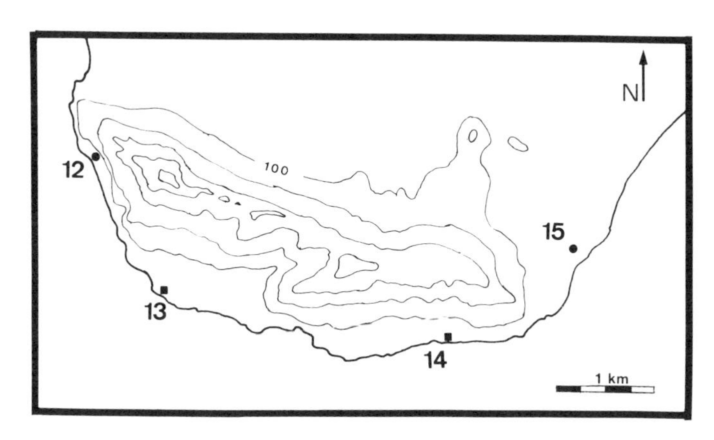

Figure 3.2. Detail map of Monte Circeo, with important Paleolithic sites. 12=Grotta Breuil; 13=Grotta Barbara; 14=Grotta del Fossellone; 15=Grotta Guattari. Contour intervals=100 m. Hollow symbols denote open-air sites; solid symbols denote caves and rockshelters.

Figure 3.3. View of Monte Circeo from the north.

been examined as part of a recent pedological study (Sevink et al. 1984, 1991). The highest elevations on the Agro Pontino are closest to the coast, where a thick series of Pleistocene marine terraces is covered in places with aeolian sands. A discontinuous cordon of low dunes, formed during the terminal Pleistocene and Holocene (Kamermans 1991:23), runs parallel to the coast just behind the present-day beach. In many areas, brackish water lagoons are found directly behind these dunes. The lowest-lying part of the graben is actually situated farther inland, at the base of the Lepini and Ausoni mountains, where it is filled with a complex series of Holocene peats and clays, overlying thick Pleistocene deposits.

The entire west-central Italian land mass has undergone massive neo-tectonic uplift throughout the Quaternary (Kamermans 1991; Sevink et al. 1984). Because of this, relative sea levels were actually higher at times during the Pleistocene than they are at present. Both the mountains and coastal plains preserve a variety of features, including terraces, wave-cut notches, and fossil beaches, left by Pleistocene high sea stands (Blanc and Segre 1953; Blanc 1957; Dai Pra et al. 1985; Sevink et al. 1982). Although complex faulting and differential uplift sometimes make it difficult to correlate these features, recent studies of amino-acid ratios in shells from fossil beach lines have permitted geologists to correlate a series of beach/terrace deposits containing a distinctive "Senegalese" (warm water) malacofauna which includes *Strombus*. Geological and radiometric (K/Ar) dating techniques indicate that these levels date to the last interglacial (oxygen-isotope stage 5e) (Hearty and Dai Pra 1986: 139). The stage 5e erosional features and beach lines are normally located at levels between 13 and 16 m above present sea level, with the variation due to localized faulting.

The oxygen-isotope stage 5e marine transgression represents the lower chronological limit on traces of human occupation in the coastal caves. A number of important sites dating to before 120,000 BP exist in the inland portions of Latium (Piperno and Segre 1982): the best-known of these include Sedia del Diavolo (Taschini 1967), Torre in Pietra (Malatesta 1978), Monte delle Gioie (Blanc 1955a), Saccopastore (Sergi 1931), and Fontana Ranuccio (Biddittu et al. 1979; Segre and Ascenzi 1984). Any significant coastal archaeological deposits predating the last interglacial were apparently destroyed by marine erosion during the high sea stands associated with isotope stage 5. Archaeological deposits at Grotta Guattari and Grotta dei Moscerini, as well as Grotta del Fossellone (Durante 1974–75), rest on so-called "Tyrrhenian" beach deposits postdating the highest sea stands. The stage 5e *Strombus* beach line is easily visible above the level of archaeological deposits at Guattari, Fossellone, and other cave sites on Monte Circeo (Durante and Settepassi 1974, 1976–77). Of the sites discussed here, only Grotta di Sant'Agostino is situated above the level of the highest (relative) stage 5e sea stands: nonetheless, deposits at the site are relatively recent (see below).

Periodic marine transgressions and regressions during the Pleistocene resulted in the formation of four major complexes of Pleistocene beach ridges on the Pontine Plain (Sevink at al. 1982, 1984). These Pleistocene soil units are exposed over large areas of the plain, except where covered by recent aeolian, colluvial, alluvial fan, and irrigation deposits (Sevink et al. 1984). The oldest terrace, with an average elevation of 25 m above current sea level, is the so-called "Latina" level, thought to date to isotope stage 15 (Sevink et al. 1984; Voorrips et al. 1991). Although exposed Pleistocene surfaces associated with the "Latina" level could in principle yield Lower Paleolithic archaeological remains, artifacts found to date belong to the Mousterian and later chrono-stratigraphic units only (Kamermans 1991:25). The next oldest ridge, the "Minturno" level, corresponds with the oxygen-isotope stage 5e high sea stand. The top of this ridge has an average elevation of about 16 m above mean sea level. The date of the succeeding beach ridge, known as the "Borgo Ermada" level, is unknown. Associated with well-developed dune and lagoonal deposits representing conditions much like those found on the coast today (Conato and Dai Pra 1980; Sevink et al. 1984), this 6-m ridge does clearly postdate the "Minturno" (stage 5e) level. Middle Paleolithic artifacts are found on some but not all of the exposed "Borgo Ermada" surfaces. It may be significant that they are most abundant in and around lagoonal deposits (Kamermans 1991:28), although it is not clear whether the association is behavioral or taphonomic. The fourth, and lowest, beach terrace, called the "Terracina" level, is of terminal Pleistocene or Holocene age, and forms part of the dune/lagoon complexes along the modern coast (Sevink et al. 1984).

Deposits of flint-rich gravels (average diameter ca. 3 cm) are exposed several meters beneath the surface of the "Minturno" beach ridges. These gravels are of indefinite, and probably varied, origins. They almost certainly represent the remains of older gravels that have been extensively reworked. Sevink (Sevink et al. 1982) believes their primary source is the Tertiary gravel deposits in the vicinity of Nettuno. Sediments from Tertiary strata near Rome, or even gravels derived directly from erosion of the Apennines and carried down coast by the sea, may also be included in them (Kamermans 1991:27). These gravel deposits are important because they were the principal, in some cases exclusive, source of raw material for Paleolithic populations in coastal Latium.

Paleoclimate and Paleogeography

Because of the ameliorating effects of the sea, climatic conditions in coastal central Italy somewhat less extreme and variable during isotope stages 5, 4, and 3 than in more continental parts of Europe. Nonetheless, general patterns of environmental change track global developments during this time period. The southern Apennines did not support large, continuous glaciers, but permanent snows and small, isolated pockets of glacial ice were present about 1750 m above sea level during the latter part of the Pleistocene (Cremaschi 1992; Federici 1980): in the absence of large glaciers, the formation and accumulation of loess was minimal (Cremaschi 1992:26). It is also clear that *Artemisia* steppe vegetation was present in peninsular Italy during colder periods.

Two long pollen cores from lake deposits supply indications of changing vegetation patterns during the Upper Pleistocene in the hills surrounding the study area. A core from Lake Vico, 50 km northwest of Rome, provides a record of vegetative changes since approximately 61,000 BP (Frank 1969). The recently drained crater lake in the Valle di Castiglione, 20 km east of Rome, furnished another pollen core 88 m in depth, extending back as far as 250,000 years (Follieri et al. 1988). Findings from the two pollen cores, summarized in Table 3.1, are quite consistent. The chronological scale shown is based on estimated rates of sedimentation, calibrated with radiocarbon dates in the most recent parts of the two sequences.

In the Lake Vico sequence, relatively dense pine and oak forest appears to have prevailed from about 61,000 (the bottom of the core) to around 55,000 years ago (pollen zone A). Between about 55,000

Table 3.1. Summary of pollen stratigraphy, Valle di Castiglione and Lake Vico cores (after Follieri et al. 1988; Frank 1969)

Estimated Age	*Lake Vico*		*Valle di Castiglione*	
	Zone	*Vegetation*	*Zone*	*Vegetation*
30 KY	D2	Gramineae, *Artemisia*, some *Pinus*, *Juniperus*	17a	Low pollen counts *Artemisia*, Gramineae, Chenopodiaceae, *Pinus*, some *Salix*, and *Quercus*
	D1	Gramineae, *Artemisia*		
40 KY	C	Gramineae, *Artemisia*, *Pinus*, some *Juniperus*, *Quercus*, and *Picea*	16b	*Artemisia*, Gramineae, Chenopodiaceae, *Pinus*
50 KY	B	*Artemisia*, Gramineae, some *Pinus*		
60 KY	A	*Pinus*, *Picea*, *Quercus*	16a	*Fagus*, *Quercus*, Gramineae, Chenopodiaceae, *Artemisia*
70 KY			15	Very low pollen counts Chenopodiaceae, *Artemisia*, *Juniperus*, *Pinus*
80 KY			14	*Quercus*, *Ulmus*, *Carpinus*, *Zelkova*
90 KY			13	Chenopodiaceae, *Pinus*, *Juniperus*
100 KY				
110 KY			12	Generally high arboreal pollen count (*Fagus*, *Ulmus*, *Quercus*, *Carpinus*, *Abies*, *Zelkova*) Brief interval with high Chenopodiaceae, *Artemisia*
120 KY				

and 46,000 years BP (zone B) the pollen record is dominated by *Artemisia* and grasses, indicating very dry, cool conditions. Arboreal pollen becomes somewhat more common in the period between 46,000 and 39,000 BP (zone C). The dominant species in pollen zone C is *Pinus*, however, which may reflect long-distance transport, while the abundance of grasses and *Artemisia* demonstrates that a more open vegetation still prevailed in the area of the lake (Frank 1969:173). The next pollen zone (D) is more than 3 m thick and probably represents an interval lasting from about 39,000 and 13,000 BP. The predominant vegetation consists of Gramineae and *Artemisia*, although pine and juniper periodically make their appearance (Frank 1969).

The Lake Castiglione pollen core preserves a record extending well before the last interglacial (Table 3.1). In this sequence, a sharp peak in arboreal pollen from a variety of deciduous trees (Zone VdC-12c) probably corresponds with isotope stage 5e. This is followed by a short interval dominated by Chenopodiaceae (zone VdC-13) in which pine and juniper are the only arboreal species, and then a return of the mixed deciduous forest conditions that characterized the previous pollen zone (VdC-14). A long series of deposits, almost devoid of pollen, dates between about 80,000 and 62,000 years ago (VdC-15): the little pollen present consists primarily of Chenopodiaceae, *Artemisia*, juniper, and pine (Follieri et al. 1988:340). The ensuing zone (VdC-16) provides evidence for a period of "weak" forest expansion, lasting until about 55,000 BP, in which arboreal pollen approaches 65% of the total count: this interval corresponds well with zone A in the Lake Vico core. Species diversity is high, with beech (*Fagus*) and deciduous oaks predominating. Just after 55,000 years ago (zone VdC-16b) most of the arboreal species drop out, leaving pine as the dominant tree species. *Artemisia* is the most common non-arboreal pollen, but grasses and Chenopodiaceae are also abundant (Follieri et al. 1988:340–41). The ensuing pollen zone (VdC-17a) is thought to span a period from roughly 38,000 to 14,000 BP. Overall, pollen is quite sparse, but the trend of declining frequencies of arboreal pollen between zones 16a and 16b continues. Peaks in frequencies of pollen from aquatic species may indicate a changing hydrologic and sedimentary environment, perhaps explaining the scarcity of pollen representing terrestrial species.

The paleoenvironmental record for the coastal lowlands is extremely sparse. The only pollen core from a low-elevation coastal area (Hunt and Eisner 1991) extends back only to about 16,000 years ago, making it too recent to be relevant to the Middle Paleolithic. Assemblages of pollen and the impressive array of macrobotanical materials from the Canale Mussolini (Canale delle Acque Alte) on the Pontine Plain are undated. These materials do suggest that relatively heavily wooded or broken forest environments prevailed in the lowland areas throughout the early part of the "Würm." The strata overlying a sandy beach layer at the site (thought to represent the stage 5e marine transgression) were assigned by the excavators to the Würm I. There appear to be two basic vegetative patterns represented by both pollen and preserved wood (Blanc et al. 1957; Taschini 1972:207). One includes white fir (*Zanichella palustris*) and *Potamogeton*, as found today in the transition between boreal and mixed temperate forest in Europe (Butzer 1971:70). The other consists mainly of hornbeam (*Carpinus betulus*) and deciduous oak (*Quercus robur*), more characteristic of temperate deciduous forests on well-drained soils (Butzer 1971:71–72). *Elephas antiquus* is associated with *both* these pollen assemblages. More cold-adapted mammoths (*Elephas trogontheri*) appear only at the top of the sequence in strata that contained no pollen but a mixture of Mousterian and Upper Paleolithic artifacts (Blanc and Segre 1953:16; Taschini 1972:208).

Although not as reliable an indicator of paleoenvironment as pollen, the species of mammals found in archaeological deposits provide some indications about the character of local ecosystems. Extreme cold-adapted taxa such as reindeer (*Rangifer tarandus*) are never found at sites in coastal Latium, or anywhere in central Italy for that matter (Sala 1990; Tagliacozzo 1992). In contrast to more northerly regions of Europe, *Equus* is relatively scarce in the Mousterian faunas from coastal west-central Italy. The general predominance of red and fallow deer (*Cervus elaphus* and *Dama dama*) and the scarcity of equids throughout the Mousterian suggest that the low-lying parts of coastal Latium were characterized by broken, forested, or parkland habitats rather than open grasslands during the first part of the Upper Pleistocene.

Ibex (*Capra ibex*) and marmot (*Marmota marmota*) are much more abundant in archaeological strata dating to after 55,000 BP (see below) (Stiner 1992, 1994; Tozzi 1970), a time period that probably corresponds to the cold, dry phase evident in the pollen record. The presence of ibex and marmot is probably more indicative of vegetation than of climatic conditions, since both species prefer steep, open, rocky slopes. Their increased abundance after 55,000 years ago is probably due to the disappearance of tree and brush cover from the precipitous talus slopes of the coastal mountains.

Major changes in sea level arguably had an effect equal to, if not greater than, either climate or vegetative shifts on the nature of environments around the caves of coastal Latium. A narrow subsurface shelf runs along much of the coastline of Tyrrhenian Italy, so that relatively minor changes in sea level, and the consequent shifting of the coastline east or west, would have had significant consequences for the topographical and ecological situations of the cave sites. Approximate distances

Table 3.2. Estimated distances between cave sites and the sea[a] (km)

	Sea Level (relative to present)					
Location	*Present*	*−10 m*	*−20 m*	*−30 m*	*−50 m*	*−100 m*
Grotta Breuil (West Circeo)	0.0	0.5	0.9	1.2	1.5	7.0
Grotta Guattari (East Circeo)	0.5	1.3	3.5	5.3	7.0	<10.0
Grotta dei Moscerini	0.0	0.5	0.8	1.3	1.7	8.5
Grotta di Sant'Agostino	1.0	1.5	1.8	2.3	2.7	8.0

[a] Data extrapolated from *Carta Geologica d'Italia* (Geological Map of Italy), Foglio 170 (Terracina, 1961) and 171 (Gaeta, 1968).

of each of the four caves from the sea according to increments of decreasing sea level are shown in Table 3.2. The figures are only relative estimates, as shifting patterns of sand transport would obviously affect the width of beaches or dunes built up along the coast. Moreover, the land was undergoing gradual uplift throughout the Pleistocene, so these estimates are, if anything, somewhat generous. As the estimates indicate, there were probably never vast coastal plains between the cave sites and the sea during the period of interest. Moreover, the most profound increase in the width of the coastal plains would occur with a drop in sea level of greater than 50 m. On the other hand, even relatively minor changes could have profound implications for how the caves and the areas around them might have been used. During the warmest parts of the Upper Pleistocene, two of the Mousterian cave sites, Grotta Breuil and Grotta dei Moscerini, were either submerged or situated directly on the sea, as they are today. This would have greatly limited their accessibility and usefulness to Mousterian foragers. Grotta Guattari was probably under water only during full interglacials, while Grotta di Sant'Agostino remained high and dry throughout the Upper Pleistocene. A drop in sea level of as little as 10 or 20 m would have placed the caves among or behind a cordon of dunes, perhaps close to lagoonal deposits like those that exist today. When seas were 50 m or more below present levels, the relative positions of the cave sites would have been much farther inland. During the coldest parts of the Upper Pleistocene each of the sites would have been separated from the sea by several kilometers of forested or parkland-covered plain. Such marked shifts in their ecological surroundings would have major impact on the foraging opportunities available in the vicinity of the various caves.

Lithic Raw Material Environment

Compared with much of the Mousterian range in Eurasia, coastal Latium is a raw material "stressed" region. The only sources of workable stone of consistent quality available in the coastal plains and the Monte Circeo area are the gravels found in fossil marine beaches associated with the "Minturno" (stage 5e) transgression. The flint pebbles are quite small. They seldom exceed 8–10 cm in maximum diameter, and the great majority are considerably smaller. Some very high-quality, fine-grained materials are found in the gravel deposits, but the overall quality of flint is highly variable. Moreover, the distribution of sources is quite discontinuous. Today, the pebble beds on the Agro Pontino are buried by late Pleistocene and Holocene sands and alluvial deposits. They are only visible in road, ditch, and quarry cuts. Pebble beaches would of course have been less deeply buried during the Pleistocene, but the density of pebbles also varies markedly from place to place within the beach terraces.

No spatial sorting of pebbles by shape has been noted to date (S. Loving, personal communication, 1992), but it does seem that the sizes of pebbles increase as one moves north and away from the coast within the Agro Pontino region (e.g., Ansuini et al. 1990–91). Pebbles collected directly from the "Minturno" level deposits are somewhat larger than those found in presumably younger sediments closer to the coast. The gravels in more recent Pleistocene and Holocene deposits may simply represent erosional remnants of the "Minturno" beds. Farther north, in the vicinity of Anzio and Nettuno, are the Tertiary gravels that may be the original source of some of the Pleistocene pebbles found on the Agro Pontino. These deposits, which are exposed along the modern coast, yield pebbles that are much larger and less internally fractured than the ones found further south.

Pebble beds are comparatively ephemeral geological entities that may be repeatedly covered with sediments and/or exposed over the course of a few millennia. As a consequence, there is no reliable means of documenting all of the potential sources that might have been accessible to Pleistocene tool makers. We do know that Grotta Guattari is located immediately adjacent to a deposit of pebbles. The isotope stage 5e terrace, which predates the archaeological deposits, is exposed immediately above the entrance to the cave (Durante and Settepassi 1976–77), and unmodified flint pebbles of various sizes are found throughout the archaeological deposits. Given the number of exposed Pleistocene beach ridges on the Agro Pontino, it is reasonable to assume that pebbles were always available within a few kilometers of the other sites on Monte Circeo, including Grotta Breuil: for example, extensive pebble beds are presently exposed about 10 m below the surface in sand quarries adjacent to the town of Sabaudia (Fig. 3.1). It is very probable that raw material would always have been somewhat more difficult to come by in the more southerly Fondi Basin, in the vicinity of Grotta dei Moscerini and Grotta di Sant'Agostino. Pleistocene gravel deposits are rare in the Fondi Basin, and those that do exist contain particularly small clasts, in the range of 10–20 mm (Sevink et al. 1984:38–39). No exposed pebble beds at all were discovered during a recent archaeological survey in the Fondi Basin (Bietti et al. 1988b). Strata 37–42 at Grotta dei Moscerini consisted of sands with some unworked flint pebbles (Kuhn 1990b; A. G. Segre, unpublished excavation notes). However, there is no information as to the sizes or abundance of these pebbles, and it is possible that they were actually brought to the site by hominids.

Nodules and tabular chunks of yellow-brown, medium-grained banded flint outcrop in lower Jurassic limestones at several locations on the eastern flank of Monte Circeo. However, the Circeo flint deposits do not appear to have been a significant source of lithic raw material at any time during the Pleistocene. This raw material has a marked tendency to fracture into angular pieces along internal fracture planes, and it is virtually impossible to remove a complete, sharp-edged flake. Assemblages from the caves of Monte Circeo often contain a few unworked or tested angular chunks of this material, but I have never seen a retouched tool made of the local flint.

There is evidence for use of raw materials other than marine pebbles during the Paleolithic in coastal Latium. Upper Paleolithic assemblages contain many artifacts with irregular, chalky cortex, demonstrating that the material was obtained directly from *in situ* sources of nodular or tabular flints. Some artifacts from Upper Paleolithic collections, and a few rare pieces from the Mousterian assemblages, preserve eroded remnants of nodular cortex, quite distinct from the smooth, finely pitted rind found on the marine pebbles. These specimens may be derived from secondary deposits of flint near the primary sources. Unfortunately, there have been no formal surveys of flint outcrops in inland Latium and Abruzzo as of this writing. One well-known source of large nodules of high-quality flint is located on Monte Genzana in the region of Abruzzo, around 100 very mountainous kilometers east of the coastal caves, but there may be other sources nearer at hand. It is widely believed that the Lepini mountains bordering the Agro Pontino contain no flint. However, Mousterian artifacts from open-air localities near Sora (Fig. 3.1) (Biddittu et al. 1967; Segre et al. 1984) are clearly not made on pebbles like those found near the coast, suggesting that there are other primary or secondary flint sources somewhere in the vicinity, probably closer than Monte Genzana.

In summary, it appears that the size and possibly the quality of raw materials increases as one moves north and east (inland) from the coastal caves. Only very small (redeposited?) pebbles are found in the most recent beach deposits along the current littoral zone. Somewhat larger pebbles can be found in the older beach lines a few kilometers farther inland.

During the Upper Pleistocene these beds were probably more extensively exposed than at present, but the gravel deposits would still have been buried in many locations. Some 40–50 km north of Monte Circeo (and almost twice that distance from Moscerini and Sant'Agostino) are the Tertiary gravel deposits around Anzio, sources of significantly larger pebbles. Finally, at more than 50–100 km distance, are the little-known primary sources of flint.

History of Archaeological Research

Finds of worked flints from the Roman *campagna* were reported as early as the mid 19th century (Ponzi 1866), but the discovery of the Saccopastore cranium in 1929 provided the first secure evidence for the presence of "Paleoanthropic" populations in the hinterlands of Rome (Sergi 1929, 1931). Systematic archaeological and geological studies of the Pleistocene of coastal Latium began in earnest in the period between the First and Second World Wars, primarily through the efforts of A. C. Blanc and colleagues from the Istituto Italiano di Paleontologia Umana (hereafter referred to as the IIPU) and the University of Rome. In 1936, this research group made an inventory of the caves and rockshelters on Monte Circeo (Blanc 1938), and initiated test excavations at Grotta del Fossellone and Grotta delle Capre (Blanc 1937a, 1938). Finds of artifacts from various open-air localities along the coast (Blanc 1937b), as well as geological, archaeological, faunal, and paleobotanical observations on Pleistocene strata uncovered during canal construction on the Agro Pontino (Blanc 1935a, 1935b, 1935c 1936), were also described.

It was a chance discovery that would bring the region to international attention. In 1939, quarry workers accidentally uncovered the entrance to Grotta Guattari and discovered the famous Neandertal skull within, galvanizing attention on the Paleolithic and particularly the Mousterian of Monte Circeo. Systematic excavation began at Grotta Guattari in the year of its discovery. Testing or more extensive excavations were initiated at other sites on Monte Circeo, including Grotta del Fossellone (Blanc and Segre 1953), soon after. In 1947–49, the IIPU undertook a systematic reconnaissance of the rocky coast between Sperlonga and Gaeta (Blanc and Segre 1947). This survey resulted in the discovery of numerous caves and rockshelters, including Grotta di Sant'Agostino and Grotta dei Moscerini. By the end of the decade both of these sites had been excavated.

During the 1960s and 1970s, the attentions of archaeologists and paleontologists working in west-central Italy shifted to older, Middle Pleistocene archaeological deposits at sites like Arche (Bidditu 1971), Fontana Ranuccio (Biddittu et al. 1979), Castel di Guido (Longo and Radmilli 1972; Longo et al. 1981), Torre in Pietra (Malatesta 1978), and Polledrara (Anzidei et al. 1989). Over the last decade, however, interest in the Middle and Upper Paleolithic has been revived. In 1983, archaeological survey teams from the Institut voor Prae- en Protohistorie of the University of Amsterdam under the direction of Dr. A. Voorrips and S. Loving began a carefully controlled surface reconnaissance of exposed Pleistocene soil surfaces on the Pontine Plain (Voorrips et al. 1991; Loving et al. 1990–91). Between 1983 and 1990, Prof. Amilcare Bietti of the University of Rome conducted excavations at of Riparo Salvini, an Epigravettian rockshelter site located in the town of Terracina (Avellino et al. 1989; Bietti 1984). In 1985, Prof. Bietti began the testing phases of a program of excavation at the site of Grotta Breuil (Bietti et al. 1988a, 1990–91). At about that time, Dr. M. Mussi, also of the University of Rome, began small-scale but carefully controlled excavations at the site of Grotta Barbara, one of the few Paleolithic cave sites on Monte Circeo not excavated during the early IIPU studies of the 1940s and 1950s (Mussi and Zampetti 1990–91).

The Pontinian and the Mousterian of Latium

While they are quite typically Mousterian in many respects, the artifact assemblages from west-central Italy also display a number of distinctive features. Early on, A. C. Blanc recognized the uniqueness of Mousterian industries of coastal Latium, naming this distinctive "littoral" Mousterian complex the *Pontinian* (Blanc 1937a:286), and the term has been retained. Soon after, excavations at

Grotta Guattari, Grotta del Fossellone and other coastal caves brought to light large assemblages which bore many similarities to Blanc's first collections from open-air localities. However, it would be more than twenty years before Dr. M. Taschini would formalize the typological and technological description of the Pontinian (Taschini 1970, 1979).

A representative sample of artifacts from several of the Pontinian Mousterian sites discussed here is shown in Figures 3.4 and 3.5. The most obvious characteristic of the industry is the small size of the flakes, cores, and retouched tools, which average about 3 cm in maximum dimension. Typologically, the collections used to define the Pontinian are characterized by an abundance of sidescrapers, which can account for 70% or more of the total tool counts (see Tozzi 1970:72; Taschini 1970, 1979). The most characteristic artifact types are simple, transverse, and pointed scrapers made on completely cortical flakes or *calotte* (e.g., Fig. 3.4, nos. 14–22). "True" (typological) Levallois flakes and/or points are comparatively scarce in Pontinian assemblages, owing in large part to the use of small-pebble raw materials. Attributes such as complex dorsal scar patterns and a cortex-free dorsal face are simply unlikely to develop when cores are very small to begin with. Although it is not particularly diagnostic, it is worth pointing out that a single irregular foliate biface was recovered from a Mousterian context at Grotta dei Moscerini (Fig. 3.5, no. 24).

It is somewhat unexpected to find that virtually all of the typical sidescraper forms found elsewhere are represented in the Pontinian—only in miniature. Unbroken, simple, transverse or *déjeté* scrapers as small as 1.5 cm long are not uncommon (e.g., Fig. 3.5, nos. 7, 8, 12, 13). Size directly effects the functionality of tools, but there is no evidence that tool forms were modified in response to the small raw materials. This observation implies that the overall form of the sidescrapers had very little to do with specific function. Instead, they are best treated as flakes with retouched edges. The number and orientation of the edges reflect variable reduction and the original shapes of the flake blanks (Dibble 1987a, 1987b; Kuhn 1992a).

Because of the strong predominance of scrapers, the Pontinian assemblages have traditionally been classed as a variant on the Quina/Charentian facies, a regional manifestation adapted to the use of small pebbles (Bietti 1980, 1982; Bordes 1984:203; Taschini 1979). There is some disagreement as to just how much of the distinctive character of the Pontinian assemblages can be directly attributed to the nature of the raw materials utilized. Some scholars (e.g., Bordes 1968:119, Piperno 1984:48; Taschini 1972:70–74) ascribe the *sizes* of the artifacts and the scarcity of Levallois technique, but not the form of the tools and the character of retouch, to the use of pebbles. Other researchers (e.g., Tozzi 1970) have argued that raw material size somehow limits options for tool manufacture to an even greater degree. The most reasonable position is that many of the distinctive characteristics that define the Pontinian simply reflect the adaptation and modification of common Middle Paleolithic stone-working techniques to an unusual and difficult raw material (see also Bietti 1990–91).

A typological breakdown of retouched tools in the assemblages used in this study appears in Tables 3.3a and b (Table 3.4 contains the definitions of the various artifact classes used). Data for Grotta Guattari and Grotta di Sant'Agostino are taken from Taschini (1979) and Tozzi (1970), respectively. For previously unpublished assemblages (from Grotta Breuil and Grotta dei Moscerini), my own counts are used. The typological uniformity of these collections is remarkable. In all cases, simple scrapers are the most common forms. *Déjeté* and transverse varieties are generally the second most abundant scraper class. Pointed, double, and other scraper forms (i.e., types 25–29) are present in small numbers, seldom exceeding 10% of any assemblage. Notches, denticulates and "Upper Paleolithic types" are generally quite scarce. In only one case do denticulates and notches combined exceed even 15% of the total retouched-tool count. It is not surprising that the Pontinian assemblages have always been considered typologically and, by implication, culturally homogeneous.

The "heartland" of the classic Pontinian Mousterian is limited to the littoral zone and the Agro Pontino. Very similar Mousterian assemblages made on small pebbles are known from coastal sites in other regions, but they are called by different names.

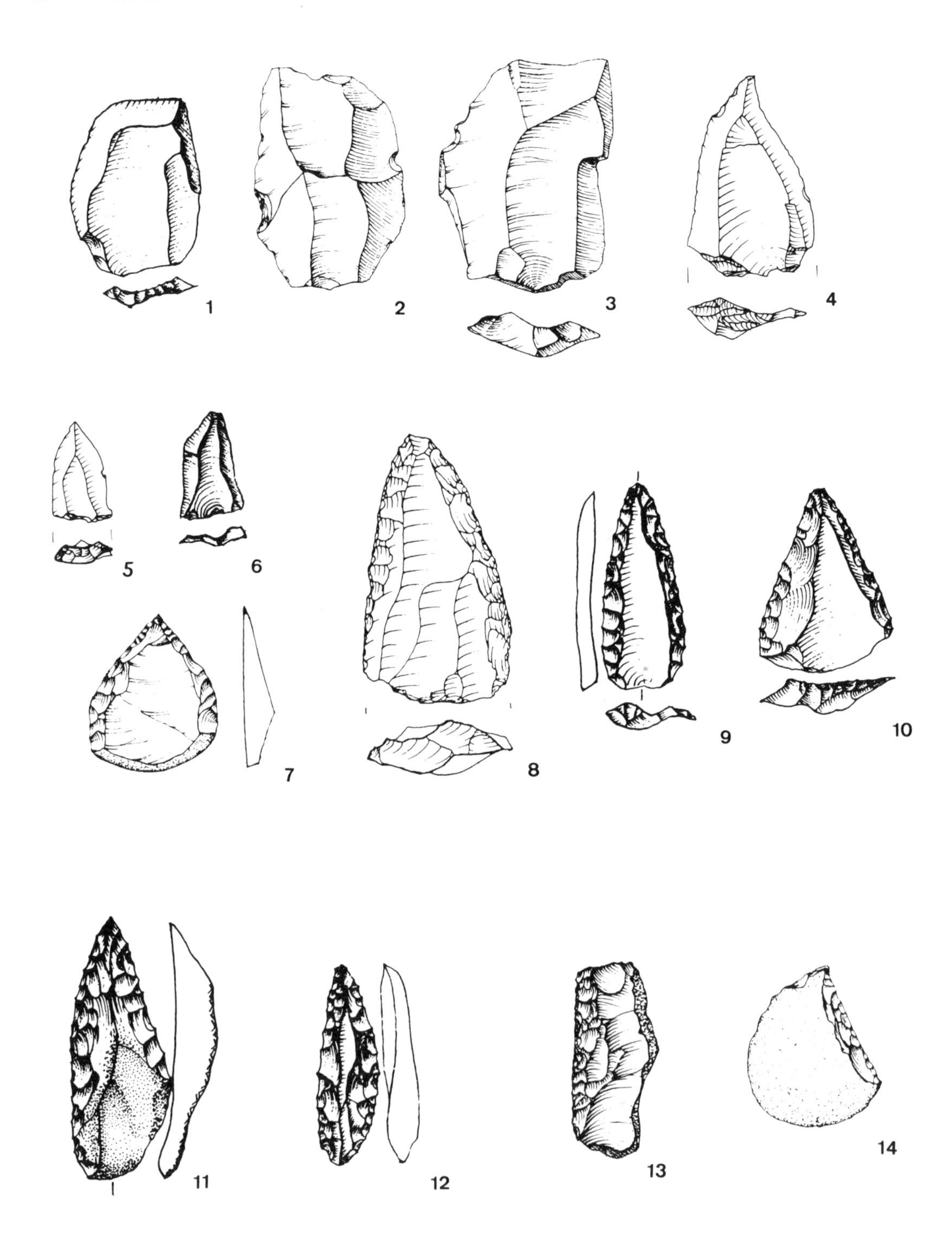

Figure 3.4. Typical artifacts from Pontinian Mousterian sites. 1–3=Levallois flakes; 4–6=Levallois points; 7–12=Mousterian points; 13–24=simple sidescrapers. (1, 6, 9–12, 16, 18–21, 24=Gr. di Sant'Agostino; 2, 3, 14, 15, 17, 23=Gr. Guattari; 4, 5, 7, 8, 13, 22=Gr. Breuil.)

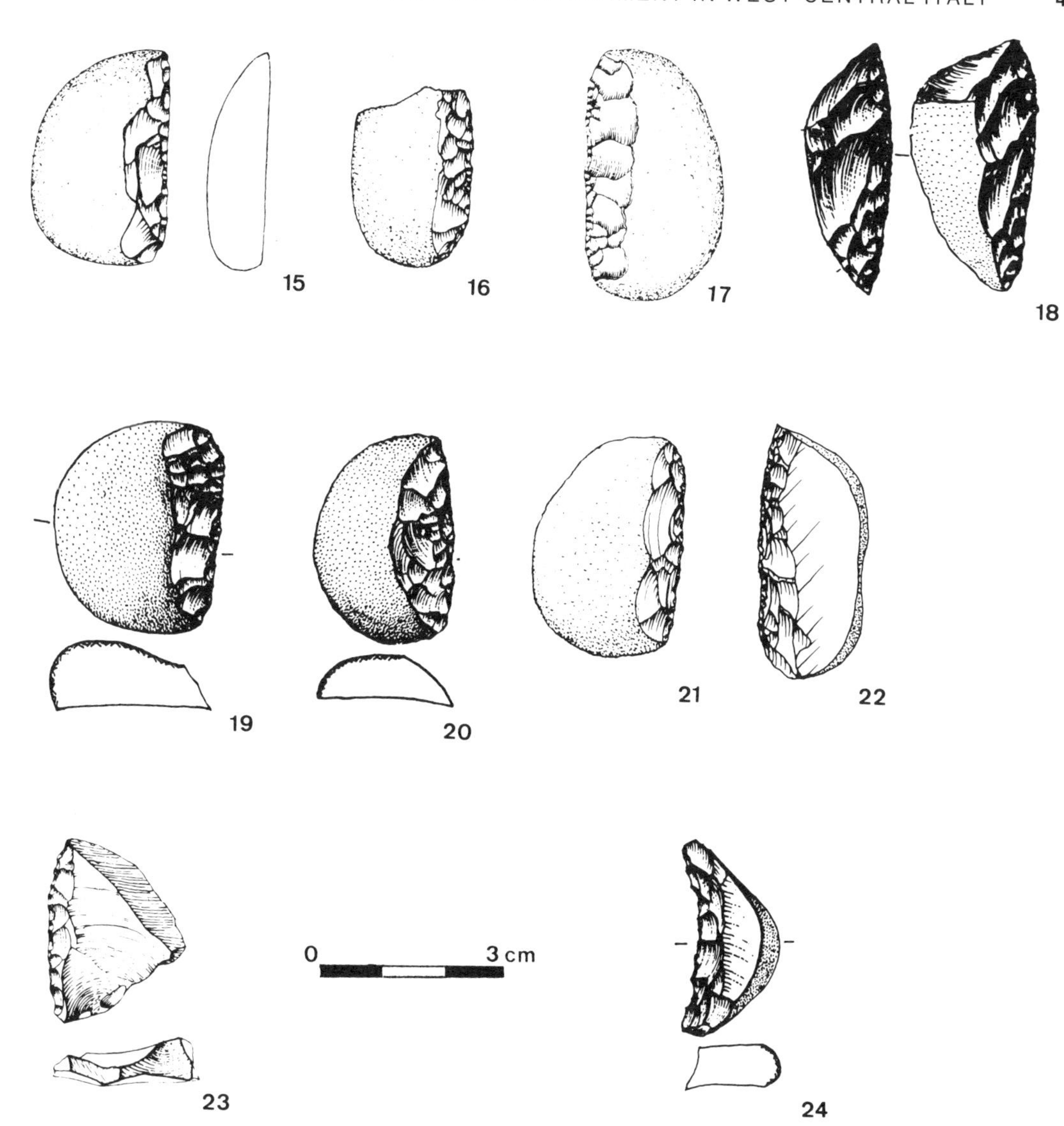

Figure 3.4. *cont.*

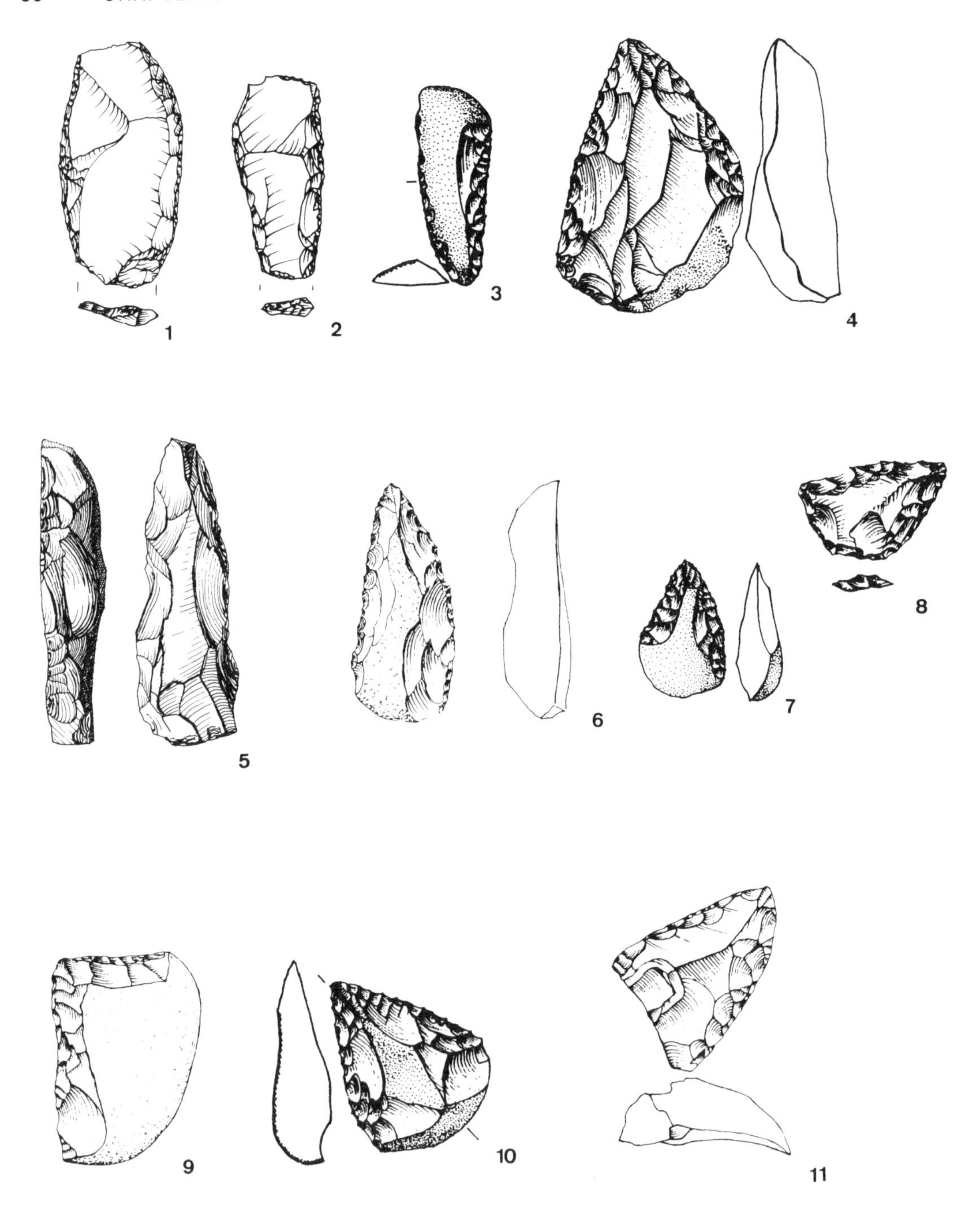

Figure 3.5. Typical artifacts from Pontinian Mousterian sites. 1–3=double sidescrapers; 4, 6, 7=convergent sidescrapers; 5=limace/convergent scraper; 8–11=*déjeté* scrapers; 12–19=transverse scrapers; 20=endscraper; 21=notch; 22, 25, 27=denticulates; 23=truncation/sidescraper; 24="laurate" biface; 26=bec; 28=Tayac point. (1, 2, 27=Gr. Breuil; 3–5, 7, 8, 10, 12–15, 17, 18, 25, 28=Gr. di Sant'Agostino; 6, 9, 11, 16, 19–23, 26=Gr. Guattari; 24=Gr. dei Moscerini.)

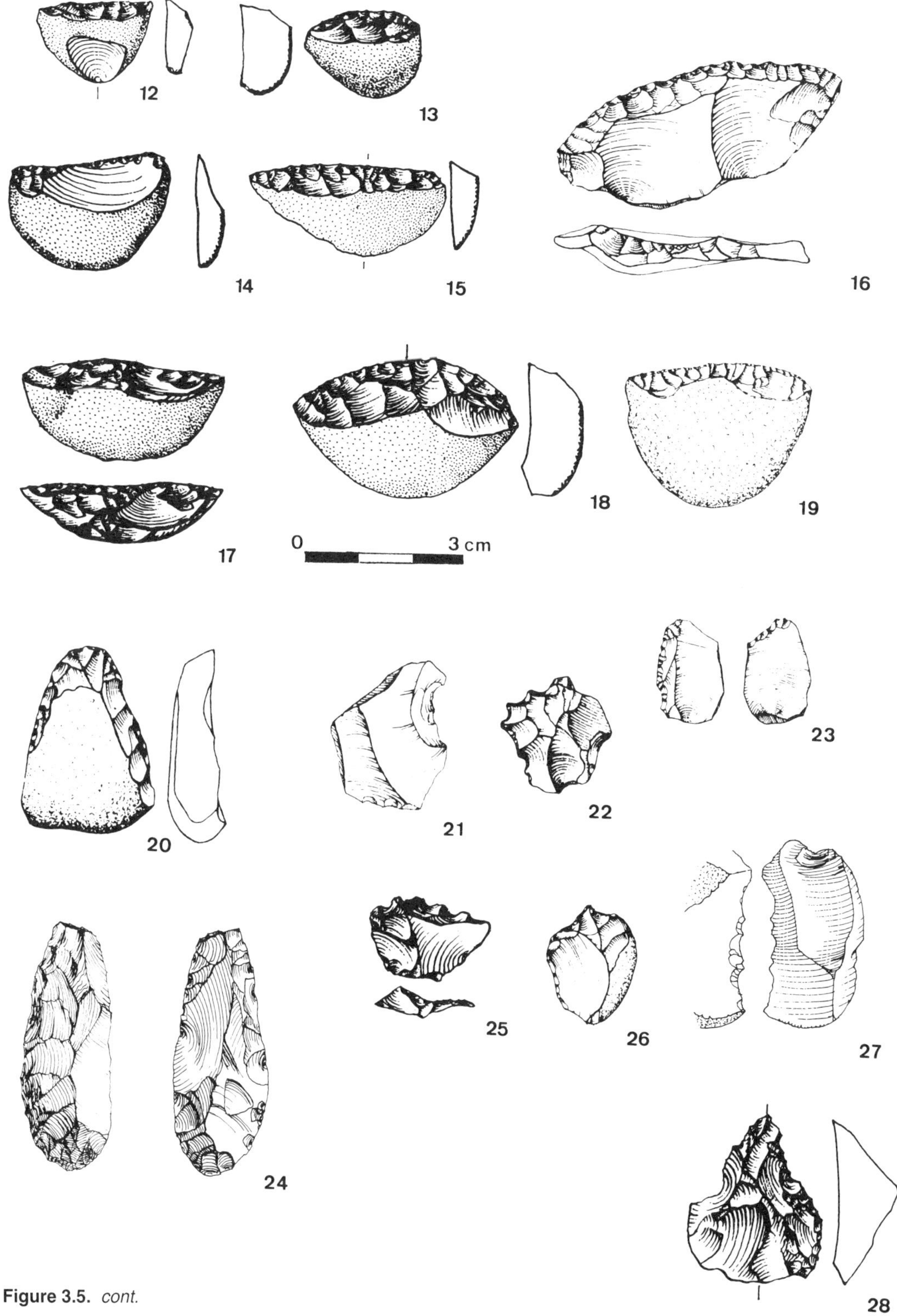

Figure 3.5. *cont.*

Table 3.3a. Typological summary of subject assemblages (frequencies)

	Simple scrapers	*Double scrapers*	*Points & cnvgt. scrapers*	*Déjeté & transv. scrapers*	*Other scrapers*	*Upper Paleo. types*	*Notched pieces*	*Dentics.*	*Misc. & diverse*	*Ret'd flakes*
Agro Pontino	59	8	14	47	5	1	2	9	2	3
Breuil										
1986–87 sample	72	10	14	47	4	6	3	19	7	21
old *sondage*	12	2	4	9	0	3	1	0	0	0
strata 3/4	22	7	5	2	0	2	5	6	0	4
Guattari										
stratum 1	25	0	3	10	3	1	6	3	3	2
stratum 2	44	5	4	24	6	11	5	10	2	5
stratum 4	84	5	12	58	17	13	5	11	6	5
stratum 5	46	4	6	15	13	2	5	6	3	3
Moscerini (strata groups)										
group 1	5	0	2	2	1	0	0	0	1	0
group 2	53	15	11	29	5	4	1	1	4	0
group 3	63	11	8	26	6	2	3	3	6	0
group 4	44	5	5	22	0	6	5	5	0	0
group 5	21	2	1	6	3	0	2	2	2[a]	0
group 6	38	3	6	21	6	0	1	0	0	0
Sant'Agostino										
level 1	257	16	26	126	6	7	28	24	1	99
level 2	59	13	7	25	2	1	8	5	0	24
level 3	32	5	5	20	2	0	1	1	0	12
level 4	13	4	5	7	2	0	1	2	0	1

[a] Misc. category in Moscerini strata group 5 includes small laurate biface.

Localities in other parts of Italy that have yielded pebble-based Mousterian industries include Monte Argentario (Segre 1959), open-air dune sites in Tuscany, and cave sites such as Grotta dei Ladroni (Bidditu 1986) and Grotta dell'Alto (Borzatti von Lowenstern 1966) on the Adriatic coast. Inland sites with similar assemblages are rare, primarily because it is only along the coast that the diminutive pebbles are found. However, "Riss"-aged assemblages made on pebbles from the Tiber river gravels at Monte delle Gioie, Sedia del Diavolo (Taschini 1967), and stratum D at Torre in Pietra (Piperno and Bidditu 1978) are sometimes referred to as "Protopontinian."

Mousterian assemblages found in the inland hills and river valleys of Latium do not closely resemble the more widely known materials from the coastal caves (Piperno 1984). The inland Mousterian assemblages generally seem to have been made on nodules or large cobbles of flint. Artifacts tend to be larger, and assemblages exhibit greater typological variation than is seen in the Pontinian. A similar phenomenon occurs in Tuscany, where the pebble Mousterian assemblages are replaced by denticulate or Levallois-dominated assemblages (e.g., Blanc 1937b; Pitti and Tozzi 1971) a few tens of kilometers from the sea. Knowledge of the Mousterian of inland Latium and neighboring Abruzzo is based largely on open-air sites and surface collections. This seems strange in light of the fact that most of the mountain ranges are calcareous, at least in the southern part of the study area, but it seems that surficial karstic systems are more extensively developed nearer the coast. The small

Table 3.3b. Typological summary of subject assemblages (percentages)

	Simple scrapers	*Double scrapers*	*Points & cnvgt. scrapers*	*Déjeté & transv. scrapers*	*Misc. scrapers*	*Upper Paleo. types*	*Notched pieces*	*Dentics.*	*Misc. & diverse*	*Ret'd flakes*
Agro Pontino	39.3	5.5	9.3	31.3	3.3	0.7	1.3	6.0	1.3	2.0
Breuil										
1986–87	35.5	4.9	6.9	23.2	2.0	3.0	1.5	9.4	3.5	10.3
old *sondage*	38.7	6.5	12.9	29.0	0.0	9.7	3.2	0.0	0.0	0.0
strata 3/4	40.7	13.0	9.3	3.7	0.0	3.7	9.3	11.1	0.0	7.4
Guattari										
stratum 1	44.6	0.0	5.4	17.9	5.4	1.8	10.7	5.4	5.4	3.6
stratum 2	37.9	4.3	3.5	20.7	5.2	9.5	4.3	8.6	1.7	4.3
stratum 4	38.9	2.3	5.6	26.9	7.9	6.0	2.3	5.1	2.8	2.3
stratum 5	44 7	3.9	5.8	14.6	12.6	1.9	4.9	5.8	2.9	2.9
Moscerini										
group 1	45.5	0.0	18.2	18.2	9.1	0.0	0.0	0.0	9.1	0.0
group 2	43.1	12.2	8.9	23.6	4.1	3.3	0.8	0.8	3.3	0.0
group 3	49.2	8.6	6.3	20.3	4.7	1.6	2.3	2.3	4.7	0.0
group 4	47.8	5.4	5.4	23.9	0.0	6.5	5.4	5.4	0.0	0.0
group 5	53.9	5.1	2.6	15.4	7.7	0.0	5.1	5.1	5.1	0.0
group 6	50.7	4.0	8.0	28.0	8.0	0.0	1.3	0.0	0.0	0.0
Sant'Agostino										
level 1	43.6	2.7	4.4	21.4	1.0	1.2	4.8	4.1	0.2	16.8
level 2	41.0	9.0	4.9	17.4	1.4	0.7	5.6	3.5	0.0	16.7
level 3	41.0	6.4	6.4	25.6	2.5	0.0	1.3	1.3	0.0	15.3
level 4	37.1	11.4	14.3	20.0	5.7	0.0	2.9	5.7	0.0	2.9

Table 3.4. Artifact classes used in Tables 3.3a and 3.3b

Class	*Artifact forms*
Simple scrapers	Bordes types 9–11
Double scrapers	Bordes types 8, 12–17
Points and convergent scrapers	Bordes types 4, 6, 7, 18–20
Déjeté and transverse scrapers	Bordes types 21–24
"Other" scrapers	Bordes types 25–29
Upper Paleolithic types	Bordes types 30–41, 44
Notched pieces	Bordes types 42, 52, 54
Denticulates	Bordes type 43
Misc. & diverse	Bordes types 62, 63
Retouched flakes	Flakes with discontinuous, light retouch. Includes Bordes types 45–50, if clearly artifactual

collections from Sora (Biddittu et al. 1967; Segre et al. 1984), located about 50 km inland in the valley of a tributary of the Liri river (Fig. 3.1), contain a large number of Mousterian points and convergent scrapers of varying sizes: these materials are reported as being of "Levallois facies." Recently reported materials from a series of high-altitude open-air localities (some above 1200 m elevation) on Monte Gennaro and Monte Pellecchia, northeast of Rome (Fig. 3.1), fit a "Typical Mousterian" typological profile, with approximately equivalent numbers of scrapers and denticulates. Levallois flakes are rare, and it appears that retouched pieces are extensively modified and heavily resharpened (Biddittu et al. 1990–91). An intriguing though regrettably small collection comes from Grotta della Cava, in the western foothills of the Lepini mountains near Sezze Romano (Fig. 3.1). Remnant deposits in a cave that had been largely destroyed through quarrying operations yielded a sample of 88 artifacts which included small tools made on flint pebbles as well as a number of larger tools, flakes, and cores made on fine-grained local limestone (Segre-Naldini 1984).

The Sites and the Samples

The following section summarizes the topographical and geological situations, basic layouts, and research histories of the four Mousterian cave sites that lie at the heart of this study. Open-air Mousterian and Upper Paleolithic localities used for comparative purposes are briefly described as well. The samples employed here, and their relationships to the original collections, are also discussed. The study samples are often subsets of the original excavated assemblages: many of the collections studied were made 40 to 50 years ago, and it was impossible to locate *every* artifact collected. This is not a serious problem, as the original assemblages were themselves samples drawn from an unknown universe and cannot be considered complete or discretely bounded entities in any real sense. In selecting samples from large collections, or in evaluating the nature of the samples made available to me for study, I have only attempted to avoid systematic biases in the kinds of variables and attributes that I have chosen to investigate.

Grotta Guattari

The best-known Mousterian site in Latium, Grotta Guattari is often considered synonymous with the Paleolithic of Monte Circeo. Guattari is situated on a gently sloping hillside a few hundred meters from the present shoreline, 10–15 m above sea level. A relatively small cave, only 15 m deep and at most 12 m wide, Guattari originally formed as a solution chamber. At its highest point, the limestone roof is less than 2 m above the most recent archaeological deposits. The floor plan is quite complex, with several small lobes or side-chambers radiating off a central chamber (Figure 3.6). Originally, the cave was accessible through two or three low, narrow openings less than 1.5 m high, which would have allowed little light to penetrate into the interior.

Grotta Guattari was discovered accidentally in February 1939 by workers excavating a trench (Blanc 1939a, 1939b). Systematic excavations at Guattari were initiated in the spring of that year, under the supervision first of A. C. Blanc and later of L. Cardini (Taschini 1979:181); some additional excavation was later carried out by A. Segre in 1950 (Taschini 1979:182). Because different individuals directed the excavations, level designations vary. Blanc excavated in stratigraphic units, while Cardini used thinner, arbitrary levels. Segre (in Taschini 1979:184) established the geological correlations between the various excavation units, dividing the deposits into seven major strata (Table 3.5). The levels or strata vary considerably in thickness in different parts of the site, and are deepest near the front of the cave (Figure 3.7). Strata 4 and 5 thin out and actually disappear deep inside the cave (Taschini 1979:182–83). The cryoclastic limestone debris on the surface and in stratum 1 led Segre to suggest that these most recent deposits dated to a very cold part of the Würm II. The presence of a "warm" fauna (including hippopotamus) is used to argue for a Würm I–II (Brörup stadial) date for strata 4 and 5 (Taschini 1979:247). The marine beach deposits in strata 6 and 7 were thought to date either to the retreat of the Riss/Würm (isotope stage 5e) high sea level or, more likely, to some later transgression.

Artifactual materials were recovered from strata

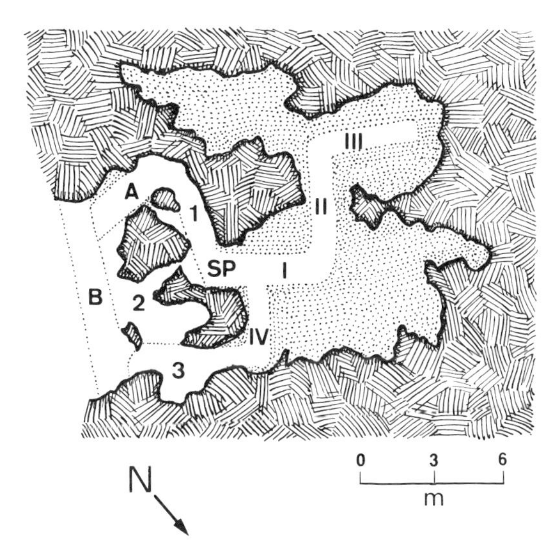

Figure 3.6. Schematic floor plan of Grotta Guattari. "Sp" indicates area known as "sotto il ponte." Roman and Arabic numerals refer to trenches (after Taschini 1979).

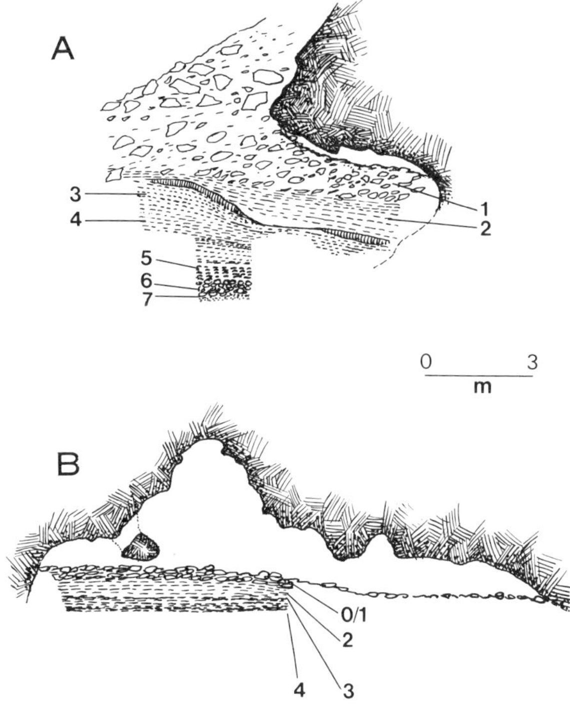

Figure 3.7. Stratigraphic sections of Grotta Guattari. A. exterior sagittal section through trench 2. B. interior sagittal section through trench I (after Taschini 1979).

Table 3.5. Description of stratigraphic units at Grotta Guattari

Stratum 1: partially cemented, concretion-covered angular blocks of limestone *eboulis*, mixed with loose brown earth. This stratum includes the surface deposits
Stratum 2: loose, "incoherent" brown earth
Stratum 3: compact brown soil, heavily cemented and containing numerous calcareous concretions and crusts
Stratum 4: partially cemented, sand fill with fossil and actual root casts
Stratum 5: brown-violet sandy soil
Stratum 6: cemented grey sand with fragments of marine molluscs
Stratum 7: marine beach conglomerate of cemented sand with whole marine shells

1, 2, 4, and 5 only. Although bones were present in profusion, no stone artifacts were found exposed on the surface. Stratum 3 and the basal layers (strata 6 and 7) were also archaeologically sterile. There has probably been some intermixing of materials from the two pairs of adjacent archaeological strata (1 and 2, 4 and 5). However, as they are separated by a sterile layer between 10 and 75 cm thick (stratum 3), strata 1 and 2 can be assumed to represent a completely independent depositional entity from strata 4 and 5. It is interesting that many bones and stone artifacts show evidence of burning, but no hearths, ash lenses, or other thermal features were noted by excavators.

After A. C. Blanc's death, Dr. M. Taschini assumed the task of describing the archaeological collections from Grotta Guattari. Dr. Taschini herself died suddenly in 1975, and Prof. A. Bietti took over the task of finishing the monograph, which finally appeared in 1979 (Taschini 1979), forty years after the original excavations. The Taschini/Bietti monograph on Guattari was the first full-scale publication of a Mousterian assemblage from the study area to employ Bordes's typological system (Bordes 1961a). The lithic materials are discussed in much greater detail below, and only a few observations need be made here. Taschini/Bietti concluded that the "essential" typology of the assemblages from all four strata at Guattari placed them closest to the French Charentian/Quina facies, and in particular the "Charentien Oriental" of Provence (Taschini 1979:247). A small difference was noted between assemblages from the upper (strata 1 and 2) and lower (strata 4 and 5) groups in terms of the faceting and Levallois indices (Taschini 1979). This represents the strongest typological contrast between *any* of the Pontinian cave assemblages included in this study.

All of the lithic artifacts from Grotta Guattari, including tools, cores, and unmodified pieces that could be located, are included in the artifact sample discussed here. The materials studied are presently housed at the Istituto Italiano di Paleontologia Umana in Rome and the Museo Pigorini in Rome. Discrepancies between artifact counts reported in this book and those reported in the original report reflect a combination of differences of opinion about how to classify individual specimens, and inevitable attrition to collections over the course of more than fifty years.

Grotta Guattari and the issue of Neandertal cannibalism

When workers first entered the newly revealed opening to Grotta Guattari, they discovered a nearly complete Neandertal cranium lying exposed atop the pavement of limestone debris which constitutes the uppermost stratum at the site. Two human mandibles were later found, one inside the cave and one in the conglomerate near the cave's entrance. The skull was described by its discoverers as having been located within a "circle of stones" near the rear of the cave (Ascenzi 1990–91; Blanc 1939a, 1939b; Piperno 1976–77). In his original description of the Guattari cranium, Sergi (1939) also observed that the *foramen magnum* of the cranium was extensively damaged and enlarged: this, he felt, showed that the brain had been removed (and presumably eaten) by hominids. Even though a number of researchers have questioned it (e.g., Chase and Dibble 1987; Gamble 1986:167), the Guattari discovery continues to be widely cited as an example of Neandertal cannibalism and attendant ritual behavior (Marshack 1989, 1990).

Recent reassessments of the cave-floor deposits, the faunal assemblages, and the cranium itself have cast doubt on assertions that Grotta Guattari was a locus of Neandertal ritual cannibalism. The skull was recently examined in detail by T. White and N. Toth, who were primarily interested in evidence of trauma or human modification. They found no cut marks or other damage patterns on the skull that could unambiguously be attributed to human actions. However, they did document a possible carnivore toothmark on the inside of the skull, near the *foramen magnum* (Toth and White 1990–91; White and Toth 1991). As for the alleged artificial "ring of stones," the skull was moved by workmen immediately after discovery, and no archaeologist ever observed its original placement within this feature (Ascenzi 1990–91). Indeed, no such stone feature currently exists, except in reconstruction at the

Museo Pigorini in Rome. Moreover, the Neandertal cranium was not alone inside the "circle." Maps show that both animal bones and limestone blocks were located inside the "ring," casting additional doubt on the supposedly special treatment of the human skull (Giacobini 1990–91).

In a monograph on the "living floor" at Grotta Guattari, M. Piperno (1972) observed that the faunal remains found on the surface included an overabundance of economically marginal elements such as skull parts and antlers. Piperno asserted that this demonstrated some sort of ritual use of the cave by Neandertals (Piperno 1976–77), and that the bones and antlers were "offerings" of some kind. In a recent taphonomic study Dr. M. Stiner (1990–91, 1991a) has shown that the faunal remains from the surface and first stratum at Grotta Guattari were extensively modified by a large, non-human carnivore, most likely spotted hyaenas (*Crocuta crocuta*). These data, along with the presence of numerous hyaena coprolites (also noted by Blanc and Segre [1953:89]) and the abundant remains of both adult and juvenile hyaenas, led Stiner to conclude that the site was used intensively as a den by these animals while the surface deposits and stratum 1 were forming. The anatomical biases in the faunal assemblage are also consistent with the contents of recent hyaena dens. Thus, the animal bones found on the surface and stratum 1 at Guattari, and perhaps the damage to the Neandertal skull itself, now appear to be attributable to the actions of non-human carnivores, rather than to human behavior, ritual or otherwise (see also Piperno and Giacobini 1990–91).

Grotta di Sant'Agostino

Grotta di Sant'Agostino is located about 45 km southeast of Monte Circeo, just north of the town of Gaeta. The site is situated around 25 m above present sea level, in the walls of a small "box canyon" that opens on a depression, possibly a former lagoon or embayment (Figure 3.8). The cave consists of a single large, roughly hemispherical chamber measuring approximately 16 m deep by 20 m wide at the entrance (Figure 3.9). The vault is around 8 m high. The cave faces south, and thanks to its

Figure 3.8. View of Grotta di Sant'Agostino from the west (courtesy of G. Barker.)

broad entrance it is well lighted throughout the day, even along the back wall. Today, there are active springs along the back wall of the cave and high up along the west wall. Thick horizontal travertine deposits are exposed along the base of the east and west walls.

Grotta di Sant'Agostino was first excavated in 1948, under the direction of E. Tongiorgi, an associate of the IIPU. One trench was placed near the mouth of the cave, adjacent to one of the travertine exposures. A second, larger trench was excavated in the rear of the cave. The deposits near the cave mouth, which extend to a depth of only 50 cm, were particularly rich in fauna and lacking lithic material. More balanced collections were obtained from the deeper (ca. 2.5 m) deposits in the main trench. The sediments in the main trench consisted

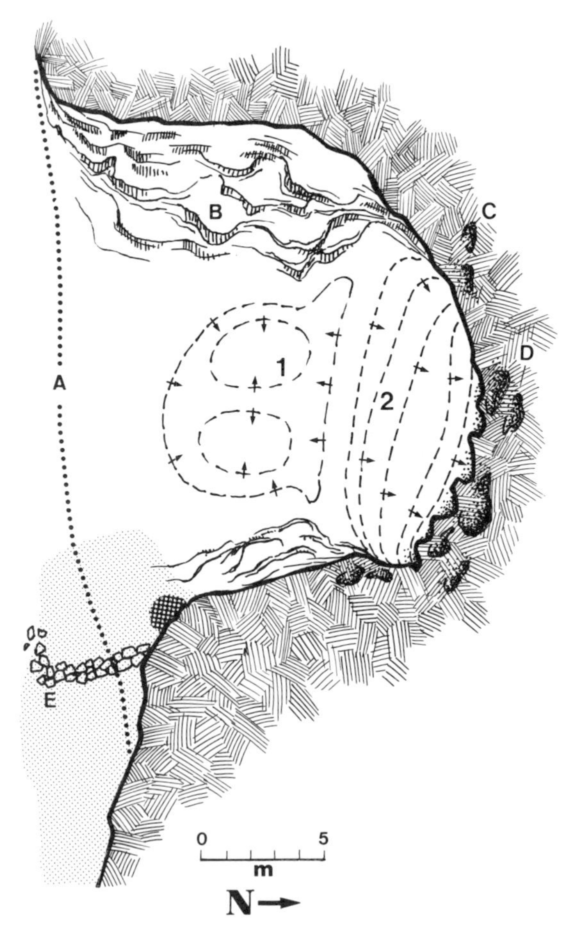

Figure 3.9. Schematic floor plan of Grotta di Sant'Agostino.

Table 3.6. Grotta di Sant'Agostino: level thickness and area

Level	*Average thickness (cm)*	*Area excavated* (m^2)
A0	30	12.40
A1	30	5.75
A2	40	5.75
A3	60	4.05
A4	90	3.00

of relatively uniform loose sands. The excavators divided them into five levels of various thicknesses and extensions (Table 3.6). Level A0 contained debris resulting from modern use of the cave as a sheep pen, and some bones of wild species, along with a mixture of both Middle and Upper Paleolithic artifacts. The remaining four levels (A1–4) yielded Middle Paleolithic artifacts and the bones of wild species exclusively. Because stratigraphic definition in the sandy deposits was poor, the excavators were unable to connect the deposits from the front of the cave with those in the main trench. Materials from the shallower excavation unit near the cave mouth were therefore given the level designation "A*x*." As at Guattari, no discrete combustion features, hearths, or major ash concentrations are mentioned in published discussions of the site.

The first analyses of the lithic materials from Sant'Agostino were published soon after the excavation (Laj-Pannocchia 1950). This work was largely technological, focusing on patterns of core reduction and flake production, and it was in many ways ahead of its time. Laj-Pannocchia recognized several distinct methods of flake production, which she illustrated both schematically and using archaeological specimens; she also attempted to sort flakes according to the order of their removal from cores. A more complete study of the lithics and fauna from Sant'Agostino was published by C. Tozzi (1970). Tozzi continued the earlier investigator's approach to studying flakes and flake blanks, and adopted Bordes's systematics in the analysis of the retouched tools. Tozzi also presented a remarkably complete summary of faunal data, including tabulations of all identifiable remains by species and anatomical element.

Grotta di Sant'Agostino yielded by far the largest artifact assemblage of any of the Mousterian sites discussed here. By Tozzi's counts, a total of 1373 retouched pieces, 712 cores, and 5792 flakes and debris were collected (Tozzi 1970). This material, except for a small segment at the Museo Pigorini in Rome, is housed at the Dipartimento delle Scienzi Archeologiche at the University of Pisa. Because of the size of the collection, it was necessary to sample some portions of it for this study. The available materials from levels A2, A3 and A4 were studied in their entirety. Because of time limitations, it was not possible to systematically study all of the 2172 flakes and pieces of debris from level A1; from this stratum, all of the cores and a random "grab" sample of approximately 52% of the retouched pieces were studied, along with a number of whole flakes that had been sorted into reduction stage categories by Laj-Pannocchia. A random sample of 184 retouched pieces from the disturbed stratum A0 was also analyzed, along with all diagnostic Upper Paleolithic implements from this stratum. A collection of retouched tools from various strata currently on display at the University of Pisa could not be included in the sample.

Grotta dei Moscerini

Grotta dei Moscerini (Gnat Cave) is one of the largest and least widely known of the coastal Mousterian cave sites in Latium. The spacious, open cave is cut into Lower Jurrassic limestones on the steep western slope of Monte Agmemone, about 5 km north of Grotta di Sant'Agostino (Fig. 3.1; Vitagliano 1984:156). It was discovered during the 1936–38 survey of the coast, and excavations were carried out during the summer of 1949. A deep trench was excavated through the entire vertical stratigraphy at the front of the cave (Figure 3.10). Another, shallower test trench was placed on the back-slope of the archaeological deposits near the rear of the cave. Excavation in the main trench followed the geological stratigraphy. The internal trench, which must have been dug in near darkness, was excavated using a combination of natural stratigraphy and artificial cuts. Because of its size and long stratigraphy, Moscerini would be an ideal candidate for re-excavation using modern techniques. Unfortunately, the cave is completely buried beneath a thick layer of rocks and boulders blasted out of the side of the mountain during construction of the coastal highway on the slope above. The thick mantle of debris does protect the site from clandestine looters, however, so that it might be excavated in the future once the logistical and civil-engineering problems are solved.

Grotta dei Moscerini contains a stratigraphy more than 8.5 m in thickness. The base of the archaeological sequence sits about 3 m above current sea level. Archaeological deposits formerly extended to within a meter or two of the cave's roof near the entrance, sloping off towards the back of the cave. The cultural strata appear to have extended more than 12 m back into the cave. Two narrow, tunnel-like passages at the extreme rear of the cave were not explored (Figure 3.11). According to notes and section drawings from the 1949 excavation, a total of 44 distinct geological stratigraphic units were recognized in the main trench (Figures 3.12a and b). Evidence for episodes of thermoclastic activity (large blocks of roof fall) is present in stratum 3, strata 8–10, strata 27–28, and strata 37–38. A wave-cut notch and several rows of holes excavated by a mollusk that lives at the interface of air and water (*Lithodomus lithophagus*) are thought to mark the maximum level of the Riss/Würm interglacial high sea stand.

Archaeological materials from Grotta dei Moscerini were not described until 35 years after the excavation (Vitagliano 1984). Typical Pontinian Mousterian artifacts are present throughout the entire sequence. Because of the large number of strata and small size of the total assemblage (approximately 328 retouched stone tools), Vitagliano made no attempt to pursue a conventional typological analysis, noting only the general similarities between the Moscerini materials and Pontinian assemblages from other sites on the coast of Latium. One of the more exceptional aspects of Grotta dei Moscerini is the presence of tools made on the shell of a marine clam (*Callista chione*). These shell tools are made by percussion flaking, rather than by grinding, and are quite similar in form and size to the stone tools from the site. Similar artifacts have been found at other Mousterian sites along the Tyrrhenian coast, including Barma Grande (Vicino 1972) in Liguria, and Grotta del Cavallo (Palma di Cesnola 1965), Grotta dei Giganti (Blanc 1958–61), and Grotta di Uluzzo C (Borzatti von Lowenstern 1965) in Puglia; but none of the other Mousterian sites in Latium yielded shell tools. Burned and broken but unworked shells of *Callista*, *Mytillus sp.*, and other marine bivalves were also recovered from most strata at Moscerini.

Because the main excavation trench at Grotta dei Moscerini covered a limited area, assemblage sizes for individual strata are small. The richest stratum (stratum 21) yielded a total of only 51 retouched stone tools and large flakes, along with additional cores and debris. For comparative purposes, strata have been combined into larger strata groups of more-or-less uniform composition. The grouping of strata was conducted as follows. There are four

Figure 3.10. Excavations in main trench at Grotta dei Moscerini (courtesy of A. Segre, IIPU).

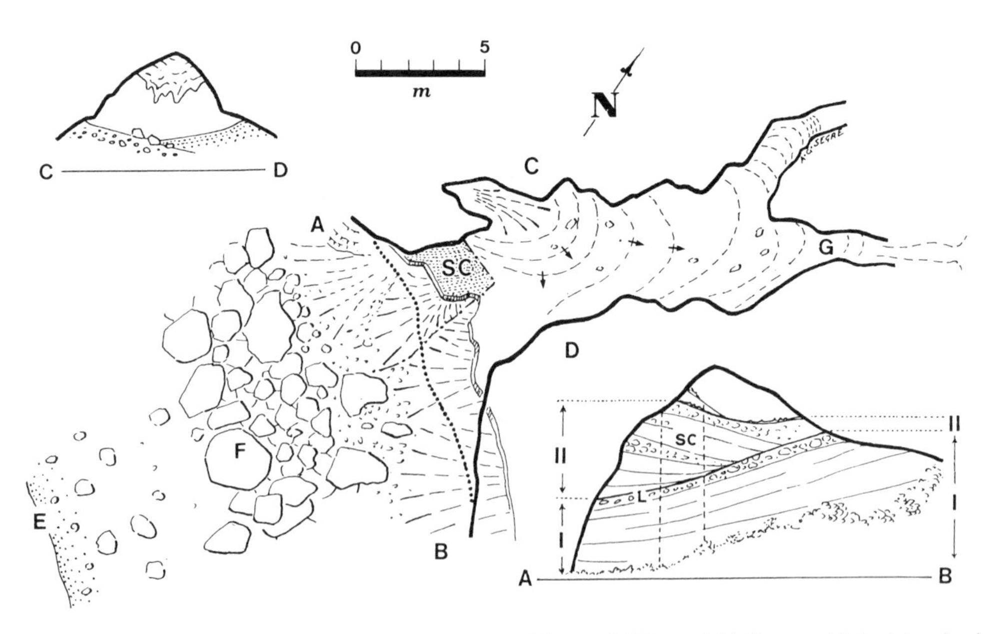

Figure 3.11. Schematic floor plan and cross-sections of Grotta dei Moscerini (after unpublished drawing by A. Segre).

Table 3.7. Grotta dei Moscerini: lumped provenience units

Strata group	*Strata*	*Type of deposits*
	MAIN TRENCH	
1	1–10	Thick, sandy layers with few stalagmites or crusts—little evidence of fire
2	11–20	Thin strata with sandy or loamy fill and few well-developed stalagmites. Numerous small ash lenses.
3	21–36	Thin layers, with alternating stalagmites and sandy fill. Extensive burned areas.
4	37–42	Thick, sandy strata, extensive burned horizons in 39 and 41, charcoal lens in upper 40. Unworked pebbles in fill.
	INTERIOR TRENCH	
5	0, 1a–1d, 2	Unknown.
6	3, 4	Unknown.

zones of relatively high artifact density in the main excavation trench: these are strata 14–18, strata 20–23, strata 31–33 and strata 37–39. Artifacts from adjacent strata within the zones of highest density were found to be technologically homogeneous. Divisions were subsequently drawn within low-density areas according to the technological similarities between low-density assemblages and adjacent high-density ones. Because they are virtually identical in composition, the concentrations in strata 20–23 and 31–33 were also combined. A similar procedure was followed with the assemblages from the arbitrary excavation levels from the interior test trench.

The stratigraphic compositions of lumped provenience units from Grotta dei Moscerini are shown in Table 3.7. Both the "rhythm" of geological change (i.e., stratum thickness) and indicators of the nature of occupation (e.g., evidence of burning) vary among the strata groups. In general, the boundaries between the lumped stratigraphic units also correspond to major changes in the density and nature of faunal materials and shifts in the frequencies of different species of marine bivalves. The latter probably reflect major changes in sea level or the nature of local marine environments. The strata groups crosscut the major thermoclastic "events" manifest in the geological stratigraphy (Figs. 3.12a and b). Even when materials from the first ten levels are combined, the combined assemblage is too small (only 11 retouched tools) to permit useful comparisons: strata group 1 (strata 1–10) is consequently not included in most of the statistical analyses.

Because more than 10 m of unexcavated deposits lie between them, it is impossible to correlate the internal and external strata at Grotta dei Moscerini; however, it is likely that the levels in the internal trench correspond with the uppermost part of the external sequence. The topmost levels in the interior trench (strata group 5, Table 3.7) yielded a fauna collected predominantly, if not exclusively, by spotted hyaenas (see below). Assuming that denning hyaenas would have preferred a relatively dark, closed space, it is likely that the interior levels were deposited when the cave opening was already restricted by archaeological deposits. In addition, the scarcity of archaeological materials in strata 1–10 suggests that hominid presence was sporadic—another possible correlate for hyaena denning. If this is the case, then the interior strata group 6 may correspond to levels near the top of exterior strata group 2 (strata 11–20) or the bottom of strata group 1 (layers 1–10).

By lumping the small assemblages from Grotta dei Moscerini assemblage size is augmented, at the obvious expense of chronological resolution. Ideally, it would be desirable to subdivide the collections in terms of the finest possible provenience unit, but the resulting assemblages would be so

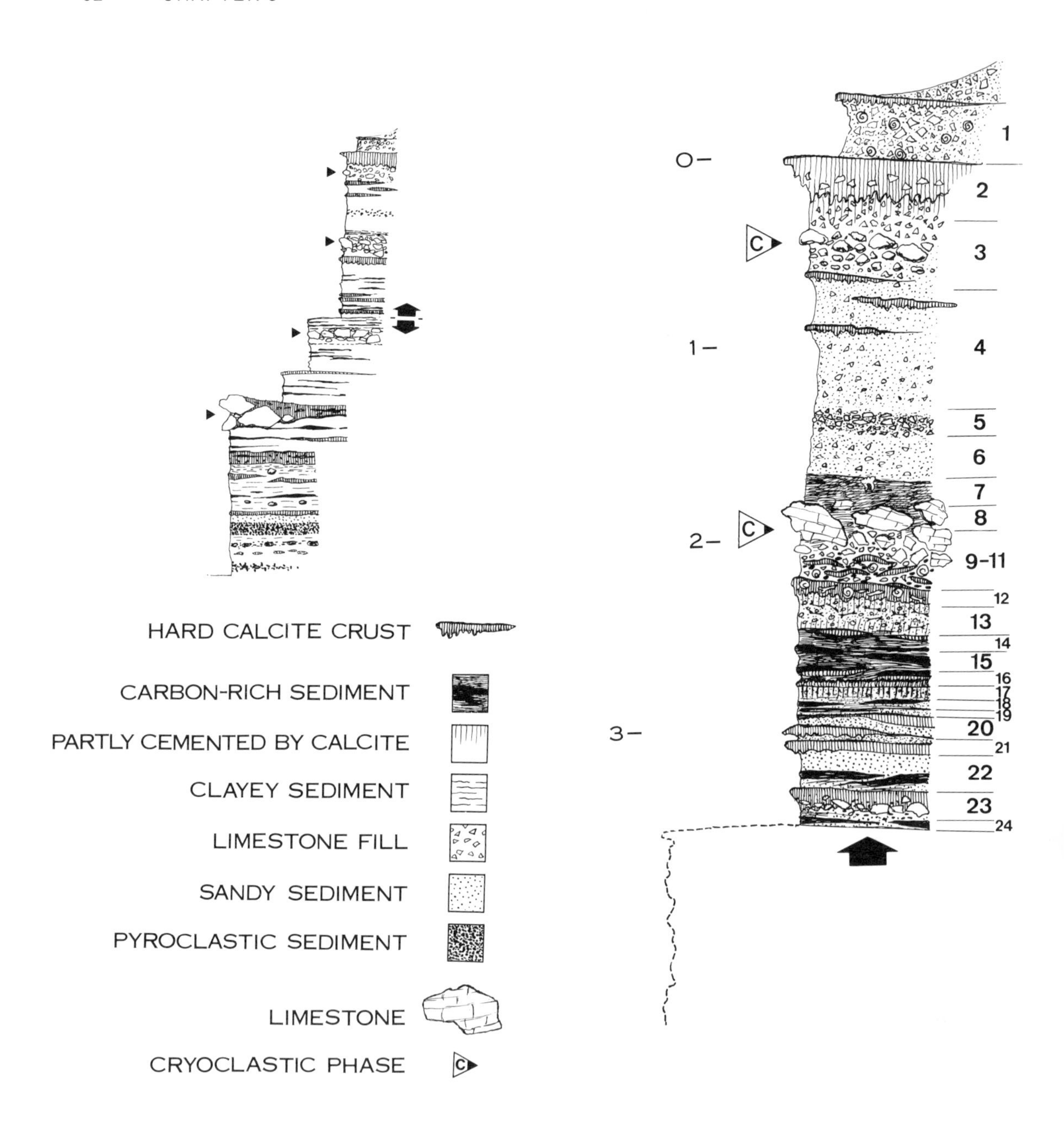

Figure 3.12a. Schematic stratigraphic section of Grotta dei Moscerini (upper part). Numbers on left indicate meters below datum; numbers on right indicate strata (after unpublished drawing by A. Segre).

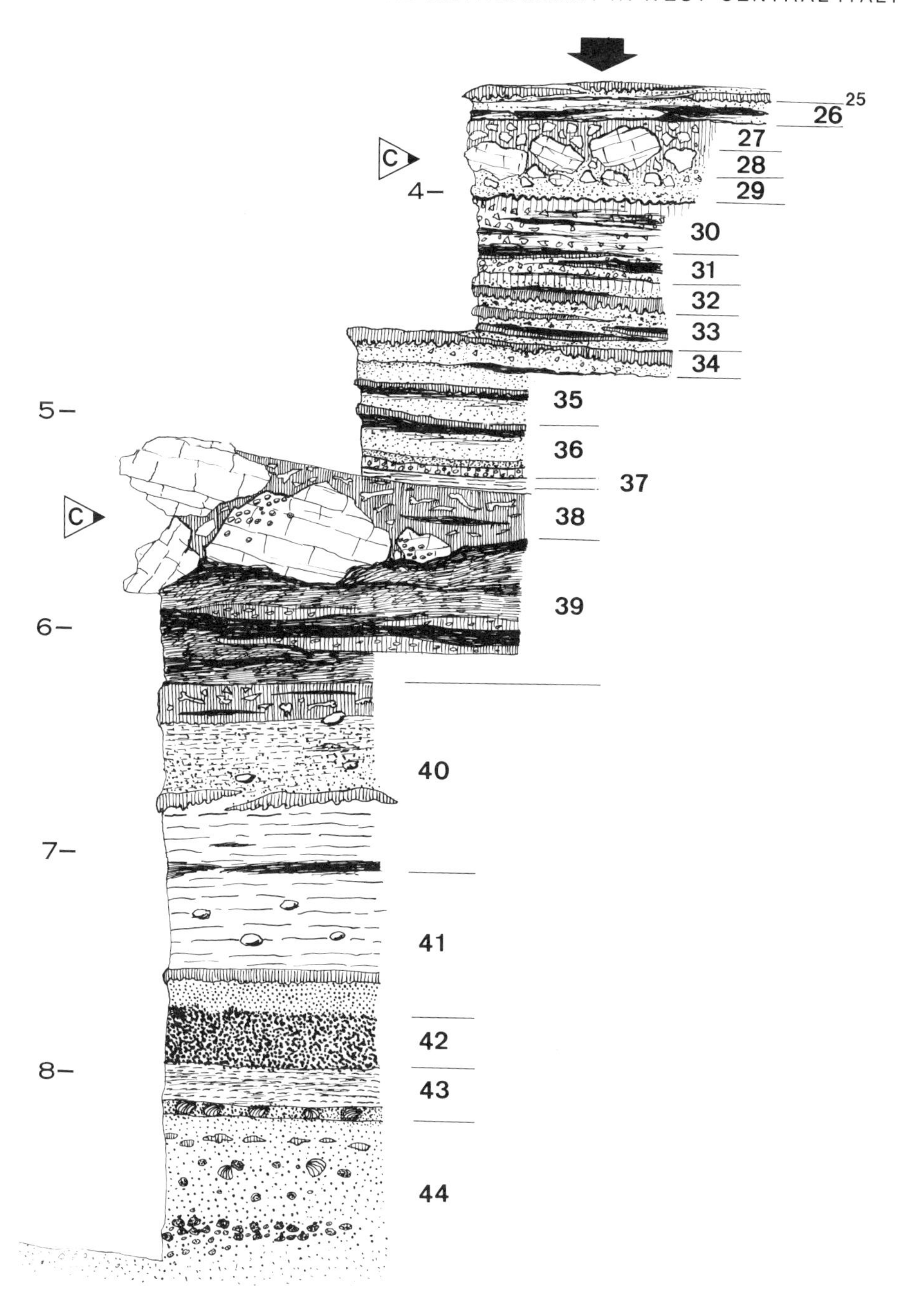

Figure 3.12b. Schematic stratigraphic section of Grotta dei Moscerini (lower part). Numbers on left indicate meters below datum; numbers on right indicate strata (after unpublished drawing by A. Segre).

small as to render comparisons statistically meaningless. On the other hand, there is no *a priori* reason to expect that changes in human behavior will correspond directly with units of geological deposition (Rigaud and Simek 1987). Combining strata is simply another arbitrary means of defining assemblage boundaries. In fact, groups of strata may not be much more internally heterogeneous than smaller stratigraphic units. Deep, finely delimited stratigraphic sequences from Mousterian sites such as Combe Grenal, Pech de L'Aze (Bordes 1972), and Le Moustier (Laville et al. 1980:174–75) often contain relatively long series of (typologically) similar assemblages in adjacent layers.

The entire recovered sample of stone artifacts from Grotta dei Moscerini was analyzed for this study. The bulk of the collection is housed at the Istituto Italiano di Paleontologia Umana, with a small sample of about 40 retouched tools and cores on display at the Museo Pigorini. The shell tools were examined but not intensively analyzed, and will not be discussed in any detail.

Grotta Breuil

Grotta Breuil is another of the cave sites discovered during the IIPU survey of Monte Circeo's seaward face. The cave is named for the Abbé H. Breuil, discoverer of the Saccopastore II cranium (Breuil and Blanc 1936) and later a participant in the excavations of Grotta del Fossellone. Grotta Breuil is located near the base of a high cliff on the western edge of the mountain, a few kilometers from Grotta Guattari (Fig. 3.2). It faces due west and opens directly onto the sea (Figure 3.13), so that it is now possible to reach the cave only by boat. The cave is approximately 10 m high and 17 m wide at its widest part, and is more than 25 m deep. The main chamber has an open plan similar to that of Grotta di Sant'Agostino, although Grotta Breuil is more elongated. A second, smaller chamber is situated high at the rear of the cave.

The intact archaeological strata at Grotta Breuil form a steep face (approximately 7 m thick) along the rear one-third of the cave (Figure 3.14; Bietti et al. 1990–91). The steep (ca. 45°) slope of the intact deposits is the result of erosion by wave action.

Figure 3.13. View of Grotta Breuil from the sea.

The deposits are protected from the sea by a series of enormous fallen limestone blocks lying near the cave mouth, but even so, waves from violent winter storms still reach the base of the stratigraphic sequence. Only about 1.5 m of intact sediments are exposed at the top of the slope. The rest of the deposit is blanketed by a layer of loose, organic-rich fill which appears to originate from the biologically aided decomposition of the parent limestone. Downslope of the exposed face of the intact stratigraphy these slope deposits are extremely rich in artifacts and bone, but above that point they contain little or no archaeological material. Based on the horizontal distribution of lithics and faunal remains, the stratigraphic relationships between soil units, and the nature of the concretions on archaeological finds, it appears that the materials in the slope deposits originated in relatively soft, unconsolidated strata exposed about 1 m below the top of the stratigraphic sequence (Figure 3.15).

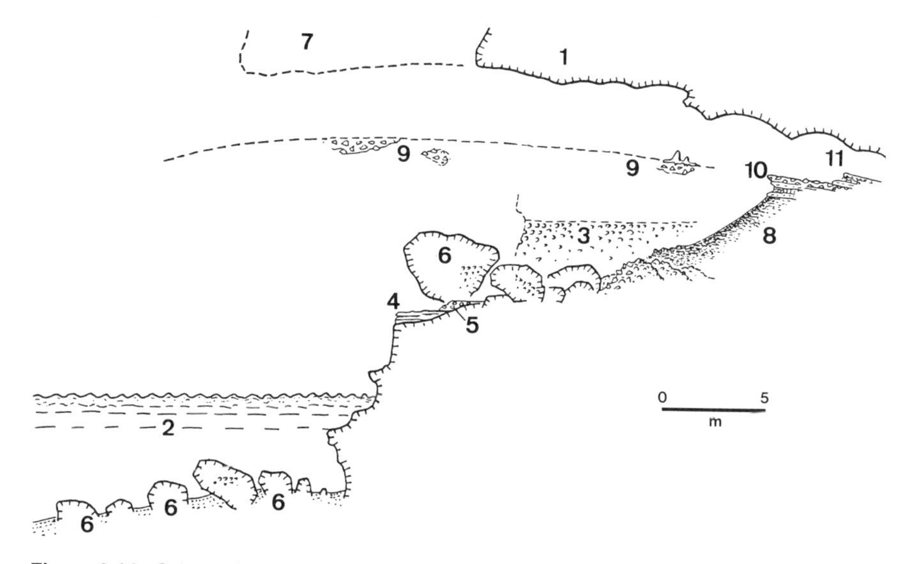

Figure 3.14. Schematic sagittal section of Grotta Breuil (after Bietti et al. 1988, from drawing by A. Segre).

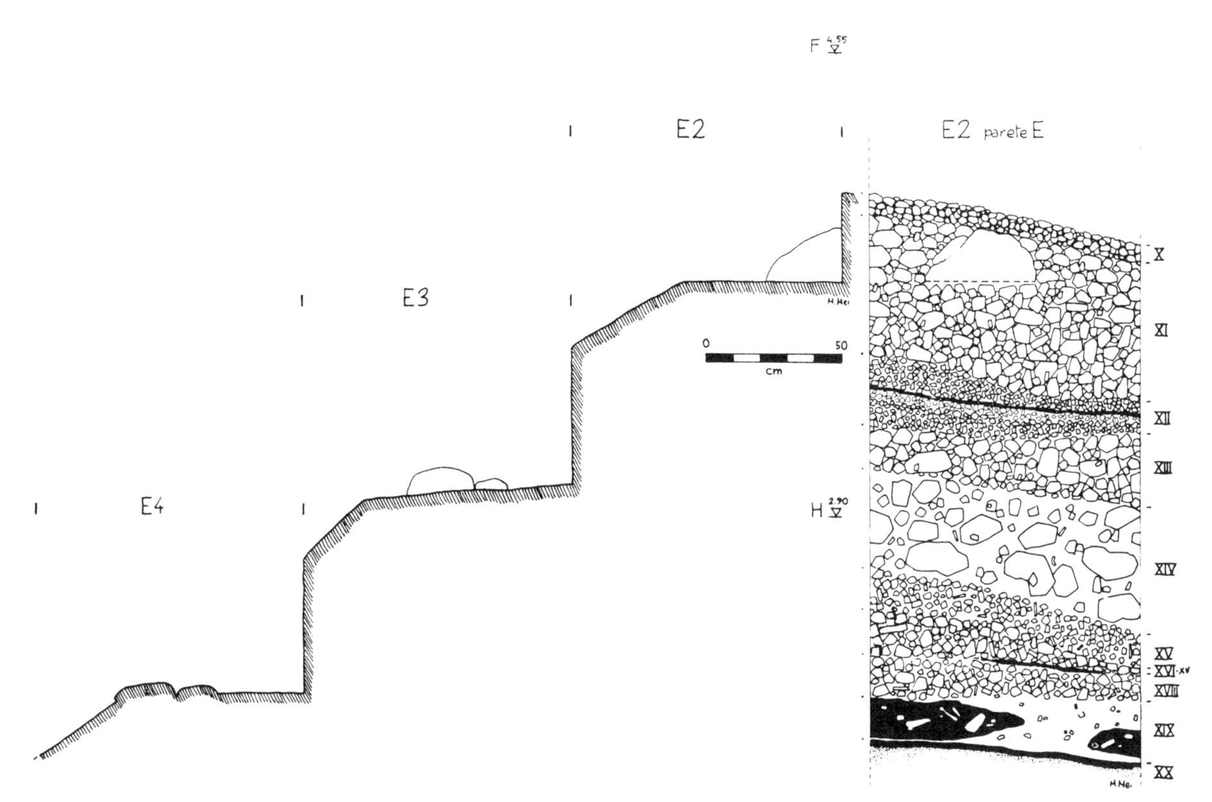

Figure 3.15. Schematic stratigraphy of Grotta Breuil.

The IIPU conducted test excavations at Grotta Breuil in the 1930s and then again briefly in the 1950s. A summary of findings was reported by Taschini (1970), who noted the presence of artifacts with a laminar "Upper Paleolithic" character, in addition to a more dominant component of typical Pontinian Mousterian materials. Because of the difficulty of access, no additional scientific study of the cave was carried out for more than 40 years. In 1985 an IIPU/University of Rome sponsored team led by Prof. A. Bietti excavated another *sondage* at Grotta Breuil, and large-scale excavations began in 1986, with the collaboration of the author and Dr. M. Stiner (Bietti et al. 1988a, 1990–91). The current excavations employ modern techniques, including three-dimensional proveniencing of finds and water screening of all sediments through 2-mm mesh.

The complex geological history of Grotta Breuil has been described in some detail elsewhere (Bietti et al. 1990–91). There are at least six major stratigraphic complexes at Grotta Breuil, four inside the cave and two cemented to the wall and cliff face outside. The two external complexes are indicated by the numerals 4 and 5 in Fig. 3.14. The oldest of these (Fig. 3.14, no. 4) consists of cemented sands containing sparse flint gravel and limestone fragments. These beach deposits correspond with traces of a high sea stand 3.2 m above current sea level, and may represent the "Würm I–II" interstadial (isotope stage 5a/4) noted elsewhere on Monte Circeo (Durante 1974–75). On top of the sands, but beneath the large blocks of roof fall, is a thin, ironized layer of cemented limestone debris (Fig. 3.14, no. 5). Traces of several high sea stands—including one about 14 m above current sea level that probably represents the "last interglacial" (stage 5e) maximum—are also visible on the cave walls.

The youngest of the four geological units inside the cave consists of the organic rich surface levels in the back of the cave, and the redeposited layer covering the slope of the deposits in which these recent sediments are mixed with debris from archaeological strata (Fig. 3.14, no. 8). The superficial layer is underlain by a series of strata approximately 4–5 m thick, consisting of alternating loose and brecciated masses of limestone fragments, locally rich in bone and chipped flint: these are the principal intact archaeological deposits within the cave. A strongly cemented layer forms a projecting shelf that extends out from the loose superficial slope deposit (Fig. 3.14, no. 10; Fig. 3.15, layer X). An archaeologically sterile stalagmite at the top of the sequence (Fig. 3.14, no. 11) has yielded a poor U/Th date of 26,000 (±12,000) years BP. Underlying this long sequence of strata are sediments of a distinctively different character. Layer XX (Fig. 3.15, layer XX) is a meter or more thick and consists of fine-grained sediments thickly shot through with dark phosphate crusts. Layer XX contains abundant lithics, but bone is almost absent: however, numerous "ghosts" of bones and occasional fragments of tooth enamel demonstrate that osseous material was present but has been dissolved. Layer XX is underlain by a thick, archaeologically sterile unit consisting of brown, strongly bedded sands (not shown in schematic stratigraphy). This basal deposit, approximately 2.5 m in thickness, sits directly atop the eroded limestone floor of the cave.

Three distinct artifact samples from Grotta Breuil are employed in this study. Two of these include material from the mobile slope deposits covering *in situ* strata. One sample (termed "Bs") comes from the old IIPU test excavations (reported in Taschini 1970). The other (called "B86–87") comes from a *sondage* excavated as part of the campaigns of 1986 and 1987 (reported by Bietti et al. 1988a). These samples are very rich in both lithics and fauna. Based on the position of the test trenches (at about the level of the numeral 8 in Fig. 3.14), it can be assumed that the archaeological materials were derived from the exposed strata below layer X (Fig. 3.14, no. 10), that is, from strata XI through XIV in Fig. 3.15. As such, these two collections sample a series of archaeological strata and are roughly equivalent to the strata groups from Grotta dei Moscerini.

The third sample from Breuil comes from strata 3 and 4, at the very summit of the sequence (Fig. 3.14, nos. 10, 11). Strata 3 and 4 consist of a series of thin stalagmitic crusts with layers of looser angular limestone fragments separating them, between 30 and 70 cm thick. Bone and Mousterian artifacts are locally abundant. Because of the difficulty of digging on the sloping face of the deposits at Grotta Breuil, excavations began simultaneously at the top and the bottom of the sequence. The most recent intact Paleolithic stratum was assigned the Arabic numeral 3 (1 and 2 are surficial deposits), and the lowermost was assigned Roman numeral XX. When the stratigraphic sequence was finally fully exposed in 1992, it became clear that strata 3 and 4 sit directly on top of stratum X (the projecting stalagmite). Thus, the assemblage from strata 3 and 4 should postdate materials found in the slope deposits, derived from strata XI–XIV, by a short interval.

It is worth noting that no signs of the Upper Paleolithic component reported by Taschini (1970) have been found during the recent excavations. Traces of Neolithic and later occupations, in the form of obsidian bladelets and some crude ceramic shards, have been found in the superficial deposits overlying stratum 3 at the top of the sequence. On the other hand, many Mousterian flakes from the uppermost strata are elongated and laminar, and these may have been mistaken for Upper Paleolithic artifacts in the sample from the first test excavations.

Open-air Paleolithic sites

Open-air locations yielding both Middle and Upper Paleolithic artifacts have long been known to exist in the Pontine Plain or "Agro Pontino," as well as in the Fondi Basin to the south (Ansuini et al. 1990–91; Bietti 1969; Bietti et al. 1988b; Blanc 1937b; Mussi 1977–82; Tozzi 1970; Zei 1970). In this study I employ data on assemblages from a number of open-air surface localities. The main Mousterian open-air sample (assemblage designation "Ap") comes from the first systematic survey of the Agro Pontino. Directed by Drs. A. Voorrips and S. Loving of the Institut voor Prae- en Protohistorie (the IPP), University of Amsterdam (Voorrips et al. 1991; Loving et al. 1990–91), the Agro Pontino survey employed a "non-site" recording strategy, oriented towards examining long-term changes in landscape use in prehistory. Extremely rigorous recording and collection procedures were followed in the IPP survey: all recognized archaeological material was collected in the field, and the presence of unmodified flint pebbles was recorded for each survey unit (normally an agricultural field).

The Pontine Plain presents a relatively unique context for the study of distributional patterns in Paleolithic surface archaeological remains. Pleistocene and Holocene soils are extremely well documented and mapped (Sevink et al. 1984, 1991). This makes it possible to identify displaced or disturbed finds of Paleolithic age through their association with more recent soil units. Historically, the "Pontine Marshes" (as the area was also known) were unsuitable for permanent agricultural habitation until they were drained in the 1930s, and the extreme scarcity of Neolithic and Bronze Age remains suggests that these conditions may have extended well back into prehistory (Koot 1991). Since the area was not subject to intensive plowing until 40 years ago, the Paleolithic surface archaeology is relatively intact compared with other parts of Europe. The frequency of recent (non-patinated) breakage and edge damage is extremely low in comparison with Paleolithic materials from parts of Italy with a long history of intensive agricultural use.

The vast majority of the artifacts which make up the Agro Pontino sample are derived from locations which would not be considered "sites" by conventional assessment. Most finds come from highly diffuse scatters or "isolated occurrences" containing fewer than ten artifacts. High artifact densities, when they occur at all, are usually encountered in association with sources of flint pebbles. A total of 194 retouched tools and unmodified Levallois flakes and 192 cores from pre-1986 Agro Pontino survey collections are included in the present study. A combination of artifact morphology and patination characteristics were employed in selecting the artifacts most likely to date to the Middle Paleolithic. Artifact types known to be unique to the Mousterian, such as Levallois flakes, discoid and Levallois cores, and sidescrapers with typical retouch, were considered to be most reliably attributable to the Middle Paleolithic. The patination on diagnostic Middle Paleolithic artifacts is both more developed and of a different color than that on Upper Paleolithic and later artifacts. Non-Levallois sidescrapers and other tools made on flakes were included in the Mousterian sample based on qualities of the patina. Since only patination criteria could be used to aid in their selection, flakes and debris lacking faceted platforms and/or dorsal scar patterns characteristic of discoid or Levallois reduction were not included in the analyzed sample. Similarly, only heavily patinated cores of typical Mousterian forms were selected for the Middle Paleolithic sample. Spatial association *was not used* to partition the sample.

Despite the care taken in its selection, the Agro Pontino sample is clearly biased. The relative frequencies of unretouched flakes from Levallois and radially worked cores are undoubtedly inflated relative to other flake types, since such artifacts could not be reliably assigned to a particular period or cultural-stratigraphic unit. The bias towards Levallois is weaker among retouched tools, since tool form and the type of retouch were also used to select the sample. The abundances of plain unretouched flakes, of flakes and blanks without faceted butts, and of Mousterian tools made on blade-like pieces are also underestimated, as these too could not be assigned to the Mousterian with any degree of confidence.

In addition to the Agro Pontino survey sample, materials from other open-air localities in coastal Latium are employed in a few analyses. The best-known of these is ***Canale Mussolini*** or ***Canale delle Acque Alte*** (Blanc 1935b, 1936; Taschini 1972) (the latter name was adopted after the Second World War, for obvious motives). Canale Mussolini is not in fact a single site, but a series of buried localities and findspots exposed in the banks of canals and wells constructed as part of the efforts to drain the swampy parts of the Pontine Plain. The various locations exhibited a rather uniform stratigraphy (Blanc 1935b; Blanc and Segre 1953) and are commonly treated as a single unit. Buried Pleistocene strata yielded not only archaeological materials but also collections of fauna and well-preserved macrobotanical materials, including large sections of tree trunks, one of which yielded a radiocarbon date of >55,000 years BP (Blanc et al. 1957). Five of the major stratigraphic units (E2, E1, D, C2, and C1) yielded artifacts of an exclusively Middle Paleolithic character, while level B2 contained a mixture of Upper Paleolithic and possible Mousterian pieces (Taschini 1972). All artifacts from the Mousterian levels in collections housed at the Istituto Italiano di Paleontologia Umana and the Museo Pigorini in Rome were restudied.

Some of the earliest reports of Mousterian materials from coastal Latium describe open-air localities on the coast near the modern town of Anzio (Blanc 1937b). Artifacts from three of these sites, ***Tor Caldara***, ***Nettuno*** (Spiaggia S. Rocco), and ***Le Grottacce*** (Vallone Carnevale) (Fig. 3.1), were restudied for this research. The materials from Tor Caldara and Nettuno are derived from a series of layers within stratified Pleistocene dune deposits (Blanc 1935d, 1937b:285–86). The artifact-bearing layers, which may represent Pleistocene beaches, contain unworked pebbles as well as Mousterian artifacts. Overlying strata yield Upper Paleolithic assemblages (Blanc 1937b). The conditions of the Mousterian artifacts vary from quite fresh to heavily rolled and abraded. Artifacts found at Le Grottacce come from the middle and upper parts of a thick stratum of reddish dune sands, which sit atop a bed of lithoid tuff containing marine fauna and thought to date to one of the later phases of the "Tyrrhenian" (isotope stage 5) (Blanc 1935b). The materials from Le Grottacce are exclusively Upper Paleolithic in character. Unambiguous "type fossils" are scarce, but both Gravettian and Aurignacian components may be represented.

Artifacts from Tor Caldara, Nettuno, and Le Grottacce, housed at the IIPU and the Museo Pigorini, were collected over a period of 20 years or more beginning in the mid 1930s. Collection procedures were not described, but the compositions of the assemblages suggest that only the most "diagnostic," and perhaps the most esthetically pleasing, pieces were retained. All of the collections studied do contain both broken artifacts and unretouched flakes and debris, but they are heavily biased towards cores and retouched tools. The assemblages from Canale Mussolini were collected over a shorter period of time, but not in a very systematic manner. Assemblages from these four open-air localities are in no way comparable to collections from excavated sites, and they are consequently used below in a restricted number of comparative contexts.

Upper Paleolithic sites and assemblages

Materials from several Upper Paleolithic sites in Latium appear in a few analyses below, mostly in discussions of raw material exploitation patterns. ***Grotta del Fossellone*** (literally "Cave of the Big Hole") is one of two sites on Monte Circeo containing substantial Upper Paleolithic strata (Fig. 3.2). This large collapsed cave was first excavated between 1936 and 1940 by the IIPU, with the assistance of the Abbé Breuil and H. Obermaier, among others (Blanc and Segre 1953:39; Vitagliano and Piperno 1990–91). Above a long Mousterian sequence, stratum 21 at Fossellone yielded a large "Typical" Aurignacian assemblage, while strata 20 and above contained less abundant material attributed to a later phase of the Upper Paleolithic, possibly the Gravettian (Blanc and Segre 1953). The so called "Circean" Aurignacian (Blanc and Segre 1953; Laplace 1966) contains many characteristic "type fossils" such as nosed and carenated scrapers and split-based bone points. The assemblage from layer 21 at Fossellone also exhibits frequent use of bipolar technique, used especially to produce thick blanks for nosed and carenated endscrapers. Typical Aurignacian prismatic blade technology was also employed, frequently in working larger non-pebble raw materials. The Middle Paleolithic assemblages from Fossellone were not available for study when data collection was in progress. A sample of more than 1200 retouched tools, cores, and unretouched flakes and blades from the Aurignacian strata was studied to provide some idea of raw material use by post-Mousterian populations in the Monte Circeo area.

Riparo Salvini is a small shelter located immediately above the town of Terracina, south of Monte Circeo (Fig. 3.1). A. Bietti has conducted extensive excavations in the shelter, which yielded a Late Epipaleolithic "final Epigravettian" assemblage (Avellino et al. 1989; Bietti et al. 1983). The deposits at Salvini exhibit very little recognizable stratigraphy, and the sediments give the appearance of a dark, organic-rich midden (Avellino et al. 1989; Bietti 1984; Bietti and Stiner 1992). A sample of burned bones from near the middle of the undifferentiated archaeological layer provided a 14C date of 12,400 ±170 years BP (Bietti 1984). A large sample of retouched and unretouched artifacts from the 1985, 1986, and 1989 excavation campaigns at Riparo Salvini was analyzed in the course of this study, again primarily as a source of information on raw material use.

Grotta Jolanda is located in the foothills east of the Agro Pontino, not far from the town of Sezze Romano and the Mousterian site Grotta della Cava (Fig. 3.1). The site was excavated in the early 1950s by the Pigorini Museum (Zei 1953), where the collections are presently housed. Like Riparo Salvini, Grotta Jolanda yielded an industry classified as "final Epigravettian" (Guerreschi 1992:214; Laplace 1964). However, on the basis of comparisons with dated assemblages from the sites of Grotta de La Cala and Grotta Polesini, it has been suggested that Grotta Jolanda belongs to a somewhat later phase than Salvini, perhaps dating to between 10,000 and 11,000 BP (Bietti et al. 1983: 340). All retouched tools and cores, along with a sample of other artifacts from the site, were examined.

The ***Cavernette Falische*** are a series of small caves and shelters located in northern Latium at the edge of a small elevated basin known as the Agro Falisco. The sites include Riparo Sambucco, Riparo Lattanzi, Caverna (or Grotta) Alta, and Caverna dell'Acqua. The first and largest excavations at the Cavernette Falische were conducted in the early part of the century (Rellini 1920); limited excavations were also carried out in the 1960s (Barra Incardona 1969). The stratigraphic context of materials from Rellini's early excavations, among the first Paleolithic digs in Latium, is somewhat ambiguous (Bietti 1990:98), but the collections are clearly all Upper Paleolithic in character. Mussi and Zampetti (1985) argue that the Cavernette Falische assemblages span principally the Gravettian period. However, the collections are highly selected, consisting entirely of large, generally whole retouched tools, cores, and blades. Because sediments were not sieved, it is likely that microlithic artifacts indicative of later Epigravettian components were missed entirely by the early excavators. A substantial number of classic carenated scrapers are also present in the collections housed at the Pigorini Museum, suggesting that an Aurignacian component is represented as well. As such, the sample of retouched tools, cores and blades from the Cavernette Falische employed in this study should be assumed to span a large proportion of, if not all of, the Upper Paleolithic period in coastal Latium.

Chronology and Absolute Dating

Before 1975, the Middle Paleolithic of coastal Latium had been dated almost solely by geological means. While geological criteria do at least provide a reliable lower age limit for Middle Paleolithic cave deposits on the coast, they are less helpful in determining either the relative or the absolute ages of individual assemblages, especially those from earlier excavations. During the past decade, Prof. H. Schwarcz has undertaken a comprehensive program of dating Middle and Upper Pleistocene sites on the Italian peninsula, including the four cave sites that form the foundation of this study.

The primary dating method used is *Electron Spin Resonance* (ESR), conducted on enamel from ungulate teeth. Uranium/thorium dating was also attempted using samples of stalagmite and marine shell. The U/Th technique, while more familiar to most researchers, is of limited utility in dating most of the cave sites, because the parent limestone in the region has a very low uranium content, and the stalagmites generally do not contain sufficient thorium to achieve an acceptable level of precision. Reliable U/Th dates were obtained only for the surface concretions at Grotta Guattari and a stalagmite underlying excavated deposits at Grotta di Sant'Agostino. The ESR technique, while theoretically sound, has several aspects which are still experimental (e.g., Grün 1988; Grün et al. 1987). One of the most pressing questions concerns rates of uranium uptake from the environment. Free uranium replaces calcium in tooth enamel and dentine; as the uranium accumulates in a specimen, it increases the level of radiation originating within the tooth itself. Since the intrinsic radiation dose rate changes over time, it is necessary to know the rates of change in order to adjust the accumulated dose for dating purposes. Two models of uranium uptake are generally considered in ESR studies. The "early uptake" model (EU) posits that the rate of uranium accumulation is quite high right after burial but decreases quickly over time. The "linear uptake" model (LU) assumes that a buried tooth picks up uranium at a more-or-less constant rate as long as it is in the ground (Grün 1988:39–40, Fig. 7; Ikeya 1978). The LU model has more often proven to be consistent with other dating techniques in the past, while the EU model generally yields what should be considered "minimum age" dates (Schwarcz and Grün 1993).

Table 3.8 and Figure 3.16 summarize the results of radiometric dating carried out on materials from the four main Mousterian cave sites to date (originally published by Schwarcz et al. 1990–91, 1991). In most cases, the results reported represent averages from multiple "runs" on one or more specimens. The standard deviations associated with the dates are generally large, so that there is considerable overlap between adjacent dates when the error terms are taken into account. However, the dates do allow at least a rough chronological ordering of assemblages. Moreover, it appears that they divide temporally into two groups. With the possible exception of M1 (Moscerini strata 1–10, with very few artifacts), the assemblages from Grotta Guattari and the external trench at Grotta dei Moscerini all appear to date to *55,000 BP or earlier*. The M2 strata (11–20) are not directly dated, but the underlying strata (25 and 26) appear to be between 67,000 and 81,000 years old. Based on the rates of sediment accumulation in the lower strata, it is very unlikely that strata 10–21 were formed more recently than 55,000 years ago. The ESR date of 51,000 for stratum 1 at Guattari is probably somewhat too recent. An unusually good U/Th date of around 56,000 BP for carbonate crusts on bones in the overlying surface deposit is probably more realistic (Schwarcz et al. 1990–91, 1991). Since the uranium-series date was obtained for the innermost layer of calcium-carbonate encrustation, the surface materials must have been deposited *before* this

Table 3.8. Dating results (all figures in years BP) (from Schwarcz et al. 1990–91)

	Technique	*Date*	*Error*
Breuil			
strata 3/4	ESR (LU)	36,600	±2700
Guattari			
surface (crust)	U/Th	51,000	±3000
stratum 1	ESR (LU)	54,200	±4100
stratum 4	ESR (LU)	71,000	±27,000
stratum 5	ESR (LU)	77,500	±9500
Moscerini			
Strata group 3			
stratum 25	ESR (LU)	79,000	N.A.
stratum 26	ESR (LU)	74,000	N.A.
stratum 33	ESR (LU)	106,000	N.A.[a]
Strata group 4			
stratum 38	ESR (LU)	101,000	N.A.
stratum 39	ESR (LU)	96,000	N.A.
Sant'Agostino			
level 0 (mixed)	ESR (LU)	32,000	±7000
level 1	ESR (LU)	43,000	±9000
level 2	ESR (LU)	53,000	±7000
level 3	ESR (LU)	54,000	±11,000

[a] Two specimens from stratum 33 gave divergent dates: one provided an LU date of 89,000 BP, the other 123,000 BP. Based on the sequence, the first is probably closer to the true date.

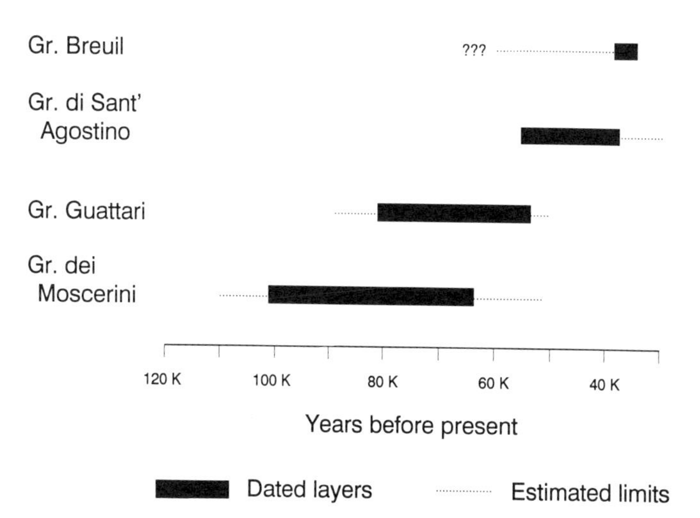

Figure 3.16. Summary of radiometric dating of Mousterian cave sites in coastal Latium.

date. Unfortunately, there is no way of knowing how long the bones on the surface were in place before crusts began to form.

Dated strata from the sites of Grotta Breuil and Grotta di Sant'Agostino appear to have been formed around *55,000 BP or later*. Two of the teeth from level 1 at Grotta di Sant'Agostino provide anomalously old dates (>110,000 BP). These particular samples may actually be derived from more ancient deposits, such as those represented by the basal stalagmite, which yielded a comparably early U/Th date (Schwarcz et al. 1990–91:54–55). The strata which are the original source of the materials in the 1986–87 sample ("B86–87") and the old *sondage* ("Bs") at Grotta Breuil have not been directly dated. However, less than 1 m of deposit separates these levels (X–XIV) from the stratum yielding the ESR date of around 37,000 BP (strata 3/4). Moreover, both geological criteria and aspects of the microfauna suggest that the bottom of the cultural sequence (several meters beneath the layers in question) postdates the beginning of isotope stage 3, about 67,000 years ago (Kotsakis 1990–91). For both these reasons, it is highly probable that the archaeological materials in the "B86–87" and "Bs" samples date to after 55,000 BP.

Hominid Fossil Associations

A number of hominid fossils have been recovered from Mousterian layers in the various cave sites discussed in this study. The best known of these is the nearly complete Neandertal cranium (Circeo I) found on the surface of Grotta Guattari (Sergi and Ascenzi 1974). Two mandibles (Circeo II and III), both attributable to *Homo sapiens neanderthalensis*, were also recovered at Guattari, one from inside the cave "a few decimeters" from the cranium, and the other from the breccia outside the cave (Manzi and Passarello 1988; Sergi 1954, 1955). Neither mandible is definitely associated with the cranium, although the one found inside the cave could belong to the same individual (Mallegni 1990–91). A fragment of human parietal recovered from the slope deposits at Grotta Breuil during the recent excavation campaigns can also be attributed to a Neandertal (Manzi and Passarello 1988; Manzi and Passarello 1990–91). Two molars from the same site are attributable to *Homo sapiens*, but it is not clear whether they represent a modern or archaic subspecies (Bietti et al. 1988a:388; Manzi and Passarello 1990–91). A fragmentary mandible from a juvenile Neandertal was also recovered from strata containing Mousterian industry at the site of Grotta del Fossellone (Blanc 1954). This specimen, considered lost for many years, has recently been relocated and is under study as of this writing (January 1994).

It is worth pointing out that the only truly diagnostic human remains associated with the subject assemblages, the Guattari specimens, date to between 52,000 and 55,000 years ago, or earlier. There are no fossils from more recent layers at Grotta Breuil and Grotta di Sant'Agostino. While the parietal fragment from Grotta Breuil seems to represent a Neandertal, its association remains ambiguous. Thus, although most of the Pontinian Mousterian assemblages were probably produced by Neandertals along the lines of the individual represented by the Guattari cranium, we cannot exclude the possibility that some other hominid form is responsible for the more recent Middle Paleolithic assemblages.

Faunal Exploitation and Subsistence

Middle Paleolithic archaeofaunas from Latium are relatively homogeneous with respect to species representation. Red deer (*Cervus elaphus*) and fallow deer (*Dama dama*) are the most common species in virtually all Mousterian faunal assemblages, a pattern continuing into the late Upper Paleolithic (Caloi and Palombo 1988, 1990–91; Stiner 1990b, 1992, 1994). The only exceptions are the fauna-poor strata 4 and 5 at Grotta Guattari, in which aurochs (*Bos primigenius*) predominates. Species such as aurochs, ibex (*Capra ibex*), roe deer (*Capreolus capreolus*), horse (*Equus caballus*), and wild boar (*Sus scrofa*) are found in most faunas, but in varying numbers. For example, ibex is quite abundant at Grotta Breuil and is present in significant quantities at Grotta di Sant'Agostino, but is practically absent from the other two cave sites. Grotta di Sant'Agostino also contains a substantial number of wild boar remains. Roe deer are relatively numerous at Sant'Agostino and Grotta dei Moscerini, second only in abundance to the two larger cervid species (Stiner 1992, 1994). A number of ungulate species, including chamois (*Rupicapra rupicapra*), wild ass (*Equus hydruntinus*), hippopotamus (*Hippopotamus amphibius*) and rhinoceros (*Dicerorhinus sp.*), are represented by a handful of bones from a few assemblages (Stiner 1990b, 1994).

Evidence derived from analyses of faunal remains from the four principal Mousterian cave sites (Guattari, Breuil, Moscerini, and Sant'Agostino) is an indispensable source of information about aspects of foraging and land use that might have influenced technological behavior. This research draws heavily on studies of archaeofaunas from Paleolithic caves in Latium conducted by Dr. M. Stiner (Stiner 1990b, 1991a, 1991b, 1992, 1994). Four facets of the faunal data—the contribution of non-human carnivores, the relative abundance of bone, the age structure of death assemblages, and body-part representation—are of particular importance to this research. Some general results pertaining to these topics are discussed below.

Non-human carnivores: presence and contribution to archaeofaunas

Many of the Middle Paleolithic faunal assemblages discussed here contain the bones of a variety of non-human carnivores. Some of the smaller carnivores such as foxes may well have been hunted by hominids as early as the late Mousterian (Stiner 1990b). On the other hand, most if not all remains of the large carnivores, principally spotted hyaena, wolf (*Canis lupus*), and brown bear (*Ursus arctos*), probably appear in the cave site deposits as the result of natural deaths and/or transport of carcass parts by other carnivores. Holocene representatives of these species commonly use caves for hibernation refuges (bears) or as dens for raising young (wolves, hyaenas). Bears frequently die during hibernation (Kurten 1976; Rogers 1981), leaving large carcasses which may remain fairly intact. Young hyaenas and wolves often die in dens, as do older animals belonging to or killed by the local group. Juvenile animals also leave behind shed deciduous teeth, a good indicator of denning behavior. Another byproduct of denning for the purpose of rearing pups—though generally not in the case of hibernation denning—are collections of bones of prey brought to the shelter to feed young.

Table 3.9 summarizes Stiner's assessment of the relative importance of hominid and non-human carnivore agents in creating faunal assemblages at the four cave sites, based on both the frequency of carnivore remains and the type of gnawing damage on bones (Stiner 1994: Table 5.24). The dominant non-human carnivore in the levels laid down before 55,000 years ago appears to have been the spotted hyaena. Two of the older assemblages, Guattari 1 and Moscerini 5, are almost completely attributable to the feeding and bone-transport behaviors of these animals. The Guattari 2 assemblage shows evidence of a greater hominid contribution, although hyaenas were still important transporters

Table 3.9. Summary of hominid and carnivore contributions to faunal assemblages (after Stiner 1994: Table 5.24)

	Dominant agent(s)		
	Primarily hominid	*Mixed, hominid-dominated*	*Primarily carnivore*
Breuil	1986–87 sample		
	strata 3/4		
Guattari	stratum 4	stratum 2[a]	stratum 1[a]
	stratum 5		
Moscerini	strata groups 2–4		strata group 1[a]
	strata group 6		strata group 5[a]
Sant'Agostino	levels 1–3		level 4[b]

[a] Dominant carnivore is spotted hyaena (*Crocuta crocuta*).
[b] Dominant carnivore is wolf (*Canis lupus*).

and modifiers of ungulate bones. The remaining faunas (Moscerini 1–4 and 6, Guattari 4 and 5) exhibit very little evidence that non-human carnivores contributed significantly to assemblage formation. Of these, Moscerini 3, 4, and 6 contain the largest numbers of bones.

The dominant non-human carnivore contributor to the post-55,000 BP assemblages was the wolf. Both cut marks and percussion fractures attributable to hominids, as well as gnawing marks made by canids, are common in faunas from all four levels at Grotta di Sant'Agostino and strata 3 and 4 at Grotta Breuil. When all things are considered, however, the fauna of level 4 at Sant'Agostino is best considered the work of wolves rather than people, while all other bone assemblages from this site, as well as all faunas from Grotta Breuil, are attributable primarily to human actions.

It may be significant that different opportunities for scheduling visits to cave sites would have been open to Mousterian hominids according to the species of carnivore that also had an interest in the same shelter. Contemporary spotted hyaena populations—and presumably Pleistocene populations as well—may occupy dens through successive seasons, and often for several years running (Ewer 1973:305; Kruuk 1972:28, 306). In contrast, wolves make intensive use of shelters for denning only in the spring, for the purpose of giving birth to and raising pups until they can keep up with the pack (Burton 1979:54; Ewer 1973:308–9). As a consequence, wolf dens are occupied continuously for no more than two or three months. It is difficult to say which group of social animals (hominids or carnivores) would have dominated the use of caves. Neandertals and carnivores probably scheduled their use of caves around one another (see Straus 1982b), rarely engaging in direct competition for space: the species that first established occupancy would have likely retained control over that space for as long as it wanted to stay. However, because wolves use dens briefly and on a highly seasonal basis, substantial hominid occupations could have alternated seasonally with wolf denning activities with little or no actual conflict over life-space. On the other hand, because hyaenas use dens year round, they are unlikely to have taken up residence in caves regularly visited by hominids. Conversely, once hyaena dens were established, even seasonal opportunities for hominid occupation would have been few and far between.

Ungulate procurement and transport by hominids

The studies of ungulate procurement patterns focus primarily on the treatment of medium-sized ungulates. The most common species in this category are red and fallow deer. The medium ungulate category may also include significant numbers of ibex, particularly in the more recent strata. There are two reasons for concentrating on the medium ungulates. First, they are the most common species in almost all faunas. Second, they exhibit the most variable treatment over time and across different assemblages. Small roe deer, as well as large animals like aurochs and horse, are also present in limited numbers in most faunas, but patterns of procurement and anatomical-part transport are relatively uniform. Horse and aurochs are more often represented by cranial elements in these faunas. Carcasses of the small roe deer tend in contrast to be relatively completely represented, regardless of time period or context (Stiner 1991b, 1994).

The discussions that follow refer only to those faunas judged, on taphonomic grounds, to be predominantly or exclusively accumulated by hominids. Assemblages in which evidence of carnivore damage, such as gnaw marks, outweighs traces of human modification (e.g., cut marks and percussion fractures), and/or faunas in which carnivore remains are especially abundant, are excluded from further consideration in this section: the excluded samples are found in the last column ("Primarily carnivore") of Table 3.9. For more detailed discussions of the taphonomic methods used, the reader is referred to other recent publications (Stiner 1990b, 1994).

Analyses of the ages of death of ungulate prey (Stiner 1990a, 1990b, 1994) reveal two contrasting patterns of mortality among the hominid-generated Mousterian faunas. One typical mortality pattern is characterized by an unusually high frequency (>30%) of relatively old animals. This is an unusual pattern because individuals of advanced age are scarce in most wild populations. Stiner (1990a) argues from extensive comparisons with modern predators, both human and non-human, that the prey assemblages with large numbers of old animals are most likely attributable to *scavenging* natural deaths or the kills of other predators, especially cursorial hunters like hyaena and wolf. The second common mortality pattern found in the Mousterian archaeofaunas of west-central Italy is more similar to a typical population's "living structure," with comparatively few old animals but many young and prime-aged individuals. Some late Mousterian faunas, as well as Upper Paleolithic and Holocene assemblages from the same region, show a tendency towards unusually high frequencies of prime-aged animals. Analogies with the mortality patterns produced by human and non-human predators in the world today suggest that the "living structure" and prime-age-dominated faunas most often occur as a result of *ambush hunting*. The tendency towards more prime animals is thought to reflect greater control over the selection of individual prey during procurement (Stiner 1990a).

The representation of anatomical elements also varies markedly among the faunal assemblages collected by Mousterian hominids. Anatomical element representation tends towards one of two patterns, exemplified in Figures 3.17a and b. In the figures, the ungulate anatomy is divided into nine basic regions; the height of each bar in the graph is equivalent to the total number of elements present for that region, divided by the expected number in a complete skeleton. In a complete ungulate carcass, all nine bars would be of equal height (see Stiner 1991b, 1994 for a more detailed discussion of the methods used). One group of Mousterian faunal assemblages from the coastal Italian sites is dominated by cranial elements, often to the near exclusion of other parts (Fig. 3.17a). Because only bony parts (maxillae, mandibles, cranial parts) and not isolated teeth were used in calculating anatomical representation, it is extremely unlikely that the head-dominated faunas are strictly products of differential preservation. Nor are these easily interpreted as highly culled assemblages containing a variety of low-utility parts (e.g., see Binford 1978a), since bones of the lower limb and foot are also strongly under-represented relative to head parts. The second group of faunal assemblages contain a much more complete range of durable anatomical parts, although cranial elements may still be somewhat over-represented (Fig. 3.17b). In these cases, under-representation of the axial skeleton does probably reflect preservational and recovery biases. It is worth noting that later Upper Paleolithic faunas from this region exhibit the same kind of essentially complete anatomical representation.

There are strong correlations between patterns of anatomical representation and age structure in these Mousterian cases. Interassemblage variation and associations between age structure and anatomical-part representation in ungulate faunas are summarized in Table 3.10. Faunas dominated by cranial elements also tend to have unexpectedly large numbers of older animals, indicative of scavenging. Faunal assemblages with more complete body-part representation are always characterized by patterns of "living structure" or prime-dominated mortality, thought to indicate ambush hunting. Different modes of game procurement and hominids' decisions about what kinds of anatomical parts to transport provide the necessary links between age structure and anatomical-part transport. The sites in question are all caves, places where ungulates seldom go of their own accord. Consequently, it is more than likely that the faunas represent animal foods transported to the cave for processing and/or consumption, and not the dregs of kill and butchery locations: in other words, the bones found in the deposits are there because hominids (or carnivores) brought them, not because hunters failed to carry them away from a kill.

The association between unexpectedly large numbers of older animals and head-dominated anatomical patterns would seem to arise from the nature of scavenging as a means of procuring food. Crania, and the rich brain tissue contained within them, are one potential source of food which frequently remains at carcasses ravaged by non-human carnivores (Binford 1981:211–16; Blumenschine 1986:36–37). Moreover, these elements retain their nutritional value even when an animal is in poor physical condition and is likely to become a candidate for scavenging (Stiner 1991b). Because hominids can break open thick cranial bones through the use of stone hammers, the brain

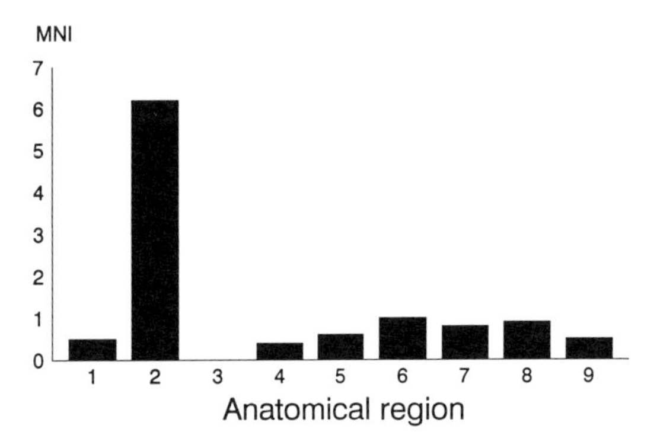

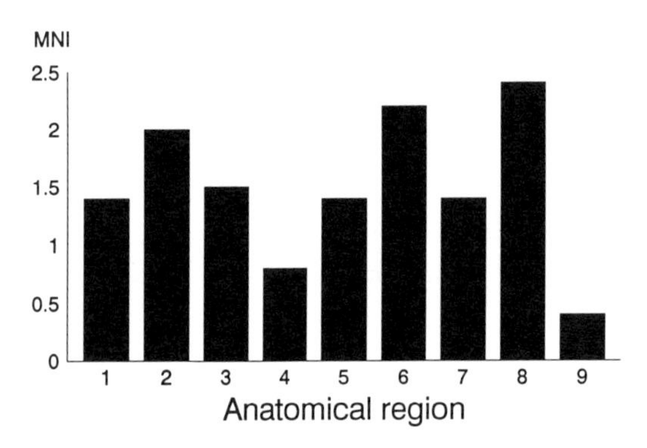

Figure 3.17. Examples of different patterns of anatomical representation in Middle Paleolithic archaeofaunas from coastal Latium. Region codes: 1=horn/antler; 2=cranium/mandible; 3=neck; 4=axial (ribs and vertebrae); 5=upper front limb; 6=lower front limb; 7=upper hind limb; 8=lower hind limb; 9=feet (after Stiner 1994: Fig. 9.6).
a. Head-dominated anatomical pattern: Gr. dei Moscerini, strata group 6 (red and fallow deer).
b. "Meaty" or ± complete anatomical pattern: Gr. Breuil, strata 3/4 (all medium ungulates).

tissues of medium and large ungulates may have been a more accessible and attractive resource for them than for certain other predator/scavengers.

In the other faunal assemblages, the presence of a wide range of meat-bearing anatomical elements is entirely consistent with the notion that humans obtained medium ungulates through some kind of ambush hunting, as suggested by the "living structure" or prime-dominated mortality patterns. Hunting prey affords early and more-or-less uninhibited access to complete carcasses of healthy animals, and provides predators with a full array of anatomical elements from which to choose. The fact that nearly all meat- and marrow-bearing anatomical parts were habitually transported to cave sites, with little or no culling of low utility parts, further suggests that the kill sites were not far away.

Table 3.10 also shows how ungulate procurement and anatomical-part transport varied over time. All the demonstrably or presumably recent Mousterian faunas from Grotta di Sant'Agostino and Grotta Breuil are characterized by "living structure" or prime-dominated mortality patterns and more-or-less complete anatomical representation. In contrast, the older Mousterian faunas from Grotta Guattari and Grotta dei Moscerini contain large proportions of older animals and are heavily dominated by cranial elements. In other words, before the 55,000 BP juncture it appears that Mousterian foragers occupying the cave sites in coastal Latium obtained game primarily by scavenging, and tended to return mostly crania and mandibular elements to caves, perhaps for further processing. Around 55,000 years ago and thereafter, the homi-

Table 3.10. Summary of age structure and anatomical representation in medium ungulates (red and fallow deer): hominid-collected faunas only (after Stiner 1994: Appendix 1, Table 11.5)

	Standardized anatomical data[a]					*Age/Classes (percent)*		
Assemblage	*Cranium + mandible*	*Neck + axial*	*Upper limbs*	*Lower limbs*	*Feet*	*Juvenile*	*Prime adult*	*Old-aged*
Breuil								
1986–87 sample	3.00	0.93	3.25	3.67	0.79	0.0	71.0	29.0
strata 3/4	2.00	2.39	3.00	4.54	0.46	55.0	45.0	0.0
Guattari								
stratum 2	2.75	0.00	0.50	0.70	0.00		N.A.	
strata 4 and 5	1.00	0.00	0.25	0.13	0.00	24.0	44.0	32.0
Moscerini								
strata group 2	1.00	0.00	0.00	0.12	0.08	0.0	43.0	57.0
strata group 3	15.00	0.00	0.00	1.75	0.17	0.0	55.0	45.0
strata group 4	7.25	0.00	0.00	0.12	0.00	22.0	22.0	56.0
strata group 6	6.25	0.08	0.75	1.58	0.33	14.0	36.0	50.0
Sant'Agostino								
level 1	3.75	0.96	4.75	4.37	0.46		(all levels combined)	
level 2	2.00	0.38	3.25	4.33	0.21	41.0	43.0	16.0
level 3	1.25	0.24	1.50	0.87	0.04			

[a] Standardized anatomical representation: total number of elements for region divided by expected number in complete skeleton (see Stiner 1990a, 1994, for definitions and explanation of calculations).

nids using these cave sites began to procure game more often by hunting, introducing a wide range of meat-bearing anatomical elements into the shelter sites. As will be shown in the following chapters, these contrasts in the exploitation of ungulate resources are associated with a number of differences in the ways stone tools were made, modified, and renewed.

It should be emphasized that inferences about hominid behavior based on the Mousterian faunal remains are greatly strengthened by combining a number of analytical perspectives. Taphonomic analyses designed to ascertain the "authorship" of each faunal assemblage are crucial precursors to drawing any conclusions about human behavior. Unless one can ascertain whether, and to what degree, hominids were actually responsible for the accumulation of bones, further conclusions rest on shaky foundations. Moreover, conclusions about modes of ungulate procurement are strengthened by the combination of both anatomical and prey age data. For example, the predominance of head parts alone would not necessarily point towards scavenging, were it not for the high frequencies of older animals in the same assemblages.

Evidence for the use of other resources

As is the case with the great majority of other Middle Paleolithic sites, the remains of medium to large terrestrial vertebrates comprise the most abundant evidence for hominid subsistence activities in the Pontinian cave sites. However, sparser remains provide tantalizing evidence of the other animal foods that played a part in the Mousterian diet. Foraging for resources other than large game seems to have been targeted largely at marine or aquatic organisms. The bones of a variety of small terrestrial species, including lagomorphs, rodents, and insectivores, were found in all four sites, but damage patterns indicate that these were introduced largely, if not entirely, by canids and other predators (Stiner 1993a, 1994).

Like many other Mediterranean coastal sites, Grotta dei Moscerini provides good evidence for human exploitation of marine bivalves. Although a wide variety of mollusks are presented in collections, three species dominate. Two, *Callista chione* and *Glycimeris sp.*, are thick-shelled, sand-dwelling clams; the other (*Mytilus galloprovincialis*) is a mussel that prefers rocky substrates. The frequencies of bottom-dwelling clams and rock-dwelling mussels vary sharply among the various levels at Moscerini, possibly indicating changes in the littoral environment associated with variable sea levels (Stiner 1993a). Many shells are fractured and burned in a manner highly suggestive of human consumption (Stiner 1993a, 1994), as well as being sometimes retouched by percussion. However, although shells are present in appreciable numbers in some levels, they are never sufficiently common to constitute a true shell midden. In fact, if only hinges are considered, clam and mussel shells are actually less abundant, on average, than stone artifacts (Stiner 1994: Table 6.12). Mousterian hominids at Grotta dei Moscerini also sometimes utilized two species of tortoise, one terrestrial (*Testudo graeca*) and one aquatic (*Emys orbicularis*). Tortoise remains include bones and carapace and plastron fragments, all of which tend to be heavily broken up. Many carapace fragments show signs of battering with a blunt object, such as a stone hammer (Stiner 1993a, 1994). Curiously, tortoise remains are confined to the interior strata at the site, where they are found in both the hominid-dominated (strata group 6) and hyaena-dominated levels (strata group 5).

It is difficult to account for the fact that evidence for hominid exploitation of marine resources is confined to Grotta dei Moscerini. Differences in preservation do not account for this, as depositional environments and the conditions of bones are quite similar at all four cave sites. Other sites do contain the occasional beachworn shell, but these are not likely to have been used as food. One possible explanation is that, because of its proximity to the sea, Grotta dei Moscerini simply afforded the easiest access to productive shell beds. Moscerini presently opens directly on the sea, and was occupied relatively early (i.e., between about 110,000 and 60,000 BP), when sea levels were relatively high. Grotta Breuil is actually slightly closer to the sea today, but the levels sampled thus far probably date to after 55,000 BP, when the sea was lower and therefore farther away. If this is so, however, it means that Mousterian hominids were not given to transporting mollusks and other aquatic foods more than a kilometer or so, since none of the caves was ever located any great distance from the coast during the period in question.

Finally, it is worth mentioning that a few isolated bones of the Mediterranean monk seal (*Monachus monachus*) were recovered at both Grotta dei Moscerini and Grotta di Sant'Agostino: moreover, at least one of the bones from Sant'Agostino bears a cut mark. The monk seal, a highly endangered marine predator, has the unique habit of retreating to coastal grottoes to give birth (Wirtz 1968). Its presence in the Mousterian deposits could be entirely fortuitous, but it is also possible that Mousterian foragers (or other predators) happened upon a female seal when she was on land just before or just after parturition. In either case, the presence of seal bones does not in any way imply the kinds of elaborate hunting strategies or technologies we have come to associate with marine mammal exploitation among modern hunter-gatherers.

New Approaches to Old Sites

Many of the collections used in this study come from excavations carried out in the 1940s and 1950s. Some archaeologists have raised objections to the use of older collections in addressing novel research issues (i.e., Rigaud and Simek 1987). Certainly, we cannot naively assume that materials excavated 40, 30, or even 20 years ago were collected using the same procedures in general use today, and in fact we know that older artifact and faunal assemblages were often highly selected. However, there is no such thing as a perfect, unbiased sample. It is a challenge in all archaeological research to identify the biases in the available observations, and to adopt an approach that circumvents or minimizes their influence. Working with older collections certainly entails a number of obstacles, but many of these can be overcome simply by how one poses questions.

The older collections employed in this study are surprisingly complete, given their vintage. Blanc, Cardini, and other members of the IIPU employed unusually rigorous collection procedures in their excavations of the 1940s and 1950s. All archaeological deposits were sieved through mesh screen (probably around 5–10 mm), and all potential artifacts were retained, including small, unmodified flakes, debris, and even unmodified pebbles. Comparisons with collections from recent excavations at Grotta Breuil, which were obtained by water screening through very fine mesh, indicate that the only major recovery bias in the older collections of lithic materials is a function of size: pieces less than about 10 mm are under-represented in the old IIPU collections. Within the sample of specimens larger than this, there does not appear to be any strong bias towards formal tools, unbroken pieces, or esthetically pleasing specimens. The impetus for this care in artifact recovery is not entirely clear. Blanc and associates did favor a technological rather than a morphological or typological approach (Taschini 1970:60), and this may have led them to collect as much of the technological material as possible. It is also my impression that Mousterian deposits along the Tyrrhenian coast exhibit relatively low artifact densities compared to some better-known Middle Paleolithic sites in France and Israel, and it may simply have been more practical to collect and curate a "complete" lithic assemblage.

In contrast to the rather advanced artifact recovery procedures, provenience designation was relatively crude in the IIPU excavations of the 1940s and 1950s. As was common practice at the time, the locations of bones and artifacts were recorded only by trench and geological (or arbitrary) level. Point proveniencing was not practiced, even for the most "important" finds. While the comparatively coarse definition of assemblages places certain restrictions on the kinds of questions that can be profitably addressed with these data, it does not prohibit useful research. Clearly, the collections of artifacts employed in this study do not correspond with single occupational events. Nor can they be assumed to be internally homogeneous with respect to the kinds of human behavior represented. Like the vast majority of Paleolithic cases, the assemblages discussed here are accumulations of debris representing tens, hundreds, or even thousands of years of human presence in the coastal caves. In other words, they are long-term samples of human behavioral *tendencies*, rather than records of *instantaneous events*.

Accumulations of materials representing dozens of human generations are certainly not appropriate for reconstructing the daily lives of prehistoric hunter gatherers in scintillating detail. On the other hand, assemblages of the type used here are entirely appropriate to addressing questions about evolutionary or adaptive processes. In fact, a sample representing a "long-term average" may be *more* suitable for addressing some kinds of evolutionary questions than are detailed records of single occupational events. After all, evolution occurs as the result of selection on variation, and understanding evolutionary processes requires understanding changing ranges of variation within populations. Thin, quasi-instantaneous archaeological occurrences may have the potential to provide a very detailed view of human behavior in the remote past, but they tell us little about variation. Moreover, the inherent inaccuracy of available dating techniques makes it virtually impossible to correlate such "snapshots" of the past with a commensurate degree of precision. Undifferentiated "palimpsest" deposits contained within thick geological layers can provide a better assessment of the ranges of behavioral options exercised by prehistoric groups over a given span of time than do the single "events" sampled in extraordinarily well-preserved, finely divided archaeological occurrences.

I by no means wish to argue that archaeologists should abandon the methodological innovations of the past two decades (point proveniencing, microstratigraphy, and micro-sedimentology), which have provided the field with surprising and important results. However, it must be recognized that thick strata preserve records of human behavior occurring over a prolonged period of time, and that these provide a useful, and necessary, *complement* to the kinds of high-resolution data provided by the application of modern data recovery and recording techniques to a few, quite possibly unique, archaeological cases. The situation is analogous to that of someone trying to reconstruct an important event using two kinds of photographic documents: a few high-resolution still photographs from which it is possible to discern many details but that cannot be assembled into a continuous record, and numerous long pieces of blurry movie film that provide a sense of action and movement but that cannot resolve individual actors or objects with any precision. If each is exploited for its strongest features, the hazy movies and the sharp but unconnected stills can together provide a more complete image of the events in question than could either kind by itself. The same is true of different varieties of archaeological record: long-term "palimpsest" deposits and thin, high-integrity sites. Not only is it uneconomical to reject older collections and thick, undifferentiated accumulations of artifacts and debris; it is tantamount to abandoning an entire perspective on the past.

4

CORE REDUCTION TECHNOLOGY

The first step in creating stone tools, once the necessary raw material has been collected, is the production of blanks from the raw stone. In European Mousterian assemblages, this initial stage generally entailed the manufacture of flakes or blades (elongated flakes) using one of a variety of procedures for preparing cores and detaching subsequent removals. Because it is (for the purpose of this study at least) the first stage of a tool's life history, blank production technology is the logical place to begin the discussion of Mousterian lithic technology in western Italy.

Two somewhat distinct schools of thought exist in the study of flake production technologies or *chaînes opératoires*; these are generally associated with French- and English-speaking researchers, respectively. The "French school," which grew out of the early technological work of François Bordes and especially of Jacques Tixier (e.g., Tixier et al. 1980), is undeniably more influential in studies of the European and Near Eastern Paleolithic. One of the foremost aspects of the French school is that core reduction is treated as a dynamic process, an evolving succession of technological choices made in response to the changing form of the core itself, rather than as an inflexible series of prescribed actions that inevitably lead to the same result. In other words, the focus is on the process of reduction rather than the products.

Not surprisingly, the most elaborate variety of Middle Paleolithic flake production technology, the Levallois "method," has received the greatest attention. The work of young French researchers, drawing on experimentation and refitting as well as on formal analysis of archaeological materials, has resulted in a complete redefinition of this most unique and characteristic facet of Mousterian lithic technology (Boëda 1986; Boëda et al. 1990; Geneste 1985). More recently, others have turned to less complex and striking, but no less widespread, methods of flake manufacture (Boëda 1991, 1993; Collina-Girard and Turq 1991; Turq 1992).

At its best, the French approach to *chaînes opératoires* can produce synthetic definitions, such as that recently proposed for the Levallois method (Boëda 1986; Boëda et al. 1990), which describe the mechanical and geometrical principles involved in one approach to "stone butchery." The Levallois method is defined not in terms of a limited range of "end products," as had been the case previously, but in terms of a set of core properties that must be achieved and sustained in order to produce these commonly recurring forms. Such a perspective enables many apparently distinct methods of flake production to be grouped together as alternative means for maintaining the requisite geometric properties of cores.

The treatment of core technology by English-speaking archaeologists, on the other hand, has been quite varied. While good use has been made of refitting studies (e.g., Phillips 1991; Volkman 1983) and experimental replication, it is safe to say that American researchers, regardless of their research orientations, have tended to rely more on statistical studies of core, flake, and blank attributes (e.g., Baumler 1988; Crew 1975). An approach relying on the collection and analysis of large quantities of metrical and character data forces one to concentrate on the by-products and products of core reduction. In addition, it tends to result in somewhat abstract and austere portrayals of the processes of flake and blank production.

In this study I employ an attribute/metric-based approach. This choice reflects both the research goals and the constraints imposed by the data base itself. The primary aim of this study is to investigate long-term evolutionary processes in terms of possible adaptive variation in stone tool manufacture and use patterns occurring over many thousands of years. The fine details of how tools were produced are of limited interest in such a context. Indeed, given the relatively coarse-grained assemblages employed here, it would be misleading to even attempt to describe in detail the precise sequences of actions that went into producing flakes, as these are likely to have varied both subtly and broadly over time. Additionally, many of the assemblages were not collected with modern techniques, and although screens were employed, recovery of the smaller fraction of lithic debris (< about 1.5 cm in maximum dimension) was far from complete. This, combined with the restricted spatial extent of some of the excavations, greatly limits the potential utility of refitting studies. Although I have undertaken limited experimental replication, mainly to check impressions of how particular attributes might be produced, this also does not play a major role in the analyses.

The primary goal of studying core reduction technology in the present context is to identify *general approaches* to flake manufacture, and to explore possible variation in their functional or economic characteristics. The first question is whether there is more than one distinct strategy of blank production represented. Core forms and attributes provide the first "reading" on variation in *chaînes opératoires* within and among assemblages. In keeping with the overall aim of the research, I have focused on only the most pronounced variation in core morphology, likely to represent truly distinct systems of flake production. The description and analyses of cores and their products are organized primarily in terms of the orientations of successive flake removals (centripetal versus parallel). Using the directions of removals as the primary criterion for dividing flake production systems runs counter to more widespread approaches, such as that based on the distinction between Levallois and non-Levallois methods. However, the directions of flake removals do appear to be strongly linked to the potential economic and functional properties of different methods of flake production, and so this system of classification is of more immediate relevance to the discussion at hand.

Cores are the residuals of more or less prolonged sequences of preparation, flake detachment, and repreparation events, and the forms of cores found in archaeological sites may not accurately reflect their entire history. The subject assemblages contain a wide variety of core forms, and is crucial to consider whether they represent separate reduction trajectories or merely stages within a single all-encompassing reduction sequence. With the materials available, this is best accomplished by comparing the metrical attributes of different kinds of cores and the flakes and tools likely to have been produced from them. Determinations of how flakes and tool blanks were produced are based on a few simple attributes, including the number, direction, and orientation of the negative scars left by previous removals, the distribution of residual cortex on the back sides of flakes and blanks, and the morphologies of striking platforms. There is an unavoidable degree of ambiguity in assigning flakes and tool blanks to a particular mode of production, and a large number of specimens cannot be reliably classified. Nonetheless, the number of "attributable" flakes and blanks is sufficient to evaluate hypotheses about whether different core forms represent independent reduction trajectories or sequential stages in a single reduction chain.

Once the basic technological variation has been documented, the next step is to look for characteristics that might have rendered alternative core reduction methods advantageous in different contexts. Three factors are of particular relevance to strategies of technological provisioning; these include the sizes, shapes, and numbers of flakes produced by different methods. Contrasts in the sizes and shapes of different kinds of flake blanks provide clues as to their functional properties and, perhaps more important, their potential for prolonged use and renewal. Differences in the productivity of alternative approaches to core reduction—measured in terms of the ratio of "attributable" flakes and tool blanks to cores—indicate which techniques might have been better suited to conditions of raw material scarcity resulting either from the distribution of flint or from human land use patterns.

The use of small pebble raw materials unquestionably constrained the options that Mousterian tool makers could exercise in turning out tools and tool blanks. Nevertheless, the Pontinian assemblages provide evidence for the use of several more or less distinct approaches to making flakes. Reliance on different approaches varies from site to site and from assemblage to assemblage. Procedures for preparing cores and detaching flakes depended to some degree on the shape of the original pebble, but the kinds of pebbles collected appear to represent "choices" made by tool makers more than simple patterns of availability. Moreover, the alternative methods of core reduction do appear to produce flakes that differ in number, shape, size, the lengths of usable sharp edges, and cortex cover. This strongly suggests, but does not prove, that alternative methods of flake manufacture could well have filled different functional or strategic roles.

Core Forms

The subject assemblages contain a (to me at least) surprising diversity of core forms. One might well predict that the use of small pebbles would have limited the choices available to flint knappers, reducing the number of practical options for producing flakes. In retrospect, raw material size is in fact probably responsible for increasing the number of apparent core types. Where raw materials are small and reduction chains short, there is little opportunity for preliminary shaping of cores. The absence of extensive dorsal preparation is shown by the fact that 85% of all retouched tools studied are made on blanks with at least some cortex, while more than 25% are completely covered with cortex on the dorsal face. In addition, larger raw materials permit artisans to alter strategies of exploitation as cores become smaller, or in response to unforeseen problems with the raw material. Although some remodeling and redirection of cores did occur, it can be shown this was not the general rule in the Pontinian assemblages. Instead, the cores—small to start with—seem to simply have dropped out of use for a variety of reasons with little or no reworking at the end. As a result, the forms of abandoned cores are more indicative of the way they were worked throughout most of their (short) use lives.

A core typology made up of about a dozen alternative forms was employed at the outset of data collection. The list was later expanded during analysis to include more than 29 distinct core types. For the sake of clarity, the final detailed classification has again been compressed into a half dozen basic classes, each with one or more variants. Much of the variation in core technology within and among the subject assemblage can be classified in terms of two broad "themes," characterized respectively by centripetal and parallel removals of successive flakes. However, as is described below, there is also considerable variation within each of these major classes.

Centripetal cores

Centripetal or *radial* core technologies are found in the great majority of Middle Paleolithic assemblages known from Europe, Africa, and the Levant (e.g., Boëda 1993; Bordes 1961a, 1968; Coles and Higgs 1969; Volman 1984), and the discoid or radial core is one of the most widely dispersed "type fossils" of the Mousterian (and Acheulean). The defining characteristic of centripetal reduction is the direction of percussion from the perimeter towards the center of the core (hence the term "centripetal"). All or part of the margin of a more or less circular core may serve as a platform for striking off flakes, either large flakes that can serve as tools or tool blanks, or flakes intended to further shape the face of detachment. Centripetally worked cores may possess either one or two faces of detachment, and platforms can vary from unprepared (cortical) to faceted *chapeau de gendarme*.

Researchers have described in great detail a variety of methods of core reduction involving centripetal detachment of flakes (e.g., Boëda 1993; Boëda et al. 1990; Crew 1975). The best known and most extensively documented of these is radial Levallois technology. Eric Boëda identifies two principal variants of radial Levallois technology. The so-called "preferential" or "lineal" Levallois method, which best corresponds to the "classic" definition of Levallois, is distinguished by the preparation of the surface of detachment for a single large flake or blade; the surface may be subsequently reshaped for one or more additional "preferential" removals; In contrast, "recurrent" Levallois is marked by detachment of a series of large flakes such that the preceding removals ready the surface for the subsequent ones, eliminating the need for extensive repreparation (Boëda 1986, 1993; Boëda et al. 1990).

The radial or centripetal scheme of detachment also occurs independently of the other characteristics that define Levallois technology. In non-Levallois radial core technology, flakes are struck from the edge towards the center of the core, often with little preparation of either the striking platform or the surface of detachment. The direction of percussion may actually tend to be tangential to the center of the piece, so that each flake carries off a portion of the core edge or platform. Characteristic products of this kind of flake production technology include pseudo-Levallois points and other triangular flakes, and sometimes side-struck pieces as well (Boëda 1993:397). When flakes are directed peripherally to the center of the core and do not carry all the way across, the face of detachment becomes more and more steeply angled with use, resulting in core forms termed "conical" (or "biconical," if flakes have been removed from two faces) (e.g., Boëda 1993:393–94; Isaac 1977:175–76).

Cores worked in a centripetal or radial fashion are the most numerous and ubiquitous forms among the subject assemblages, dominating most collections (Table 4.1). Radially prepared cores average between 2.8 and 3.7 cm in diameter. Except for their small sizes, however, they resemble specimens from other Mousterian assemblages (Figure 4.1). Three major variants are distinguished here: unifacial, bifacial, and "Levallois." The final term refers only to specimens preserving evidence of a single large removal, corresponding to the definition of "preferential" or "linear" Levallois. Among radially prepared cores, specimens with one surface of detachment (unifacial) outnumber the bifacial form by a ratio of almost five to one (Table 4.1). Many specimens termed bifacial in fact have just one or two large invasive detachments on the secondary face. It was originally postulated that the detachment of flakes from both faces of a core was a technique for extending its utility, and therefore could serve as an indicator of more extreme reduction. In the study sample, however, bifacial, radially prepared cores are in fact no smaller than the unifacial variety. The bifacial forms may even tend to be somewhat thicker (Table 4.2), indicating that they are, if anything, less completely consumed. Bifacial reduction thus may have been used primarily on larger pieces of raw material which could support successful detachment of usable flakes from both faces of the core.

Only a very small percentage of centripetally worked cores were classified as classical "preferential" Levallois cores by virtue of the presence of a single flake scar covering 50% or more of one face (e.g., Fig. 4.1, nos. 7, 11). The cores so classified appear to represent cases in which a large flake was removed late in reduction, without subsequent repreparation and removals. Although the numbers are too small to permit statistically supported conclusions, there does not appear to be significant inter-assemblage variation in the frequencies of specimens with a single preferential removal scar.

Unprepared cortical, plain (single facet), faceted, and even *chapeau de gendarme* platforms are all commonly found on centripetally worked cores. In fact, most cores appear to have produced flakes with more than one kind of butt (Table 4.3). Platform faceting is clearly not exclusive to cores classified with a single preferential flake scar. Platform faceting produces stronger, more uniform platforms, and experimental research has shown that this fosters the production of flakes that are large relative to the size of the core (Dibble 1981). Given that cores were quite small to start with, the high frequency of faceted platforms on Pontinian centripetal cores of all types is not surprising. There is some evidence that the frequency of platform preparation on radially worked cores increased with progressive reduction. More extensively consumed cores exhibit a higher frequency of platform faceting than do lightly worked pieces (Table 4.4). On the other hand, almost half of even the most casually worked cores have at least one faceted platform, suggesting that preparation began quite early in the process of working a core. On a few very heavily exploited centripetal cores, the entire prepared platform was removed and flakes were detached by percussion from the residual cortex on the "dorsal" surface (note the persistence of exclusively cortical platforms among heavily reduced pieces in the final column of Table 4.4). The final removals from such cores would exhibit cortical

Table 4.1. Core frequencies (percent)

			Centripetal			*Parallel*					
	Tested pebble	*Chopper/ ch. tool*	*Radial Levallois*	*Unifacial radial*	*Bifacial radial*	*Prepared platform*	*Pseudo-prismatic*	*Globular*	*Bipolar*	*Misc.*	*n*
Agro Pontino	**0.5**	**13.2**	**1.0**	**60.9**	**14.1**	**8.3**	**1.1**	**0.0**	**1.0**	**0.0**	**192**
Grotta Breuil	**6.7**	**7.6**	**1.3**	**31.1**	**5.9**	**13.3**	**19.3**	**2.5**	**2.5**	**10.1**	**202**
1986–87 sample	3.9	6.5	0.0	35.1	13.0	9.1	20.8	0.0	6.5	5.2	77
old *sondage*	18.2	9.1	0.0	36.4	0.0	0.0	18.2	18.2	0.0	0.0	11
Strata 3/4	8.8	5.3	1.8	21.9	0.9	21.1	21.1	2.6	0.0	16.7	114
Grotta Guattari	**22.8**	**22.3**	**0.9**	**34.2**	**6.0**	**0.6**	**6.0**	**4.1**	**0.0**	**3.1**	**319**
Stratum 1	6.9	41.4	3.5	17.2	3.5	3.5	17.2	3.5	0.0	3.5	29
Stratum 2	19.3	25.0	1.1	26.1	9.1	1.1	9.1	8.0	0.0	1.1	88
Stratum 4	20.2	21.1	1.0	44.2	6.7	0.0	2.9	1.9	0.0	1.9	104
Stratum 5	33.7	15.3	0.0	35.7	3.1	0.0	3.1	3.1	0.0	6.1	98
Grotta dei Moscerini	**7.5**	**17.2**	**0.0**	**45.2**	**11.8**	**1.1**	**8.6**	**4.3**	**3.2**	**1.1**	**93**
Strata group 2	3.5	24.1	0.0	51.7	6.9	3.5	6.9	0.0	3.5	0.0	29
Strata group 3	8.3	8.3	0.0	37.5	20.8	0.0	8.3	8.3	4.2	4.2	24
Strata group 4	4.9	19.5	2.4	46.3	7.3	0.0	9.8	4.9	2.4	2.4	41
Strata group 5	0.0	18.2	0.0	72.7	9.1	0.0	0.0	0.0	0.0	0.0	11
Strata group 6	13.6	13.6	0.0	36.4	44.4	9.0	13.5	13.5	0.8	3.0	22
Grotta di Sant'Agostino	**0.4**	**6.6**	**4.7**	**41.0**	**8.2**	**14.8**	**18.8**	**0.4**	**3.1**	**2.0**	**256**
Level 1	0.8	6.1	7.5	44.4	9.0	13.5	13.5	0.8	3.0	1.5	133
Level 2	0.0	7.0	0.0	33.7	5.8	23.3	25.6	0.0	2.3	2.3	86
Level 3	0.0	12.0	4.0	44.0	0.0	0.0	28.0	0.0	8.0	4.0	25
Level 4	0.0	0.0	8.3	50.0	33.3	0.0	8.3	0.0	0.0	0.0	12

Table 4.2. Sizes of unbroken unifacial and bifacial centripetal cores (all sites)

	Maximum dimension (mm)				
Core category	*Mean*	*sd*	*n*	*t-value*	*p*
Unifacial	32.7	7.4	342	–0.097	.922
Bifacial	37.7	7.5	61		

	Thickness				
Core category	*Mean*	*sd*	*n*	*t-value*	*p*
Unifacial	12.2	4.4	342	–2.629	.009
Bifacial	13.9	5.3	61		

Table 4.3. Treatment of striking platforms on centripetal cores

	Centripetal core type (percent)		
Treatment of striking platforms	*Levallois*	*Unifacial*	*Bifacial*
cortical	0.0	19.1	1.0
plain (one facet)	5.0	2.0	2.1
faceted	5.0	1.5	1.0
cortical + plain	5.0	21.3	12.5
cortical + plain + faceted	40.0	38.3	40.6
plain + faceted	45.0	17.8	42.7

Table 4.4. Platform treatment and reduction, centripetal cores (all types combined)

	Extent of Reduction (percent)		
Treatment of striking platforms	*Light*	*Moderate*	*Heavy*
cortical	19.6	17.5	9.3
plain (one facet)	4.4	2.1	1.6
faceted	2.2	1.5	1.6
cortical + plain	28.3	21.3	11.6
cortical + plain + faceted	19.6	36.8	51.9
plain + faceted	26.1	20.8	24.0
At least part of platform faceted	47.9	59.1	77.5

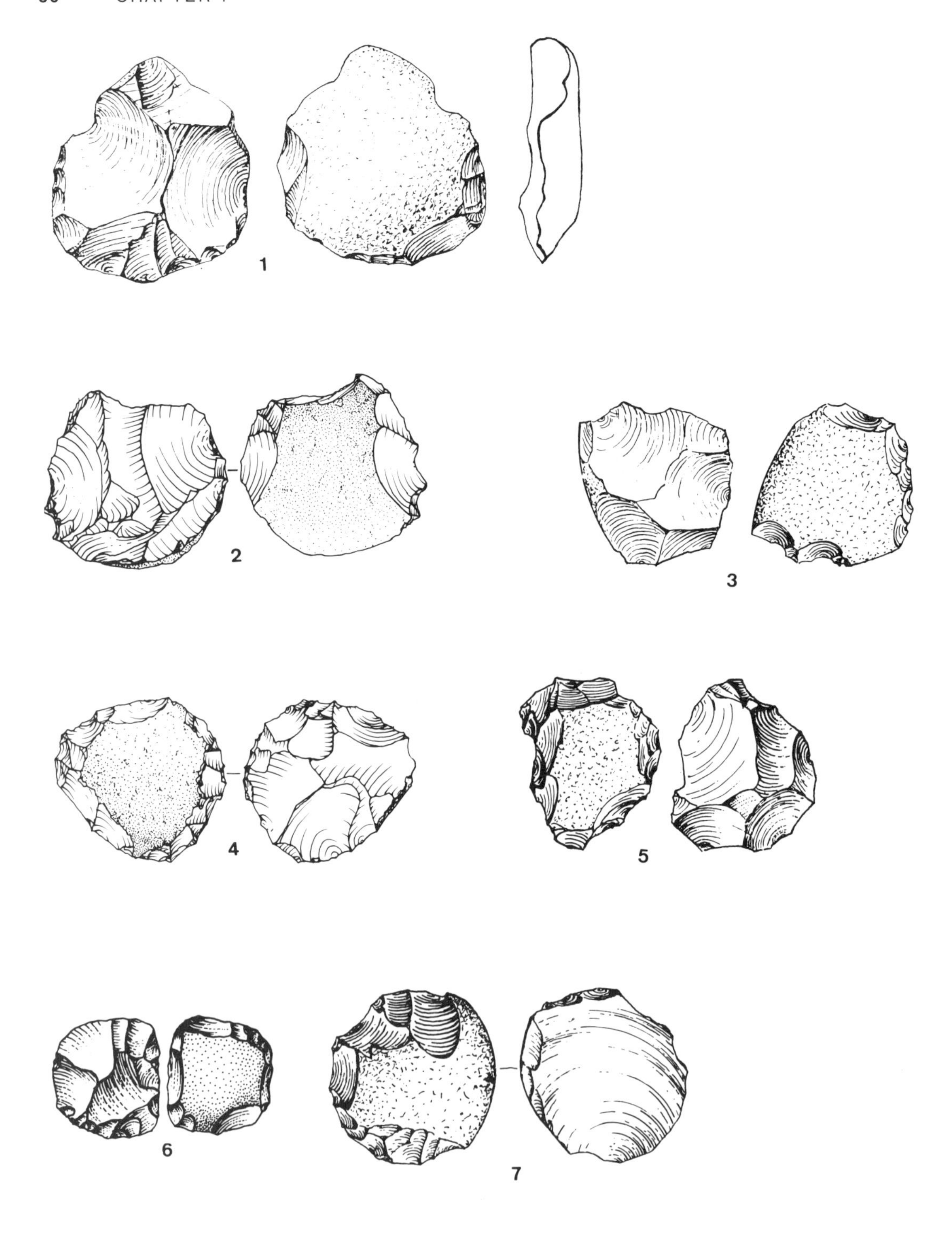

Figure 4.1. Centripetal or radially worked cores from Pontinian Mousterian sites. (1, 12, 14=Gr. Guattari; 2, 4, 8, 9, 13=Gr. Breuil; 3, 5–7, 10, 11=Gr. di Sant'Agostino.)

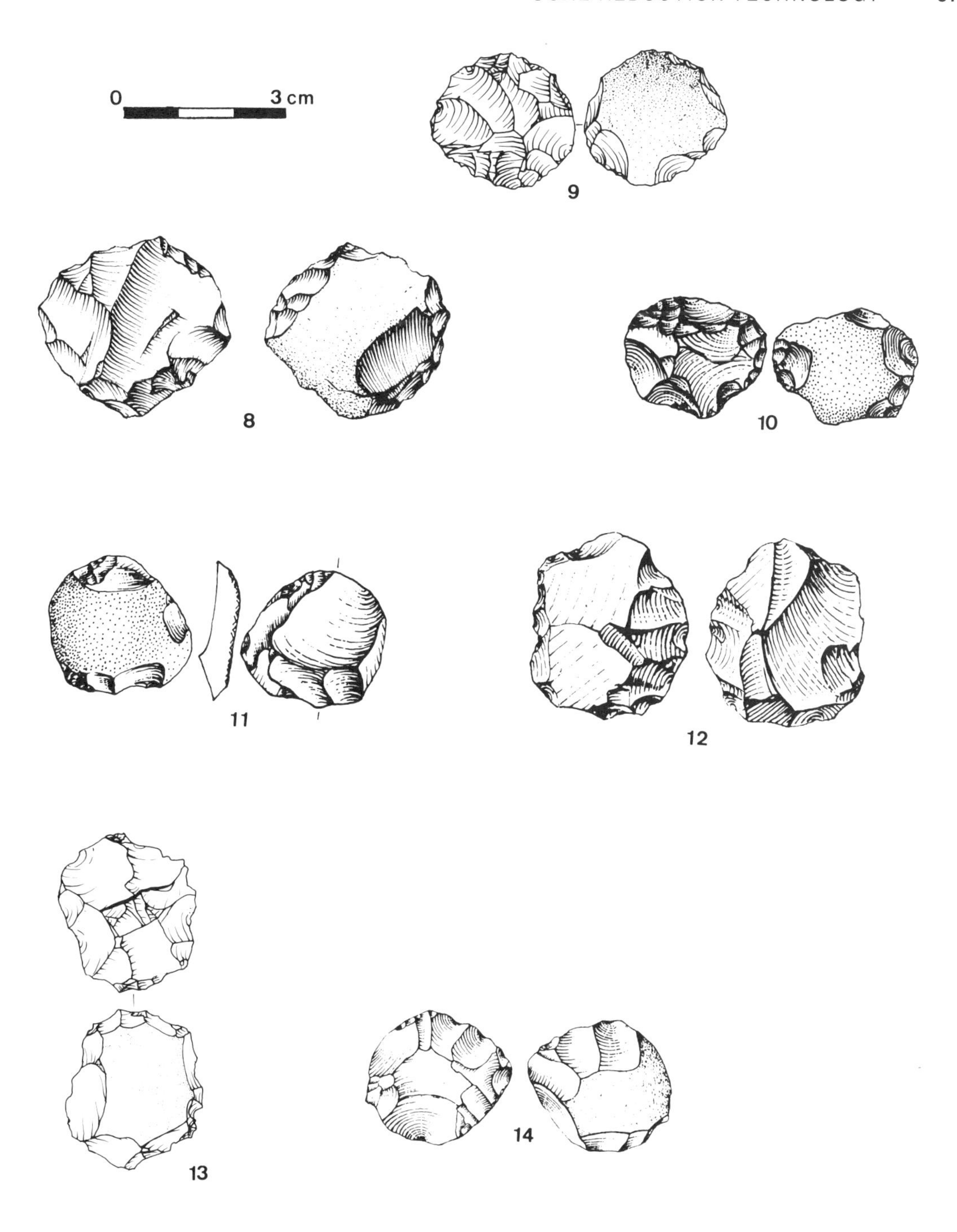

Figure 4.1. *cont.*

butts along with extensive dorsal preparation and evidence for many prior removals, normally a somewhat unusual combination of attributes.

Parallel core technology: cores with one or two platforms

The second major *schema* of blank production found among the cores from the subject assemblages involves detachment of a series of flakes in parallel along a single axis or, much less frequently, along two axes of a core. In this general mode of core reduction, a relatively short, straight striking platform was created at one (or both) ends of a pebble, and flakes were struck off from it (or them), most frequently along the long axis of the piece. The most familiar example of parallel preparation and detachment of flakes is prismatic blade technology, variants of which are found in a great variety of contexts worldwide. While not traditionally considered typical of Mousterian industries, technologies employing single and double platform cores and producing elongated blade-like pieces are relatively common in Middle Paleolithic and even earlier contexts throughout Eurasia and Africa (e.g., Conard 1990; Révillion 1993; Van Peer 1991). Variants of recurrent Levallois technology involving removal of parallel or convergent flakes from a single platform are arguably the most widespread methods of blank production throughout the Middle Paleolithic of the Near East Meignen and Bar-Yosef 1988).

One characteristic of blank production schemes utilizing single and opposed platform cores is that each removal creates a pair of more or less parallel ridges. As with "recurrent" radial Levallois technology, each flake "sets up" subsequent removals, reducing the amount of repreparation of the face of detachment between successive detachments. Still, while the general blueprint for producing flakes and blanks may entail a series of parallel removals, all working of the core does not necessarily take place along a single axis. In the manufacture of true prismatic blades, for example, blows are often directed perpendicular to the axis of detachment in order to create, realign, or strengthen a ridge, yielding the classic *lame à crête* (Tixier et al. 1980: 82–83). In unidirectional Levallois core technology, preparation of the face of detachment may entail trimming the lateral and distal edges of the core in order to achieve the required convex shape (e.g., Boëda et al. 1990: Fig. 4). In referring to parallel, as opposed to centripetal/radial, core technology, emphasis is therefore placed on the general plan for striking off large flakes and potential tool blanks: it is not implied that *all* work on the core is in parallel from one or two directions.

In the subject assemblages, there are two main variations on the general theme of working cores by detaching a series of flakes in parallel. One is represented by a type of core, termed *pseudo-prismatic*, which resembles a somewhat rudimentary version of a prismatic blade or bladelet core. Pseudo-prismatic cores typically possess a single plain, unprepared platform or, less commonly, two (opposed) unfaceted platforms at the end(s) of a pebble (Figure 4.2). The single platform variant accounts for almost 82% of the sample as a whole: although the varieties were assigned to separate classes during data collection, they were later recombined because their products are nearly indistinguishable. While they are never the dominant form, pseudo-prismatic cores are found in most assemblages (Table 4.1). They are most abundant in assemblages from Grotta Breuil and Grotta di Sant'Agostino, where their presence has been noted in previous studies (Laj-Pannocchia 1950: Fig. 5; Tozzi 1970: Table 6).

Although the pseudo-prismatic cores superficially resemble blade and bladelet cores found in Upper Paleolithic sites in coastal Latium, the respective *chaînes opératoires* appear to have differed in a number of important respects. The basic procedure in the Mousterian case (broadly schematized in Figure 4.3) appears to have begun with the removal of a large cortical flake from one end of a—usually elongated—pebble. The resulting oblique face was used directly as a platform for detaching flakes. There was little or no prior preparation of the face of detachment. The "crested blade" technique (see Tixier et al. 1980) does not seem to have been employed to direct and control the first removals from pseudo-prismatic cores, and no distinctive *lames à crête* were recovered from Mousterian deposits. In some cases where the scars from the initial removals are preserved, the first flakes

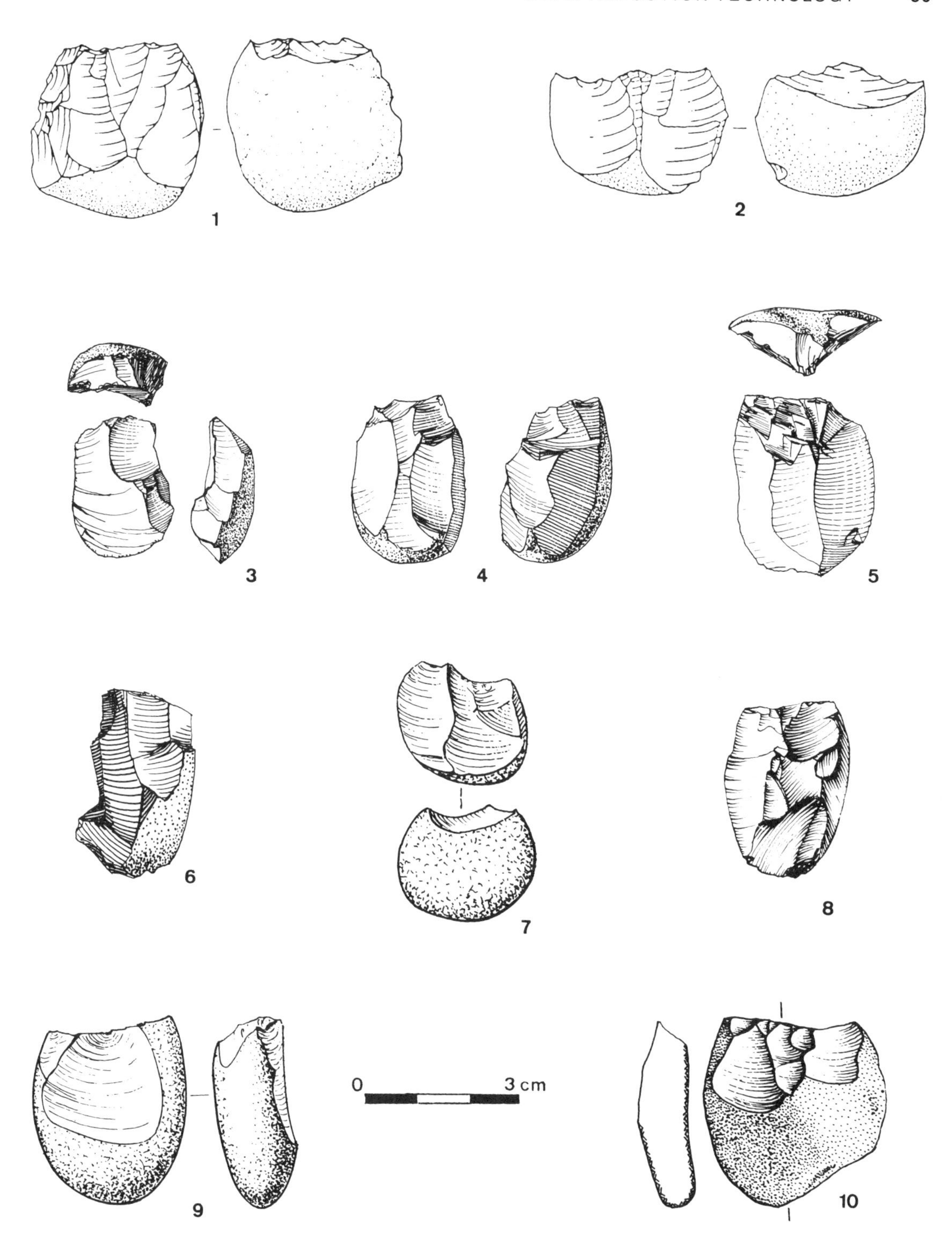

Figure 4.2. Pseudo-prismatic cores from Pontinian Mousterian sites. (1–6, 8=Gr. Breuil; 7, 9, 10=Gr. di Sant'Agostino.)

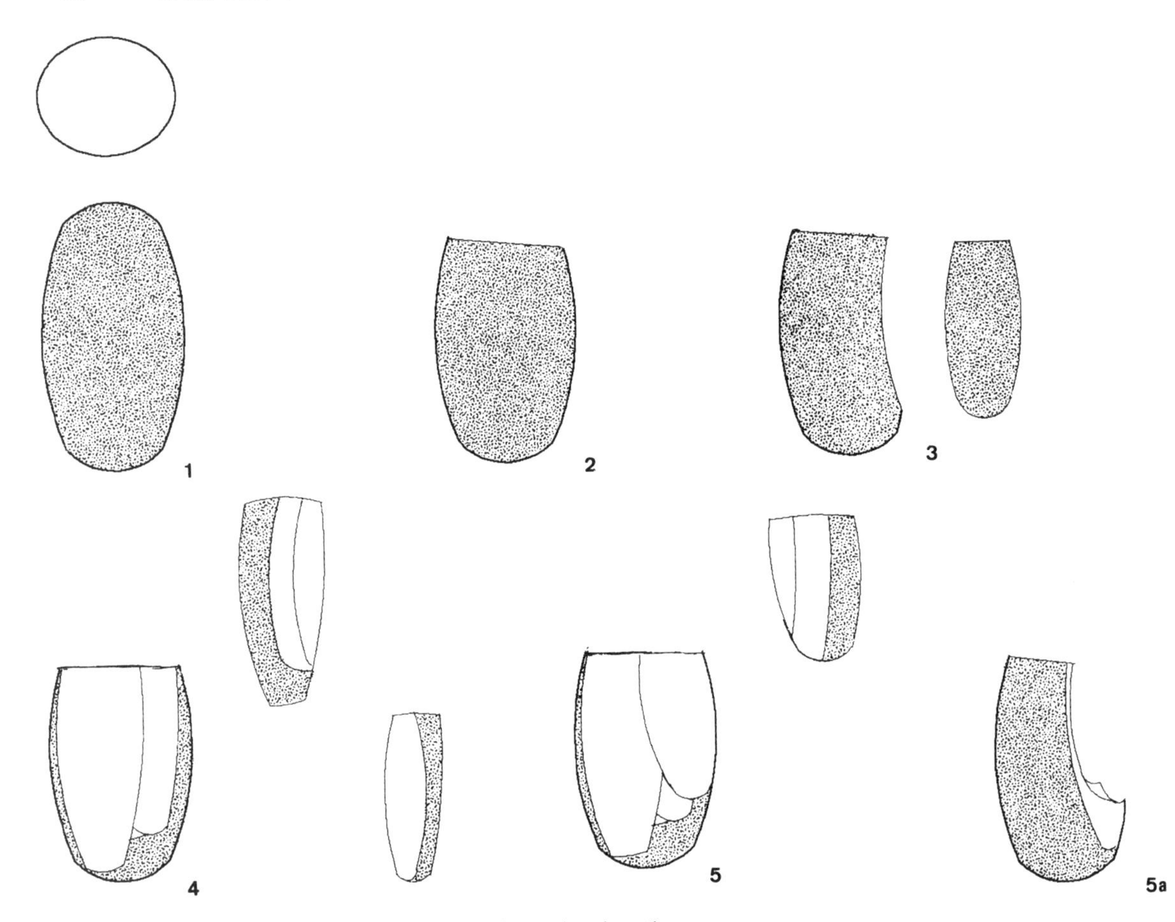

Figure 4.3. Schematized illustration of reduction of pseudo-prismatic cores.

were struck so as to follow natural ridges on the pebble surface. In other cases, however, the first flakes were detached directly in the center of a relatively flat surface (Fig. 4.2, nos. 9, 10). Many of the blanks detached from pseudo-prismatic cores would have retained a strip of cortex along the end and one margin, giving them a form approximating a naturally backed knife (Type 38 in Bordes's 1961 typology).

Flakes were normally removed from less than 50% of the surface of pseudo-prismatic cores. In a *very* few cases there are two distinct faces of flake removal on either side of the core. Quite frequently, the face of detachment on the abandoned core is flat or even slightly concave, rather than convex as in a true prismatic core. The premature "hinging" off of flakes was a frequent problem in the exploitation of pseudo-prismatic cores, and some abandoned specimens preserve a series of steps along the face of detachment at the end opposite the platform (Fig. 4.2, nos. 4, 6, 8). These distal steps were rarely if ever removed by lateral retouch. Instead, the distal end of the core may have been reshaped by intentionally striking off an *outrepassé* flake that carried it away: typical "overpassed" flakes are numerous in some assemblages. Another response to the development of multiple hinges on pseudo-prismatic cores may have been to remove the end with a strong perpendicular blow, creating a second platform.

The pseudo-prismatic cores also commonly bear evidence for platform renewal, in the form of small, flat flakes removed from the striking platform by a blow directed at an edge lateral to the face of detachment. Unlike "platform tablets" of the type associated with true prismatic blade technologies, these flakes rarely carried away the entire platform. This may be an effect of raw material size, as completely eliminating the platform would have quickly rendered a core too short for use.

Raw material size alone cannot explain why the crested blade technique was not used in the working of pseudo-prismatic cores. Small crested blades derived from bladelet cores made on pebbles are quite common in late Upper Paleolithic assemblages such as Riparo Salvini (Avellino et al. 1989). Instead, the absence of the crested ridge technique in the Middle Paleolithic assemblages may indicate that there was no attempt on the part of Mousterian tool makers to maximize the elongation of flakes or blades. Rather, the tendency to exploit flat or lightly concave flaking surfaces suggests that the Italian Mousterian flint workers sought to create relatively wide—though still somewhat elongated—flakes, rather than true blades. Pieces bearing parallel dorsal scars, the kinds of flakes likely to have come from pseudo-prismatic cores, were most frequently retouched along one margin to create simple sidescrapers (Kuhn 1992a): had they been made much longer and thinner, it is unlikely that they could have been successfully retouched and manipulated.

Another variety of core found in the Pontinian assemblages at first appears to represent something of a hybrid between the centripetal Levallois and pseudo-prismatic types, but in fact forms a distinct class. As with pseudo-prismatic cores, flakes were struck from one or two (opposed) platforms, so that their forms and dorsal scar patterns are reminiscent of those found on the products of pseudo-prismatic cores. At the same time, the cores themselves tend to be flat, and the striking platforms are faceted, more like those of radial Levallois specimens. The term *prepared platform cores* is used to refer to this group of nuclei (Figure 4.4). The term is somewhat imprecise, in that many radially worked cores also have prepared or faceted striking platforms: however, a more precise designation, such as "prepared platform parallel cores," would be somewhat unwieldy. I emphasize that in this context the term "prepared platform cores" refers specifically to specimens from which flakes were detached in parallel, rather than radially, from one or two retouched striking platforms. Prepared platform cores were first recognized as technologically distinct by Laj-Pannocchia in her early study of the collections from Grotta di Sant'Agostino (Laj-Pannocchia 1950; see also Tozzi 1970: Table 6). Taschini (1970:73) also described cores with one or two intentionally faceted platforms in the assemblage from early *sondages* at Grotta Breuil. In fact, the majority of the artifacts falling into the "prepared platform" group come from these two sites.

The prepared platform cores are overwhelmingly unifacial. Compared with pseudo-prismatic cores a relatively large proportion (31.25% overall) have two or more striking platforms. When two platforms are present, they are almost always opposed. Cores with three platforms are also known; the third and final platform is normally oriented perpendicularly to the first two, and often has only a single broad flake detached from it. Although cores with three platforms may resemble centripetal or disk cores superficially, they can be distinguished based on a number of other characteristics and were classified as the prepared platform variety whenever recognized as such.

The sequence (or sequences) of technological acts that eventually led up to the prepared platform cores is not entirely clear. It does appear that the operation began with the preparation of a short, straight, faceted platform on one edge or end of a pebble. Platform faceting was an early rather than late step in manufacture (Figure 4.5): this is indicated by the existence of completely cortex-covered flakes with faceted butts. If a natural ridge was available, the first flake or flakes produced were struck off along it. The dominant scheme of sequential removals appears to have been parallel rather than convergent: although some cores appear to show convergent flake scar patterns (e.g., Fig. 4.4, nos. 1, 2) very few flakes and tool blanks do. As with pseudo-prismatic cores, many of the flakes produced would have retained a strip of cortex on one margin, giving them the *couteau à dos naturel* form.

Unlike pseudo-prismatic cores, prepared platform cores often show signs of peripheral or marginal flaking to shape the surface of detachment (Fig. 4.5). Lateral and distal preparation appears to have been used occasionally to set up the initial removals. It was more commonly employed to

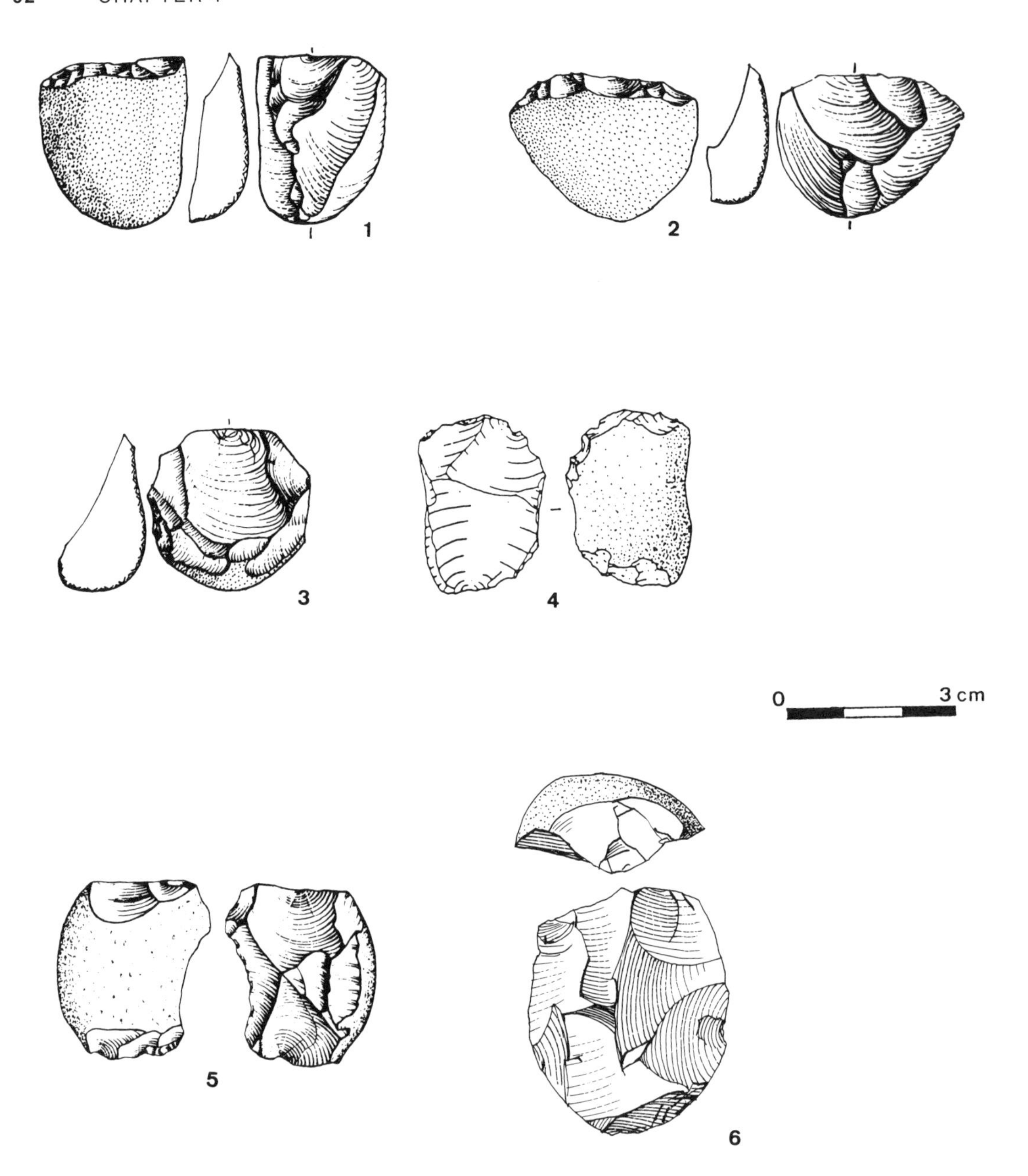

Figure 4.4. Prepared platform cores from Pontinian Mousterian sites. (1–3, 9=Gr. di Sant'Agostino; 4, 6–8, 10=Gr. Breuil; 5=Gr. Guattari.)

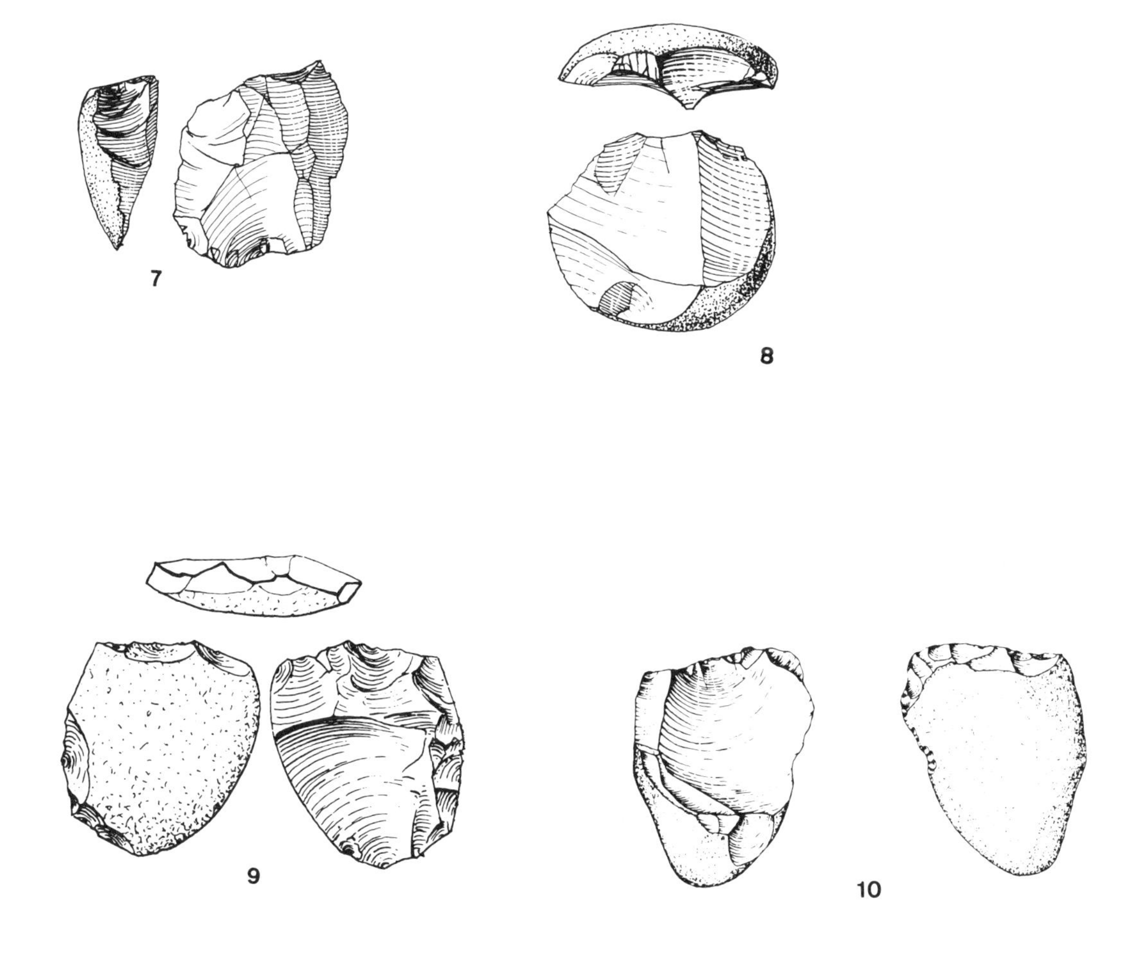

Figure 4.4. *cont.*

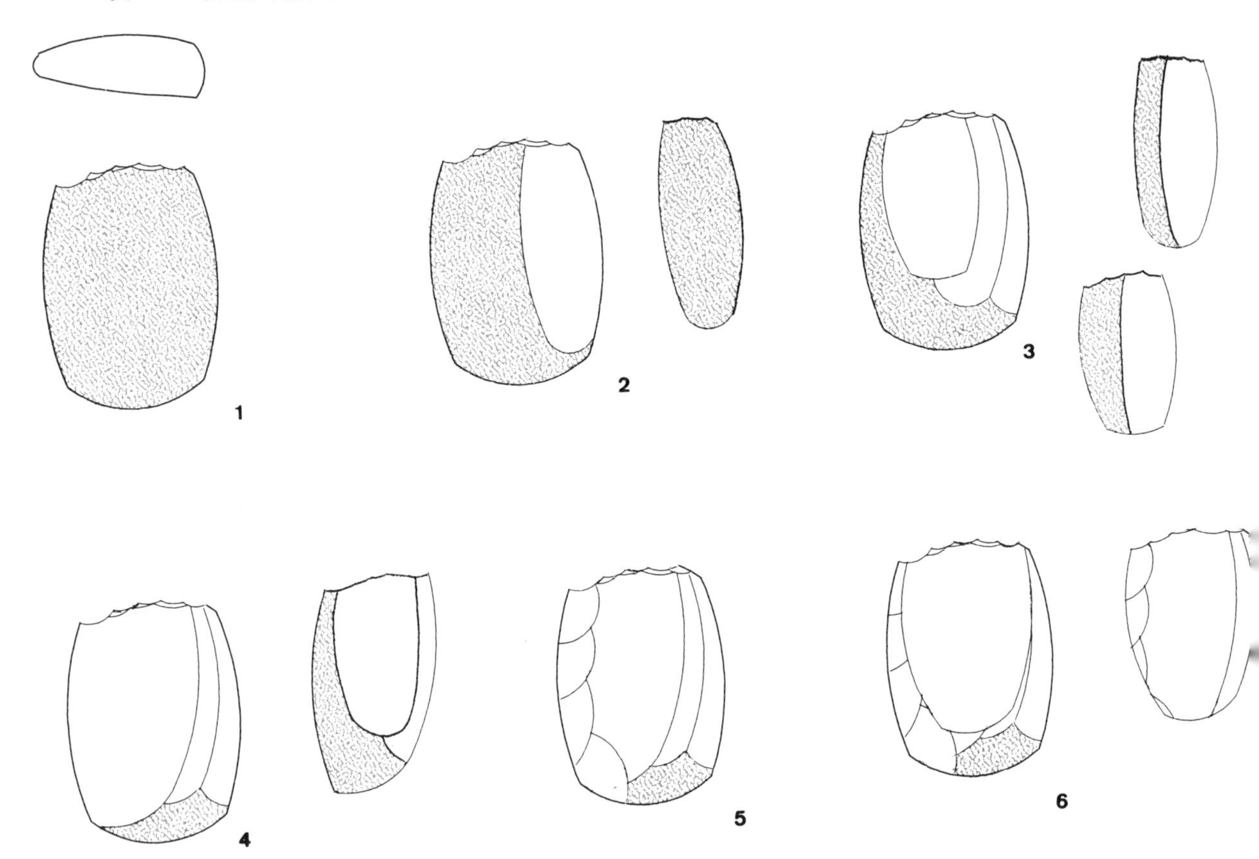

Figure 4.5. Schematized illustration of reduction of prepared platform core.

reprepare the face of the core after removal of one or more large flakes (Fig. 4.4, no. 6; Fig. 4.5). Lateral preparation can generally be identified by the small size of the removals perpendicular or oblique to the normal axis of flake detachment, and by the absence of platform preparation for these small flakes. Prepared platform cores were infrequently prepared on the end opposite the platform. As a consequence extensively worked specimens are quite often wedge-shaped in cross section (Fig. 4.4, no. 3).

Other core forms

Centripetal, pseudo-prismatic, and prepared platform cores encompass much of the identifiable variation in core morphology in the Pontinian Mousterian assemblages. Even so, these are not the only kinds of nuclei present. Approximately one-third of the pieces used as cores do not fall into any of these categories. By and large, these remaining specimens do not appear to represent identifiable reduction chains with distinctive products, and are therefore somewhat peripheral to this discussion. Nonetheless they merit brief description.

As Table 4.1 indicates, there are only a small number of *globular cores* in the Pontinian Mousterian assemblages. Globular cores are defined by the presence of three or more platforms of percussion that are not oriented in the same plane. True globular cores and their more extreme variant, so-called polyhedrons, may well represent a distinctive kind of technology. The scarcity of globular cores may be at least partly a function of raw material forms. The pebbles found in the study area tend to be either flattened or cylindrical in form: the subspherical pebbles necessary for the manufacture of cores with multiple platforms in different planes

Table 4.5. Frequencies of tested pebbles, chopper/chopping tools, and formal cores (percent)

	Tested pebbles	*Choppers and chopping tools*	*Formal cores*	*n*
Grotta Breuil	7.0	7.0	86.0	273
Grotta Guattari	22.9	22.3	54.8	319
Grotta dei Moscerini	7.5	17.0	75.5	94
Grotta di Sant'Agostino	0.4	6.6	93.0	256

Table 4.6. Maximum dimensions of unbroken chopper/chopping tools and formal cores

	Core category	*Maximum dimension (mm)*				
		Mean	*sd*	*n*	*t-value*	*p*
Grotta Guattari	Ch./ch. tools	40.7	10.0	67	2.425	.016
	Formal cores	37.2	8.7	111		
Grotta dei Moscerini	Ch./ch. tools	35.8	17.0	13	−2.148	.037
	Formal cores	28.7	6.4	36		
Grotta di Sant'Agostino	Ch./ch. tools	35.2	7.0	16	−5.368	<.001
	Formal cores	28.1	4.9	196		
Grotta Breuil	Ch./ch. tools	34.6	6.7	16	−1.154	.250
	Formal cores	32.7	6.3	180		
Agro Pontino	Ch./ch. tools	39.8	6.7	24	−3.081	.002
	Formal cores	34.7	7.5	122		

are difficult to find. Because they are scarce, and because it is nearly impossible to systematically differentiate their products from flakes produced by other means, globular cores are not considered further.

All Mousterian assemblages from coastal Latium contain pebbles with a few marginal removals on one side or one end, and little or no platform preparation (Figure 4.6). These pieces are commonly classed as *choppers and chopping tools* (Bordes's types 59–61), in typologically oriented studies. Taschini considered these diminutive "core tools" to be typological markers of the Pontinian Mousterian (Taschini 1972, 1979). I tend to believe that the chopper/chopping tools represent partially worked cores abandoned for some reason at an early stage of reduction, rather than intentionally shaped implements. They are quite small for core tools, with an average length of just over 37 mm, and could not have been applied with much force, although they could have served as irregular-edged scraping and cutting implements. Several lines of evidence at least support the idea that the choppers and chopping tools actually represent lightly worked cores. First, along with *tested pebbles* (pebbles with a few unsystematic removals), choppers and chopping tools are most abundant at Grotta Guattari, the only site which has a source of raw materials right at hand (Table 4.5). Second, in three of the four site assemblages together with the Agro Pontino sample, chopper/chopping tools are appreciably larger than more extensively worked formal cores (prepared platform, radial, globular, etc.), consistent with the notion that the latter are more heavily reduced versions of the chopper/chopping tools (Table 4.6). On a final, impressionistic note, the forms of choppers and chopping tools tend to echo the most common "formal" core types. Where radially worked cores are most abundant (Grotta Guattari and Grotta dei Moscerini), the choppers are generally circular in plan, with strongly concave edges. At Grotta Breuil and Grotta di

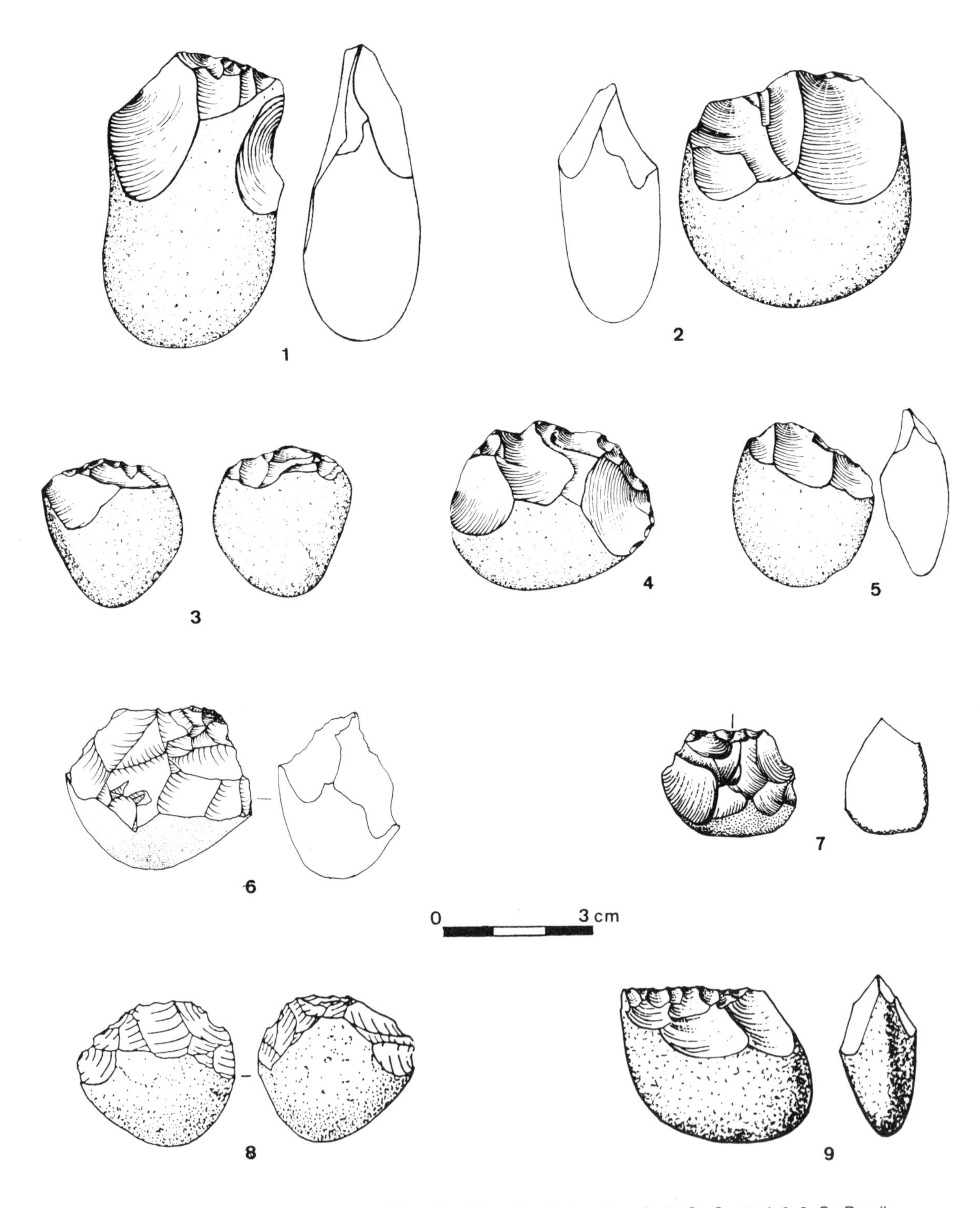

Figure 4.6. Small choppers and chopping tools from Pontinian Mousterian sites. (1–5=Gr. Guattari; 6, 8=Gr. Breuil; 7, 9=Gr. di Sant'Agostino.)

Sant'Agostino, where single and double platform cores are more common, the choppers are more elongated, with short, straight edges. In any event, while many of these "core tools" did yield flakes of usable size, their products would be in no way distinguishable from the flakes produced during early-stage reduction of other core forms.

Bipolar technique

Bipolar or hammer-and-anvil technique is frequently encountered in archaeological cases where raw materials are small, and west-central Italy is no exception. Classic bipolar cores or *pièces esquillées* are relatively abundant in the late Epigravettian assemblages from Riparo Salvini (Avellino et al. 1989), and in the Aurignacian of Grotta del Fossellone (Blanc and Segre 1953; Laplace 1966). The *pièce esquillée* is also a type fossil of early Upper Paleolithic Uluzzian assemblages from the Italian peninsula (Gioia 1988, 1990). Although the importance of bipolar technology in the Pontinian Mousterian was recognized early on (Blanc 1937b; Laj-Pannocchia 1950), bipolar cores with typical wedge-shaped platforms are quite scarce, constituting only 2% of the core sample. Instead, Mousterian tool makers appear to have employed a type of bipolar percussion which rarely leaves a residual core.

Figure 4.7 is a schematic illustration of the inferred technique, reconstructed experimentally. The Pontinian technique of bipolar percussion is not uncontrolled pebble-bashing producing randomly shaped splinters. Using the small but high-quality flint pebbles found in coastal Latium, pebble shape can be exploited to control the products of percussion quite precisely. In the Mousterian, bipolar technique was most frequently employed to split flat or lenticular pebbles by longitudinal percussion directed at one edge: similar pebble-splitting techniques have been documented in prehistoric assemblages from North America (e.g., Ball 1987). If there were no hidden flaws in the raw material, striking an anvil-supported pebble on its edge often yielded two cortical flakes with a surface area approximately that of the original pebble. Sometimes the pebble failed to split, or did not split evenly in half. If the first flake removed was small, or if the fracture did not carry all the way through, the pebble might be rotated and struck again on another part of the margin in an effort to cleave it. These split pebbles were usually retouched directly, but in some cases one half of the pebble was further worked into a radial or other core form.

Although this technique of pebble-splitting does not usually leave a residual core, it does produce distinctive traces on the flakes produced (Figure 4.8). Split pebbles occasionally preserve two bulbs of percussion on opposing ends of the ventral face (Fig. 4.8, nos. 5, 7, 9), but this is a comparatively rare occurrence, as the distal end has often been removed by retouch (e.g., Fig. 4.8, nos. 3, 8) or fracture. Moreover, experiments suggest that the opposing bulb may not even be created if the anvil support is fairly soft (e.g., wood as opposed to stone). A better marker of bipolar reduction in the study assemblages is flat or concave "sheared" bulbs of percussion with strongly marked concen-

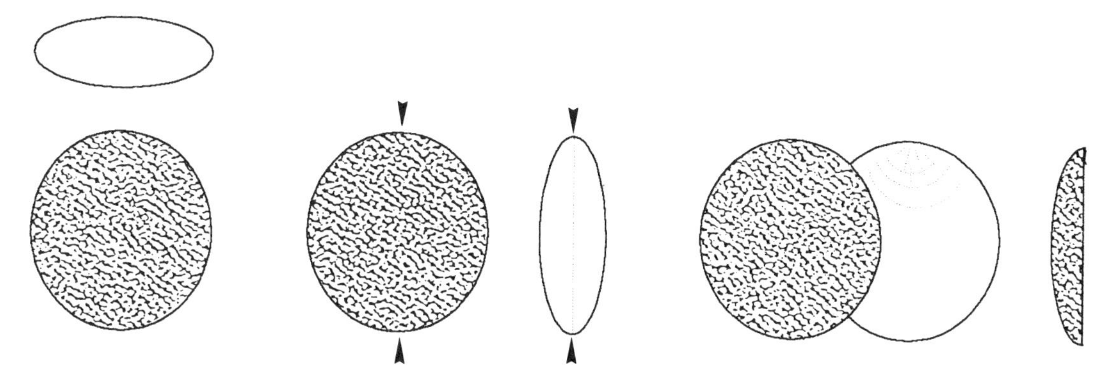

Figure 4.7. Schematized illustration of splitting of pebbles by bipolar technique.

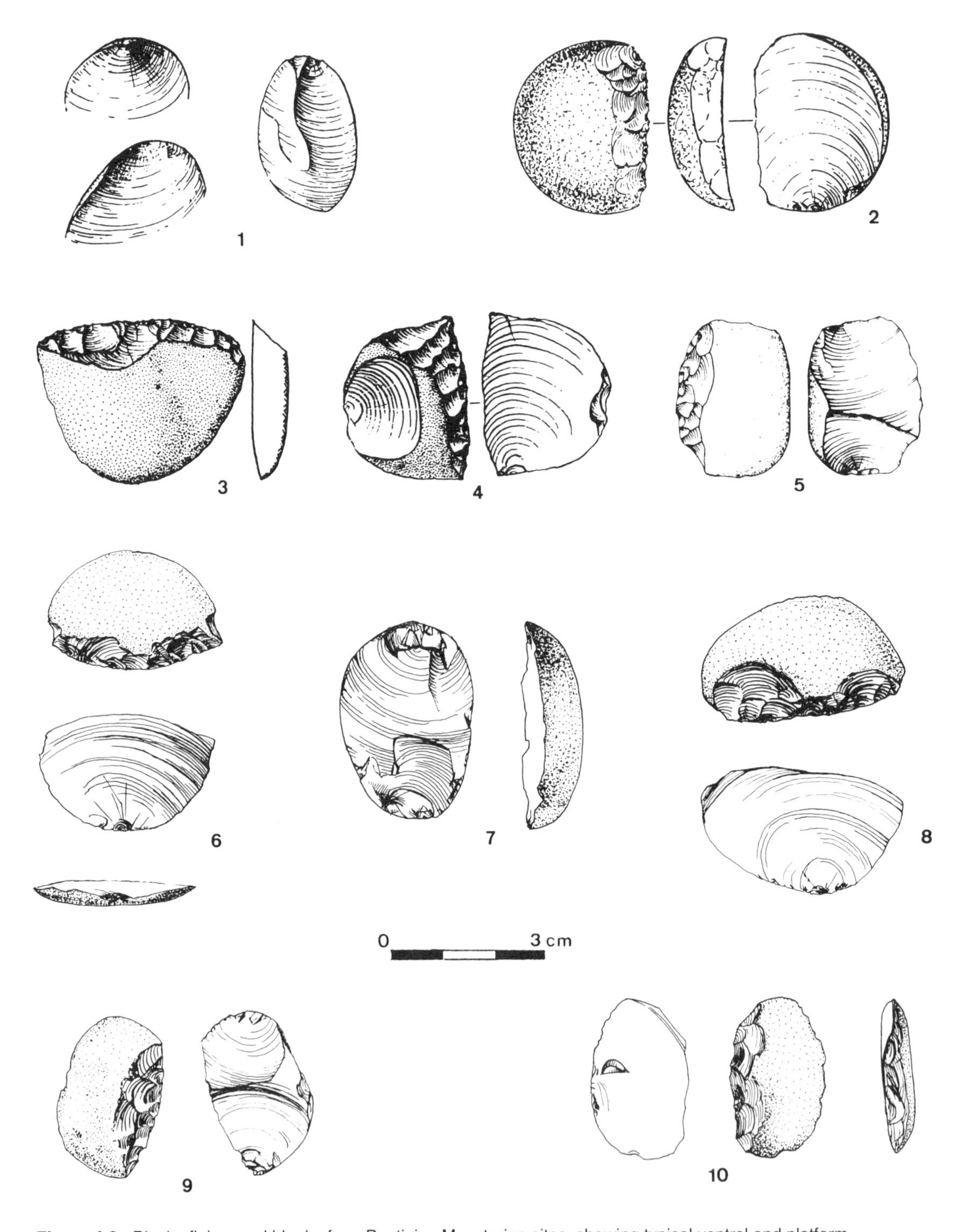

Figure 4.8. Bipolar flakes and blanks from Pontinian Mousterian sites, showing typical ventral and platform attributes. (1–4=Gr. di Sant'Agostino; 5–8=Gr. Guattari; 9–10=Gr. dei Moscerini.)

tric ripples on the ventral surface, and small semicircular zones of crushing at the point of impact (Fig. 4.8, nos. 1, 2, 4, 6, 8, 10). Since the most predictable results are obtained by striking directly on the edge of a lenticular pebble, the dorsal and ventral surfaces of split pebbles intersect to form an acute angle, and distinct, measurable platforms are not present on the resulting flakes.

Flake Production: One Method or Many?

I argue that the variety of different core forms present in the Pontinian assemblages attests to a comparable, if not equal, variety of approaches to making flakes and blanks for retouched tools. While this assertion may at first seem obvious to some, it remains to be demonstrated. An alternative viewpoint holds that there are actually fewer distinct *chaînes opératoires* than there are core forms. Taking Dibble's propositions about sequences of Mousterian scraper transformation and reduction (1987a, 1987b, 1988) as a guide, one could argue that at least some different types of cores actually represent sequential stages in one or more complex core reduction sequences. Both experimental work and refitting studies conducted on archaeological materials have demonstrated that cores can and often do change form over their use lives (e.g., Baumler 1988; Volkman 1983). Even where the actual sequence of removals cannot be replicated, there is often reason to believe that core forms were altered radically during exploitation. Baumler (1988) has made a strong case for radical transformation of core forms during reduction at the Mousterian site of Zobiste, in the former Yugoslavia. In this Mousterian assemblage, most of the flakes and tool blanks appear to have been produced from unidirectional cores using parallel preparation and detachment, while the cores are almost exclusively small, with radial preparation. Discarded cores are smaller than the majority of flakes and retouched pieces, so it seems probable that the original uni- or bidirectional parallel Levallois nuclei were transformed into radial cores, perhaps as a means of extracting one or two final flakes.

Bietti has proposed a model similar to Baumler's, arguing that the variety of core forms found at Grotta Breuil represents successive stages in a single reduction process, rather than independent approaches to flake production. The "transformational" model applies specifically to two assemblages, one from strata 3/4, also discussed here, and a collection from stratum XX not analyzed in this study (Bietti and Grimaldi 1990–91; Bietti et al. 1991): the latter is not included in the study sample because it is undated and lacks associated fauna, so important to this research. One of the two hypothetical sequences of core transformation proposed by Bietti et al., which are quite similar to one another, is shown in Figure 4.9. As in the case cited above, Bietti et al. consider centripetal or radial cores to be the ultimate form. Cores are thought to start out in a variety of shapes (chopper/chopping tools, prepared platform, pseudo-prismatic, etc.), presumably in response to the shapes of the pebbles on which they are made. A chopper/chopping tool is thought to have been gradually worked around more and more of the edge until it acquired a disk-like form. The assumption is that a unidirectional parallel (prepared platform or pseudo-prismatic) core had a second platform added, and then a third on one side. Eventually, if reduction proceeded far enough, the core was worked completely around the margin, taking on a radial form. The implication of these propositions is of course that a single basic method of blank production, rather than many, is represented in the diverse core assemblages from Grotta Breuil, and perhaps in the Pontinian Mousterian more generally.

Because the raw materials used to make the Pontinian Mousterian artifacts are very small, it is somewhat difficult to imagine that core forms would have changed very much over the course of reduction. Nonetheless, it is possible to evaluate the opposing models of core reduction in a systematic fashion. If the scenarios proposed by Bietti et al. are valid, the more reduced core forms should be smaller—more extensively consumed—than those types thought to represent an earlier stage in the chain. Specifically, centripetal or radial cores should be smaller than the pseudo-prismatic and prepared platform varieties. In contrast, if different

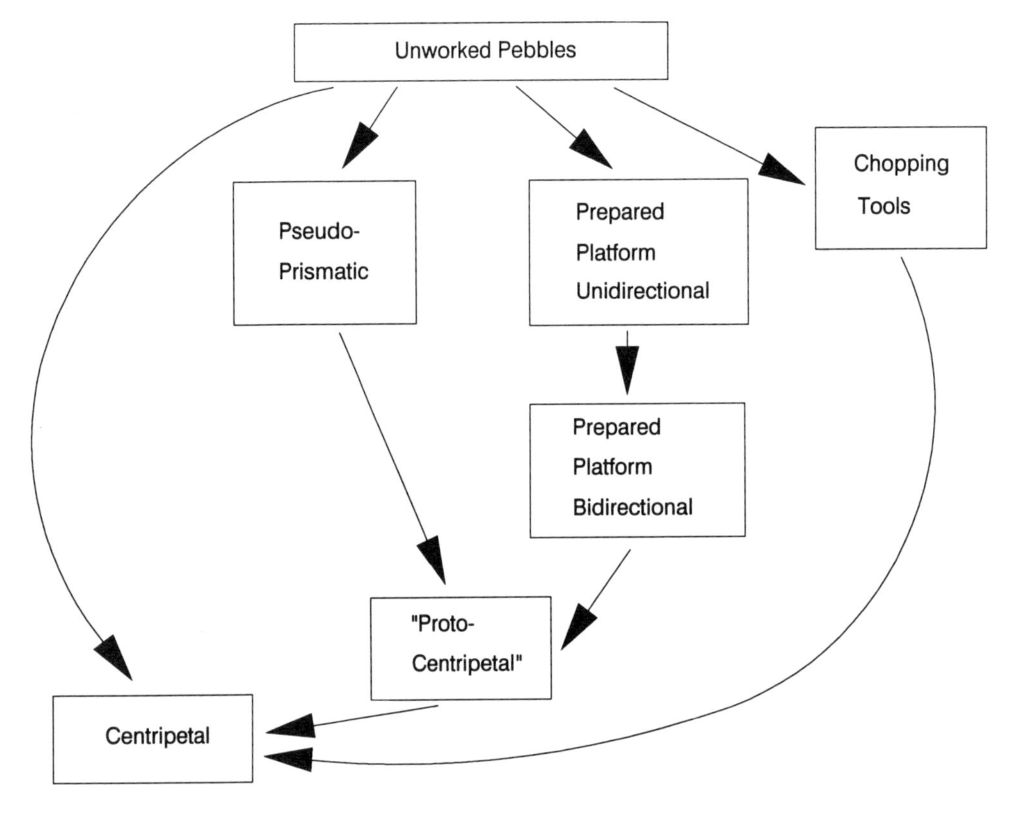

Figure 4.9. Hypothetical sequence of core reduction and transformation proposed by Bietti et al. (1991).

core forms represent independent reduction trajectories, all core forms should be reduced to approximately the same degree, and there should be no systematic differences in their sizes. The average sizes of cores from the Pontinian Mousterian clearly favor the latter alternative. As Table 4.7 indicates, cores worked in a parallel fashion from one or two platforms are *not* significantly larger than centripetal nuclei in any of the four sites. Similarly, in those individual assemblages with sufficiently large numbers of unbroken cores for comparison, all differences in core size are also statistically non-significant (for the sake of economy, data from other assemblages are not presented). This includes Breuil strata 3/4, discussed by Bietti et al. It is not even certain that cores with two platforms are more reduced than cores with just one. There is no size difference between them for the sample as a whole (Table 4.8). In one smaller sub-sample (Grotta di Sant'Agostino) single platform pseudo-prismatic cores are significantly larger than the dual platform variant, but in the other (Grotta Breuil 3/4) the opposite is true.

One could deny that the overall similarities in core sizes actually disconfirmed the "transformational" model of core reduction by arguing that all cores were simply reduced until they could be worked no more (e.g., see Barton 1990). However, the uniformity of summary statistics across formal categories does not mean that all cores were worked down to the same size and abandoned. In fact, each of the mean values shown in Table 4.7 has a substantial standard deviation associated with it, and, as Figure 4.10 illustrates, within any category there are many specimens considerably larger or smaller than the mean. What these data show is that shapes of cores have little relationship to variation in the extent of reduction.

The sizes of *products* of core reduction may provide a somewhat clearer picture of the relationships between different core forms. Baumler was able to demonstrate in the Zobiste material that flakes and

Table 4.7. Maximum dimensions of unbroken cores (mm)

		All centripetal	*Pseudo-prismatic*	*Prepared platform*
Grotta Breuil	mean	33.0	32.6	33.1
	sd	6.4	6.6	6.4
	n	81	48	30
Grotta Breuil, strata 3/4	mean	34.4	33.5	33.7
	sd	6.8	6.2	5.6
	n	22	24	23
Grotta Guattari	mean	36.6	35.7	—
	sd	7.7	11.8	
	n	67	11	(2)
Grotta dei Moscerini	mean	29.3	25.7	—
	sd	6.8	1.8	
	n	25	6	(1)
Grotta di Sant'Agostino	mean	27.9	27.6	28.7
	sd	4.4	7.0	3.7
	n	109	43	32
Agro Pontino	mean	34.9	—	32.9
	sd	7.8		6.0
	n	102	(0)	16

Table 4.8. Maximum dimensions of unbroken pseudo-prismatic and prepared platform cores, according to the number of platforms

		Maximum dimension (mm)				
	Category	*Mean*	*sd*	*n*	*t-value*	*p*
All sites	Prepared platform—1 platform	30.3	5.1	33	−1.711	.09
	Prepared platform—2 platforms	33.2	6.2	15		
	Pseudo-prismatic—1 platform	29.9	6.8	76	0.341	.73
	Pseudo-prismatic—2 platforms	30.7	10.9	16		
Grotta di Sant'Agostino	Prepared platform—1 platform	28.8	3.5	16	0.284	.78
	Prepared platform—2 platforms	28.1	3.8	3		
	Pseudo-prismatic—1 platform	27.6	5.0	37	2.854	*.01*
	Pseudo-prismatic—2 platforms	20.3	2.8	4		
Grotta Breuil, strata 3/4	Prepared platform—1 platform	29.4	5.4	5	2.388	*.04*
	Prepared platform—2 platforms	36.8	4.3	5		
	Pseudo-prismatic—1 platform	32.7	5.4	6	0.340	.74
	Pseudo-prismatic—2 platforms	34.0	7.1	4		

(Note: sufficient sample sizes available only for Grotta di Sant'Agostino and Grotta Breuil, strata 3/4)

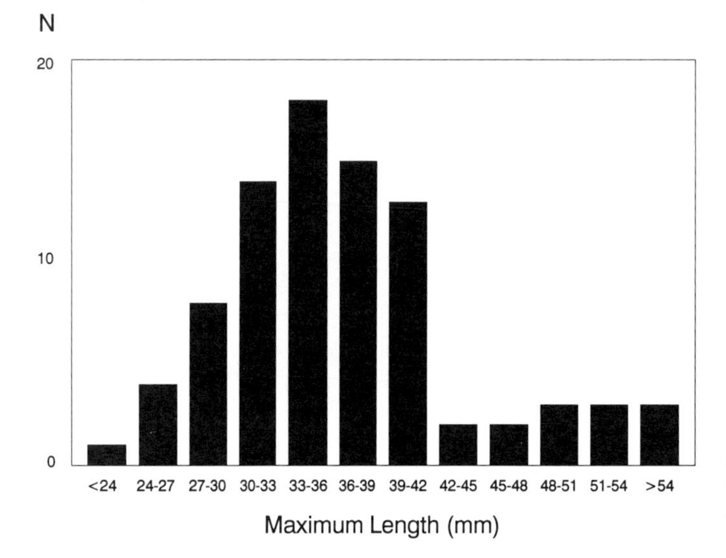

Figure 4.10. Maximum lengths of unbroken centripetal cores from Grotta Guattari.

tools with parallel dorsal scar patterns, as would be produced from cores with one or two platforms, were significantly longer than specimens with radial dorsal scar patterns, as would come from centripetal cores (Baumler 1988). This suggested to him that pieces with parallel dorsal scars were produced at an earlier stage in the exploitation of a piece of raw material, very much in keeping with the notion that core forms changed over the course of reduction. Table 4.9 contains data on the maximum lengths of unbroken flakes and retouched pieces from the Pontinian assemblages with parallel and radial dorsal scar patterns: the former are subdivided according to the platform type, corresponding to the two major varieties of single or double platform nucleus. As is clear from the table, there is no evidence that flakes and tools with parallel dorsal scar patterns are *consistently* longer than artifacts with radial scars, either in the sample as a whole or in assemblages from individual sites. Looking at sites as wholes, the only statistically significant difference goes against the model of core transformation: at Grotta dei Moscerini flakes and tools with radial dorsal scar patterns are *larger* than pieces with parallel dorsal scars ($t=-2.41$, df=145, $p=0.017$). Of the 17 individual assemblages (data not shown), in only one (Grotta Breuil strata 3/4) are specimens with parallel dorsal scar patterns significantly larger than specimens with radial scars ($t=-2.88$, df=108, $p=0.005$). It is not clear what this single difference means, however: with 17 paired comparisons, there is about a 10% chance of encountering a statistically significant difference at the 0.005 level, even with random data.

The data presented in Tables 4.8 and 4.9 clearly do not support the proposition that the shapes of cores, and the forms of their products, reflect different stages in a single protracted sequence of core preparation and exploitation in the Pontinian Mousterian assemblages. There is no doubt that cores do change over the course of reduction. Moreover, this kind of model does appear to apply quite well to a number of archaeological cases, including situations in which the general scheme of flake production shifted from parallel to centripetal or radial. However, the "transformational" model does not hold true for the assemblages at hand, at least as regards the major core types and the majority of assemblages. Instead, the subject assemblages contain the debris of a number of independent trajectories of core reduction.

I argue that at least four distinct approaches to flake production are represented in the Pontinian Mousterian cases discussed here (Figure 4.11). One was bipolar or hammer-and-anvil technique, which yielded split pebbles or similar flakes that could be retouched directly or could be further exploited by other methods. A second general method followed the centripetal scheme of core exploitation,

Table 4.9. Maximum dimensions of unbroken flakes and retouched tools (mm)

		Parallel preparation			*Centripetal*	
		All	*Plain platform*	*Faceted platform*	*All*	*Bipolar*
All sites	mean	28.9	27.4	29.5	29.5	27.3
	sd	7.8	7.7	7.7	8.1	5.9
	n	521	214	245	582	932
Grotta Breuil	mean	29.0	26.9	30.6	27.6	27.9
	sd	8.4	8.0	8.7	6.7	6.7
	n	172	78	71	160	155
Grotta Breuil, strata 3/4	mean	32.8	31.6	33.5	28.0	30.8
	sd	9.4	11.4	8.9	7.2	7.7
	n	62	20	36	48	13
Grotta Guattari	mean	32.6	29.9	33.8	32.8	28.7
	sd	8.9	7.1	10.4	10.4	5.7
	n	52	19	23	90	215
Grotta dei Moscerini	mean	27.1	26.3	27.7	30.8	26.1
	sd	7.8	9.4	5.6	9.1	6.0
	n	51	28	19	96	131
Grotta di Sant'Agostino	mean	27.9	27.8	28.1	28.3	26.5
	sd	6.3	6.8	6.1	6.7	5.2
	n	230	87	126	211	412

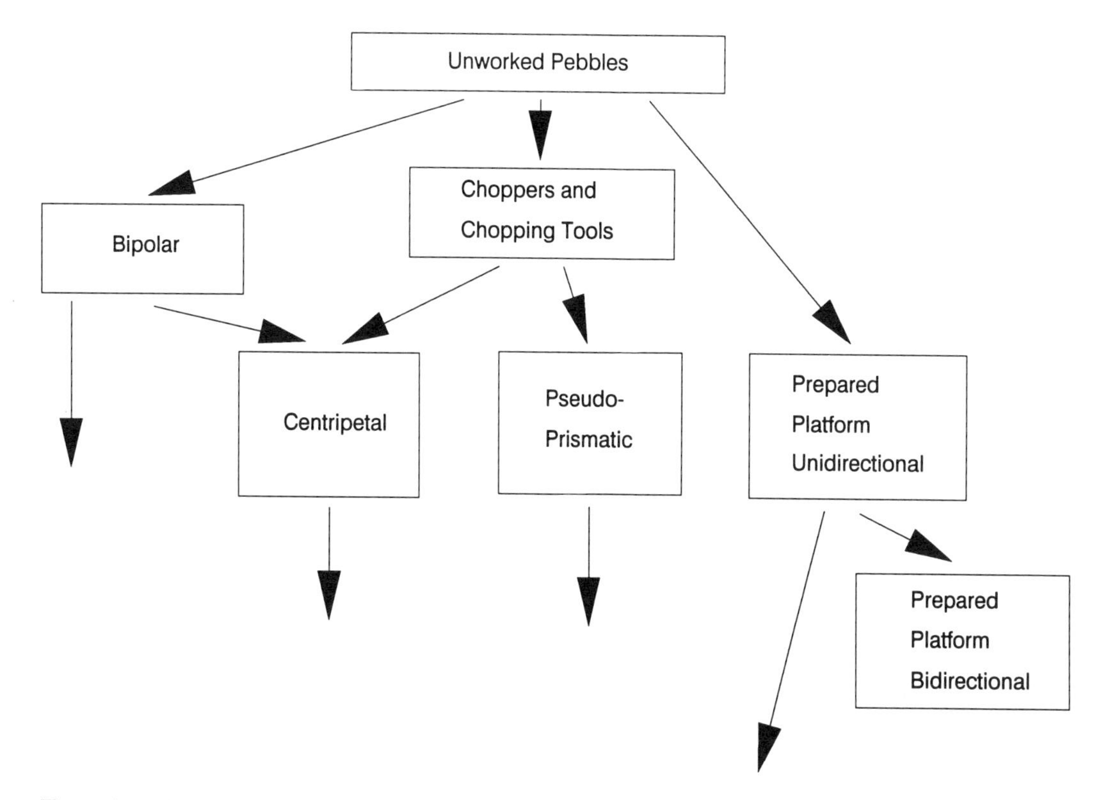

Figure 4.11. Proposed independent pathways of core reduction.

beginning from a chopper/chopping-tool-like form, and progressing to a classic disk or Levallois core. The third trajectory led to the production of pseudo-prismatic cores: here again, nuclei probably passed through a stage at which they resembled elongated choppers. The final method of flake production involved the use of so-called prepared platform cores. These probably began life with a single platform: additional striking platforms were apparently added as the opportunity arose or as they became necessary, but this did not occur as an inevitable outcome of progressive reduction. Although the residual core began to resemble a discoid as a third (or even a fourth) platform was added, there is no evidence that the majority of disk cores were produced in such a manner, even in those assemblages (from Grotta Breuil and Grotta di Sant'Agostino) where there is greatest evidence for use of the two methods of parallel flake production.

The materials from the uppermost levels at Grotta Breuil (strata 3/4) represent the one *possible* departure from the general technological pattern. Differences in the lengths of flakes and retouched tools from Breuil strata 3/4 (Table 4.9) would support a model of core transformation, while the sizes of different core forms are not consistent with such a model (Table 4.7). Either the technology of this most recent assemblage was unlike that of the other Mousterian cases, perhaps resembling the scheme proposed by Bietti et al. (1991), or this single exception from the trend represents a statistical fluke. The archaeological materials from Grotta Breuil strata 3/4 possess other unusual characteristics (discussed below), and it is entirely possible that core technology was somewhat distinctive as well. Final resolution of this question will require more detailed technological analyses (e.g., refitting studies) that lie beyond the scope of the present research. For the present, however, the flake production technology represented in the most recent deposits at Grotta Breuil will be considered a possible exception to the general pattern found in the other Pontinian Mousterian assemblages.

It is of course possible to describe methods of flake production in much greater detail than simply the general scheme of detachment (centripetal or parallel). Researchers interested in *chaînes opératoires* have identified an ample, and indeed often bewildering, array of different approaches to making flakes and tool blanks; at times it seems that every new site and new sample yields a novel variation on some technological theme. Because refitting and extensive replication were not undertaken for this research, I cannot provide a "blow-by-blow" description of the hypothetical core reduction trajectories. Nonetheless, it is possible to hazard some more general comparisons.

Given the fact that the data base includes a variety of assemblages that span many tens of thousands of years, it is unlikely that the numerous radial or centripetal cores in the Pontinian assemblages represent a single uniform *chaîne opératoire* homogeneous in every detail. Some general observations are possible, however. Not many of the products of the radial cores can be classified as classic Levallois flakes, because they tend to have relatively few dorsal scars and because so many preserve traces of cortex. However, this is almost certainly a function of the particular raw material used. On the other hand, pseudo-Levallois points and similar products, characteristic of non-Levallois radial technology (Boëda 1993), are even more rare. Turning to the cores themselves, the great majority of radially prepared specimens are relatively flat with gently convex (rounded) rather than steeply angled or "conical" faces of detachment. Moreover, striking platforms are frequently steeply angled and faceted. The general morphology of the cores is thus consistent with Levallois technology. Furthermore, the scarcity of pieces with evidence for preferential removals suggests that the "recurrent" variety was predominant. On the other hand, Levallois technology by definition involves a single face of detachment, with the other face of the core being used for preparation of the platform (Boëda 1986, 1993). The presence of a significant number of bifacially worked pieces consequently suggests that something like non-Levallois radial core technology was also sometimes used.

Despite their name, *pseudo-prismatic cores* do not appear to represent simply a kind of rudimentary or incipient prismatic blade core technology. The tendency to create and exploit flat rather than strongly convex faces of detachment, and the absence of the crested blade technique, suggest that Mousterian tool makers sought a somewhat different range of products than did Upper Paleolithic technologists. Rather than seeking to create long, thin, very regular blanks (i.e., blades and bladelets), the Middle Paleolithic technique seems to have been aimed at the production of a series of somewhat wider flakes, perhaps with the potential for being worked into sidescrapers. The blank production technology represented by the pseudo-prismatic cores most resembles the basic method often illustrated for creating *couteaux à dos naturel* (e.g., see Bordes 1968).

The *prepared platform* cores provide an interesting comparative case, which highlights the difference between a typological and a technological definition of Levallois. The prepared platform cores are somewhat reminiscent of so-called "recurrent unipolar or bipolar Levallois" technologies known from Mousterian industries in Europe and the Near East (Boëda et al. 1990; Meignen and Bar-Yosef 1988). The prepared platform cores possess at least three of the five fundamental characteristics of the "new definition" of Levallois technology (see Bar-Yosef et al. 1992:544): first, they have two intersecting surfaces, one used for detachment of flakes or blades and the other used for platform preparation; second, removals are essentially parallel to the plane of intersection between the two surfaces; third, platforms are prepared (faceted). Practical experience in knapping small flint pebbles from coastal Latium suggest that hard hammer percussion (the fourth requirement in the "new definition") was used in core reduction. The final criterion in that definition—distal and lateral convexity of the core—is difficult to assess because most of the specimens available for study were intentionally abandoned and probably possessed attributes (such as concave or flat faces of detachment) that rendered them unfit for further use. On the other hand, it is difficult to imagine how these cores could have ever yielded flakes unless their surfaces had been convex at one time.

Whereas the prepared platform cores themselves might coincide fairly well with one definition of Levallois technology, their *products* are unlikely to be classified as such. They would generally have faceted butts, but the majority—more than 75% of pieces with dorsal scar patterns consistent with a parallel scheme of flaking—preserve dorsal cortex, and most have only a few dorsal scars. Thus, in this case, where small raw materials render it impractical to extensively prepare a flaking surface and remove all cortex, we seem to be presented with a technological system with Levallois production yielding products that are predominantly non-Levallois in form. In sum, the geometry of the prepared platform parallel cores is consistent with unipolar, recurrent Levallois technology, but the nature of the raw materials used prevented products from resembling classic Levallois flakes or blades.

While the distinctions between different kinds of core technology may be quite real, representing quite different ways of dealing with the mechanics of stone fracture, we should be cautious in interpreting the apparent technological similarities between Pontinian and other Mousterian assemblages. There are, after all, a finite number of practical options to pursue in flaking stone, and there will inevitably be convergence in approaches to core reduction. Moreover, the amount of control one has over core form is—roughly speaking—inversely proportional to the size of the raw material and the duration of the reduction chain. Where raw materials are large, stone workers are free to trim and shape cores almost at will to produce a chosen range of end products. In the case of the Pontinian, the limitations and opportunities imposed by the shapes of the pebbles on which cores were made had a lot to do with determining how flakes could be manufactured and what the abandoned cores look like.

There is in the subject assemblages a fairly strong relationship between core form and pebble form, summarized in Tables 4.10a and b. The first table (4.10a) is a frequency cross-tabulation of core forms against pebble shapes for the sample as a whole, using only those specimens for which the form of the original pebble could be identified. The second table (4.10b) shows the standardized residuals, calculated as $(O-E)/\sqrt{E}$ (where O=Observed, E=Expected). A positive value for this residual indicates that a particular cell contains more than the expected number of cases, whereas a negative value indicates the opposite. Extreme values of the residual (positive or negative) for significant core forms are shown in heavy type. While the tables

Table 4.10a. Relationship between core form and pebble form (frequencies)

	Core blank						
Core type	*Flat pebble*	*Cylindrical pebble*	*Spherical pebble*	*Blocky pebble*	*Tabular piece*	*Large flake*	*n*
Tested pebble	48	26	7	2	3	0	86
Chopper/chopping tool	87	34	12	2	0	0	135
Centripetal (all types)	255	28	25	6	2	4	320
Prepared platform	48	10	2	0	0	2	62
Pseudo-prismatic	28	22	18	4	3	0	75
Globular	1	1	4	1	1	1	9
Bipolar	7	6	0	0	0	1	14
Misc.	6	7	5	0	0	0	18
Total Determinable							719

Table 4.10b. Relationship between core form and pebble form: standardized residuals $(O - E) / \sqrt{E}$

	Core blank					
Core type	*Flat pebble*	*Cylindrical pebble*	*Spherical pebble*	*Blocky pebble*	*Tabular piece*	*Large flake*
Tested pebble	−1.24	2.49	−0.59	0.15	1.85	−0.98
Chopper/chopping tool	−0.33	1.76	−0.46	−0.49	−1.30	−1.23
Centripetal (all types)	**2.83**	**−4.10**	−1.31	−0.26	−1.00	0.23
Prepared platform	1.03	−0.46	−1.71	−1.14	−0.88	1.58
Pseudo-prismatic	**−3.12**	**2.15**	**3.76**	**1.95**	**2.13**	−0.91
Globular	**−2.04**	−0.52	**3.23**	**1.87**	**2.64**	**2.85**
Bipolar	−0.77	2.10	−1.19	−0.54	−0.42	2.14
Misc.	−1.74	1.99	2.35	−0.61	−0.47	−0.45

show that any core form *could* be made on virtually any type of pebble, they also confirm the fact that the shapes of pebbles did influence the ways they were exploited. Centripetal reduction was executed most frequently using flat or split pebbles, and rarely on other forms. Pseudo-prismatic cores most often began as spherical, angular/chunky, or elongated pebbles, and rarely as flat ones. Like radial cores, prepared platform cores were preferentially made on flat pieces, although there is a more even distribution across other pebble forms.

These data strongly imply that Mousterian stone workers in the study area took advantage of the opportunities offered by differently shaped pebbles, adapting their tactics of manufacture to the peculiarities of the raw materials. Only a few utilizable flakes could have been obtained from a small flint pebble, making it necessary to produce pieces of appropriate shapes and sizes right from the outset. Even though the small size of most pebbles would have prohibited preparation of the core, particularly the face of detachment, the shapes of flakes could be controlled to some degree by selecting pebbles that already possessed some of the attributes of more extensively prepared cores. For example, the lateral and distal convexity that helps define Levallois technology could have been obtained by choosing lenticular pieces of raw material, instead of through extensive chipping of the face of detachment. This would help explain why many radial or prepared platform cores are highly reminiscent of Levallois technology, while relatively few flakes and retouched tools can be classified as Levallois products.

Explaining Variation in Core Technology

Links between core forms and pebble shapes may tell us something about the overall organization of flake production in the study area, but alone they do not explain either the variety of approaches to flake production or variation in the frequency with which different approaches were employed. Table 4.1 shows that while most assemblages contain a variety of core forms, the abundances of different types vary widely. Because the frequencies of different core forms vary from site to site, it is natural to ask whether variation in the general approaches to core reduction is a simple response to the ranges of pebbles available in different locations, or whether it represents more complex (and interesting) choices by prehistoric technologists. In other words, was the frequency of pseudo-prismatic cores determined by the local abundance of elongated or spherical pebbles, or were pebbles preferentially selected in order to facilitate a desired mode of flake production?

It is possible to conduct a simple test for potential effects that variation in the forms of locally available pebble raw materials might have on core technologies. As discussed above, the pebbles found throughout coastal Latium probably have the same geological origins, and all have been extensively reworked by marine action. The sizes and shapes of clasts found at any particular location are determined by factors such as local terrestrial hydrology, the morphology of the littoral zone, and the ages of the nearest gravel deposits, rather than by differences in the origins of the flint. There have been no systematic assays of variation in pebble sizes and shapes within the study area. Given the changeable nature of the local landforms and of the raw material sources themselves, the value of such a study would be limited: after all, it would be virtually impossible to determine whether the pebble exposures sampled were the same ones available to and exploited by Mousterian hominids. However, we can pose the question more simply. The closer two sites are to each other, the more likely it is that prehistoric tool makers at both sites would have used the same or similar raw material sources. If the forms of available pebbles determined the kinds of core forms made on them, then sites located close together should yield more similar ranges of core forms than do distant sites.

The influence of geographical proximity on similarities in the forms of cores found in different sites is addressed in Table 4.11 and Figure 4.12. Table 4.11 shows two measures of the "distances" between assemblages from the respective sites. The first is simple (straight-line) geographic distance. Grotta Guattari and Grotta Breuil are situated within 5 km of one another on Monte Circeo; Grotta dei Moscerini and Grotta di Sant'Agostino are also located near each other, but around 40 km to the south of the other two sites. The second distance measure is essentially an inverse of Robinson's index of agreement (Robinson 1951), calculated as the summed differences in the percentages of different core types found in each pair of sites. The index varies from 0 to 200: the lower the score, the more similar the proportions of core forms in a particular pair of assemblages. Because the variables of interest are geographical and not behavioral, the core assemblages are lumped by site. Finer-grained patterns of inter-assemblage variation are not of interest in this case.

Table 4.11 Geographical and technological "distances" between sites

Site pairing	*Geographical distance (km)*	*Core assemblage "distance"* [a]
Guattari–Moscerini	35	45.9
Guattari–Sant'Agostino	40	85.8
Guattari–Breuil	5	70.7
Moscerini–Sant'Agostino	5	60.4
Moscerini–Breuil	40	63.9
Sant'Agostino–Breuil	45	36.5

[a] Core assemblage "distance" calculated as inverse Robinson-Brainerd Index of similarity: distance = summed difference between percentages of each core type for a pair of assemblages.

If there were a direct link between the kinds of cores Mousterian tool makers produced and the forms of pebbles available nearest to the cave sites they occupied, there would be a strong correlation between geographical and technological distance. Instead, there appears to be virtually no rela-

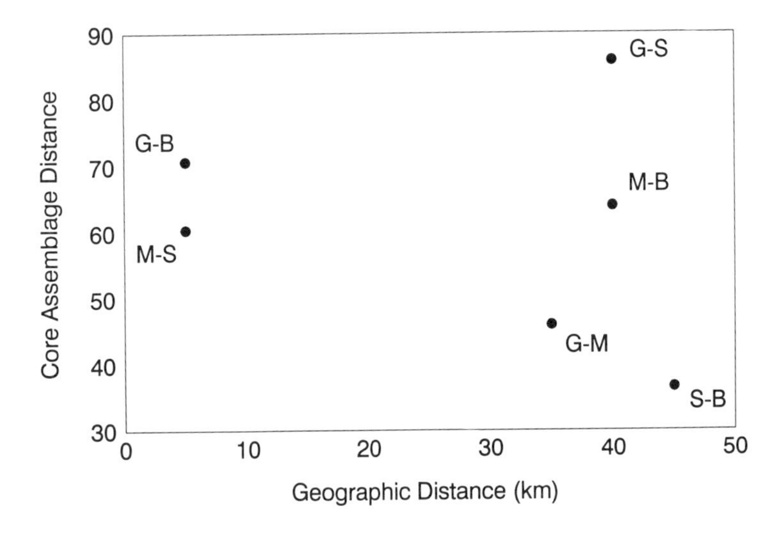

Figure 4.12. Plot of geographic and technological "distances" between pairs of sites. (G=Gr. Guattari, B=Gr. Breuil, M=Gr. dei Moscerini, S=Gr. di Sant'Agostino.)

tionship at all between these two variables. Two of the most distant pairs of sites (Sant'Agostino–Breuil, Guattari–Moscerini) exhibit the greatest overall similarity in core forms, while the core assemblages from the two most proximate site pairs (Guattari–Breuil and Moscerini–Sant'Agostino) are quite dissimilar. The relationship between geographical proximity and technological similarity is portrayed graphically in Fig. 4.12. If similarities in core forms were strongly determined by local pebble shapes, we would expect the site pairs to form a line running from lower left to upper right in the graph. Quite clearly, this is not the case.

The lack of correlation between technological and geographic distances presents a strong argument that, while the forms of pebbles played an important role in Pontinian Mousterian technology, little if any inter-site and inter-assemblage variation in core form is explicable in terms of localized variation in pebble shape. Rather, it seems that Mousterian tool makers selected different pebble forms as supports for different kinds of cores (see Van Peer 1991:138 for a similar observation about a very different region). This leaves open the more interesting question of what other factors might have influenced the choices made by Middle Paleolithic populations about how to produce flakes and tool blanks.

Functional and economic properties of different approaches to core reduction

There is pronounced variation in the frequencies of different core forms among the four Pontinian Mousterian cave sites. As is apparent from Table 4.1, *inter*-site variation is considerably more strongly marked than *intra*-site variation. The most obvious differences between sites involve the frequencies of centripetal and parallel exploitation of cores. All of the core assemblages from Guattari and Moscerini are strongly dominated by radial forms, while the Breuil and Sant'Agostino collections contain many more of the pseudo-prismatic and prepared platform core types. This inter-site variation also represents change over time, in that the dated assemblages from Breuil and Sant'Agostino are somewhat younger than the assemblages from Guattari and Moscerini, all of which were deposited before 50,000 to 55,000 years ago. In other words, it would seem that parallel reduction schemes partially supplanted centripetal methods of flake manufacture during the Mousterian in the study area.

In the second chapter I observed that two major interpretations have been offered for variation in core technology, one citing the influence of postulated cultural traditions and the other referring to functional and/or economic factors. The fact that flake production technology varies over time, as well as from site to site, could be cited in support of either interpretation. It is worth noting in this regard that patterns of inter-site variation and/or changes through time involve simple frequency shifts. With the possible exception of the most recent assemblage (from strata 3/4 at Grotta Breuil), there is no evidence for the appearance of radically new ways of working stone—only changes in the degree of emphasis on one or two alternative methods. In the long run, however, it is virtually impossible to come up with an independent test of an argument about ethnicity or "technological traditions" in the remote past. Functional hypotheses, on the other hand, can be addressed, at least indirectly, first by examining the formal and economic properties of alternative approaches to flake and blank production, and then by investigating whether or not conditions existed in the past that would have rendered any of those properties particularly advantageous. The following section describes some of the attributes that might have rendered each method of core reduction more or less suitable or useful in a particular context.

Flake and blank forms

A question of immediate interest concerns whether the alternative methods of core reduction yielded flakes with distinctive properties that might have influenced their utility or functionality. In order to investigate this possibility it is necessary to assign flakes and retouched tool blanks to a particular mode of production. Short of refitting every flake and core, there is no way of knowing precisely which flakes were produced from which kinds of cores. Flakes and blanks can be probabilistically attributed to different core types based on the orientations of the flake scars on their dorsal surfaces. Table 4.12 and Figure 4.13 show the basic typology of dorsal scar patterns used in this study; Figures 4.14 and 4.15 contain archaeological specimens corresponding to the two major classes of flakes and blanks. Flakes and tool blanks with scars originating *longitudinally*, at the proximal and/or distal ends only (Fig. 4.13, a–d; Fig. 4.14), are most likely to have come from cores with parallel preparation and one or two opposed platforms. Pieces with parallel dorsal scars can be further broken down according to whether the platform is plain or faceted, corresponding to pseudo-prismatic and prepared platform core types respectively. Flakes and tool blanks bearing dorsal scars that originate from several different directions (Fig. 4.13, e–h; Fig. 4.15) are most likely to come from centripetal cores. Bipolar technology represents a special case, as it can be identified most securely *only* on the basis of the characteristic split bulb of percussion. As a consequence, only pieces preserving the proximal end could be classified as bipolar, which is not true for the other categories.

This flake and blank typology is not without ambiguity. Flakes with essentially parallel, longitudinally oriented dorsal scars could sometimes be produced from centripetal cores, while pieces with more complex, multidirectional scar patterns could have been produced after repreparation of prepared platform cores. In order to increase reliability, only relatively large flakes (>15 mm) with more than one dorsal scar are used. Larger flakes would have carried off more of the face of the core, and are thus more likely to accurately reflect the overall pattern of preparation. Even so, some flakes and blanks cannot be reliably assigned to any particular mode of production, so an "unattributable" category was created for these ambiguous artifacts. Types of artifact treated as non-diagnostic or unattributable include flakes and blanks with "unreadable" dorsal scars, pieces with 100% dorsal cortex cover (no dorsal scars—e.g., Fig. 4.13, l), and cores and pebbles that have been retouched directly (Figure 4.16, nos. 7–11). Although the manufacture technique may seem obvious in individual cases, flakes with a single flake scar (Fig. 4.13, i, j; Fig. 4.16, nos. 1–4) were considered too ambiguous to classify, and for the sake of security all were placed in the "unattributable" category. Specimens with dorsal scars originating from the lateral edges only (e.g., Fig. 4.13, k; Fig. 4.16, nos. 5, 6) were also assigned to this class.

It might seem that artifacts with previous removals originating on one side only would be produced most frequently from centripetal or radial cores, but in reality they could derive from preparation or

Table 4.12. Flake and blank classification scheme

	Parallel preparation				
	Pseudo-prismatic	*Prepared platform*	*Centripetal preparation*	*Bipolar reduction*	*Unattributable*
Dorsal scar patterns	proximal and/or distal origins (>1 scar)		multi-directional (>1 scar)	N.A.	<2 scars, "unreadable," lateral origins
Platform	plain (single facet)	dihedral or faceted	all types	split bulb	all types

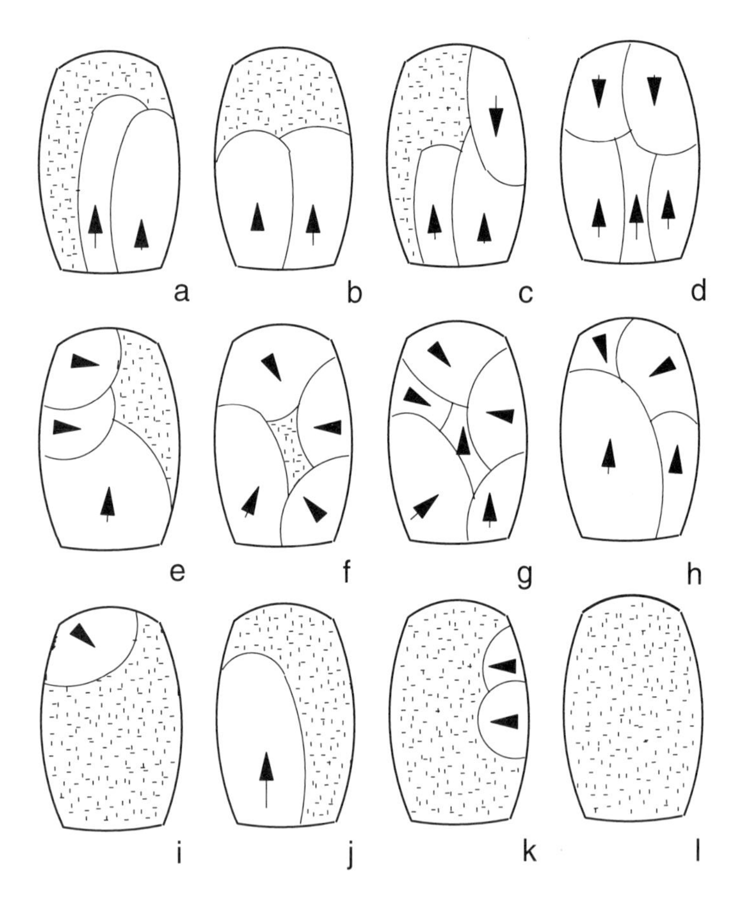

Figure 4.13. Schematic representation of typical dorsal scar patterns.

repreparation of a variety of core types (e.g., see Fig. 4.5 above), as well as from rotation of pebbles when attempting to split them with hammer-and-anvil technique. In fact, the frequencies of flakes with laterally originating scars show no significant correlation with the abundance of any particular core form across assemblages, suggesting that they were indeed by-products of several different reduction processes. I also emphasize that the unattributable pieces could come from virtually any mode of flake production: they do not necessarily represent either informal or undocumented types of core technologies.

Functional characteristics of flakes and blanks

Arguably the most important functionally linked characteristics of flakes and tool blanks are related to size. The absolute size of a tool directly influences its utility. For the majority of Middle Paleolithic retouched artifact forms—scraping and cutting implements—we can probably assume that larger tools were more effective tools, at least within reasonable limits. Artifact dimensions also have important implications for the *durability* of retouched tools, in that larger pieces have a greater potential for resharpening or renewal. Both effectiveness and potential for renewal would have been especially vital factors in the Pontinian Mousterian, since the use of small raw materials greatly limited the maximum sizes of artifacts in the first place.

As was observed previously (Table 4.9), flakes and blanks from single or double platform cores are on average about the same lengths as pieces from radial cores. The data presented in Table 4.9 do show that bipolar flakes are often somewhat smaller in their maximum dimensions than the other types. In the assemblages from Guattari, Moscerini, and Sant'Agostino, flakes and tool blanks with centripetal scar patterns are significantly larger than specimens produced by bipolar percussion. At Guattari and Sant'Agostino, pieces with parallel dorsal scars are also significantly larger than bipolar split pebbles (Table 4.13). In all of the cases, the differences amount to no more than a few millimeters, but with such small artifacts any size advantage is potentially significant.

More consequential than size differences are contrasts in the *shape* (plan-view proportions) of flakes resulting from different modes of core reduction. Table 4.14 shows the ratios of maximum length to maximum width for different classes of flake: only unmodified specimens are included, since retouch can modify the proportions of an artifact. Flakes with parallel dorsal scar patterns tend to be longer and narrower than flakes produced from either centripetal cores or bipolar percussion. The differences in proportion are not particularly impressive when each class is taken as a whole, but the contrasts become more striking if overall size is taken into consideration. The smallest pieces in each category are very similar in shape, while the differences among relatively large pieces are quite apparent (Figure 4.17). Significantly, the greatest contrasts in shape are found in the size range of flakes (>30 mm) most frequently selected for retouch (Table 4.15).

Differences in plan-view proportions have potentially important implications for the durability or potential for renewal of different classes of flake blank. In the Mousterian in general and the Pontinian assemblages in particular, tools tend to be retouched along the long edge of the blank (Kuhn 1992a). The number of times a tool edge can be resharpened is thus limited by its width. Narrow, blade-like pieces have less potential for renewal than wider, more "stubby" blanks of the same length. Pebble sizes would have constrained absolute flake length, so the only practical option for making longer-lasting tools would have been to produce broader tool blanks. If the renewability of stone tools were an important strategic consideration, radial cores and bipolar/hammer-on-anvil technique would have been the most suitable options for flake production, as they yield blanks that are comparatively wide relative to overall length, especially in those size ranges preferred for further modification.

Other characteristics of blank forms are more difficult to quantify but no less important at a functional level. One of these is the quantity (Table 4.16) and the distribution of dorsal cortex on different forms of flake/blank. Cortex cover plays a large role in determining the form of a flake's edges, and hence its utility for different purposes. Flakes with

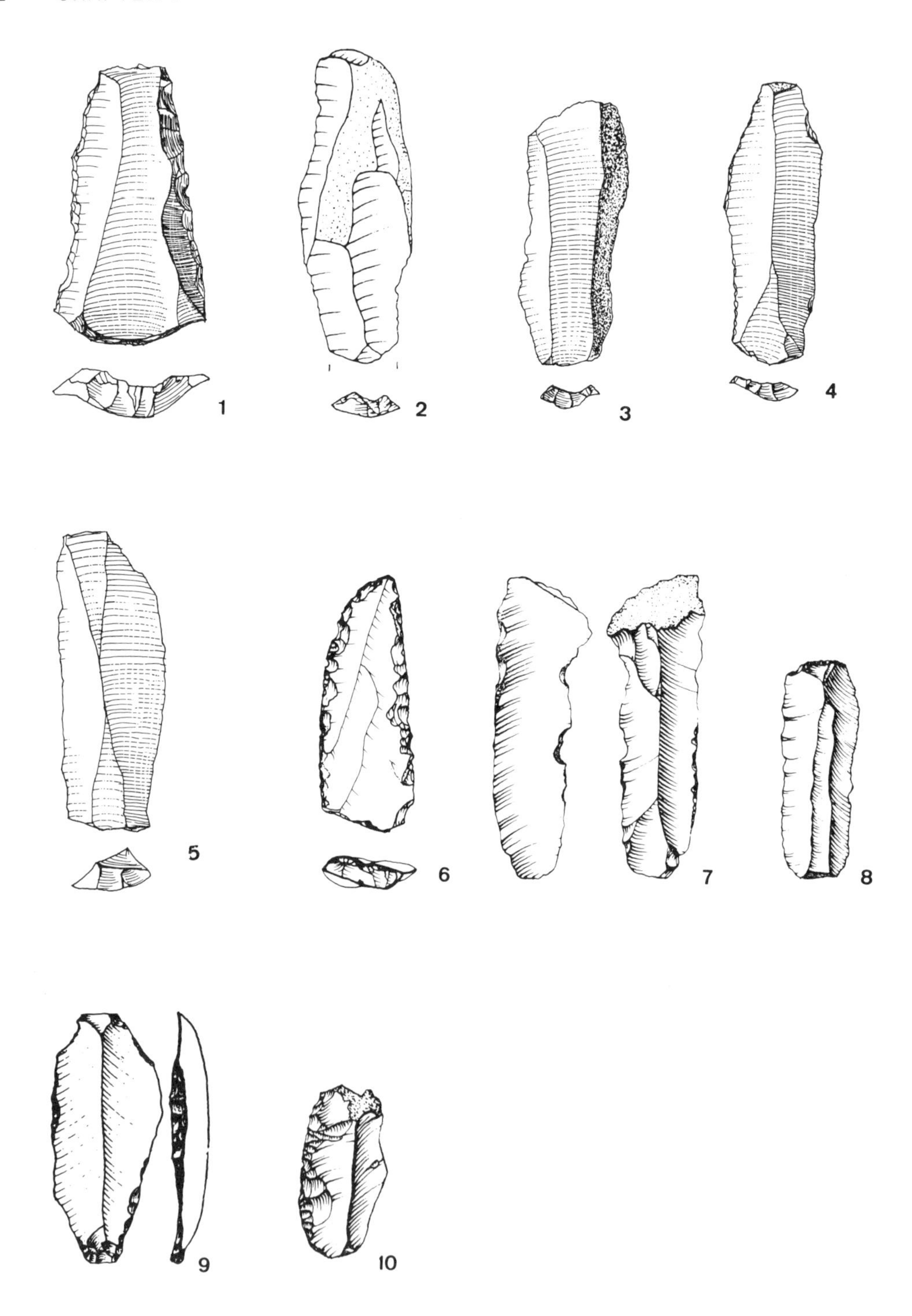

Figure 4.14. Examples of flakes and retouched tools attributed to parallel core preparation and reduction. (1–8=Gr. Breuil; 9–19=Gr. di Sant'Agostino.)

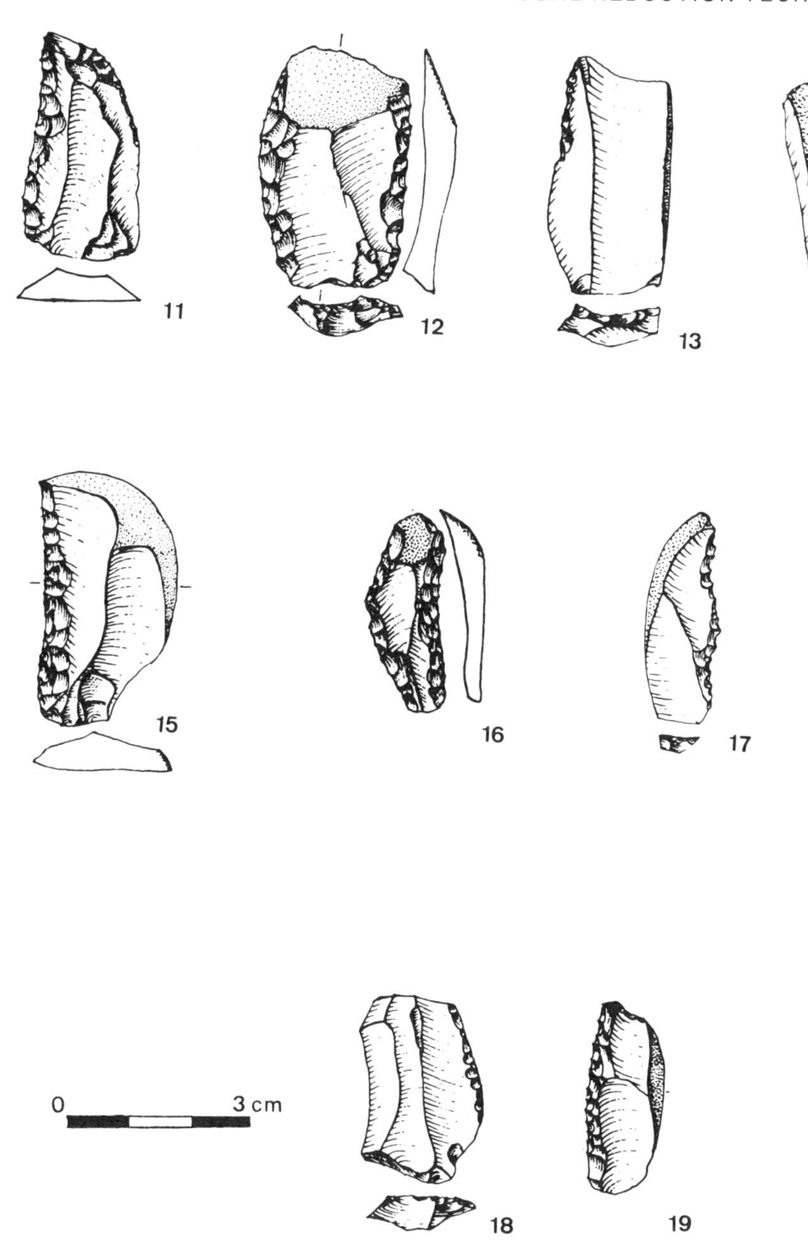

Figure 4.14. *cont.*

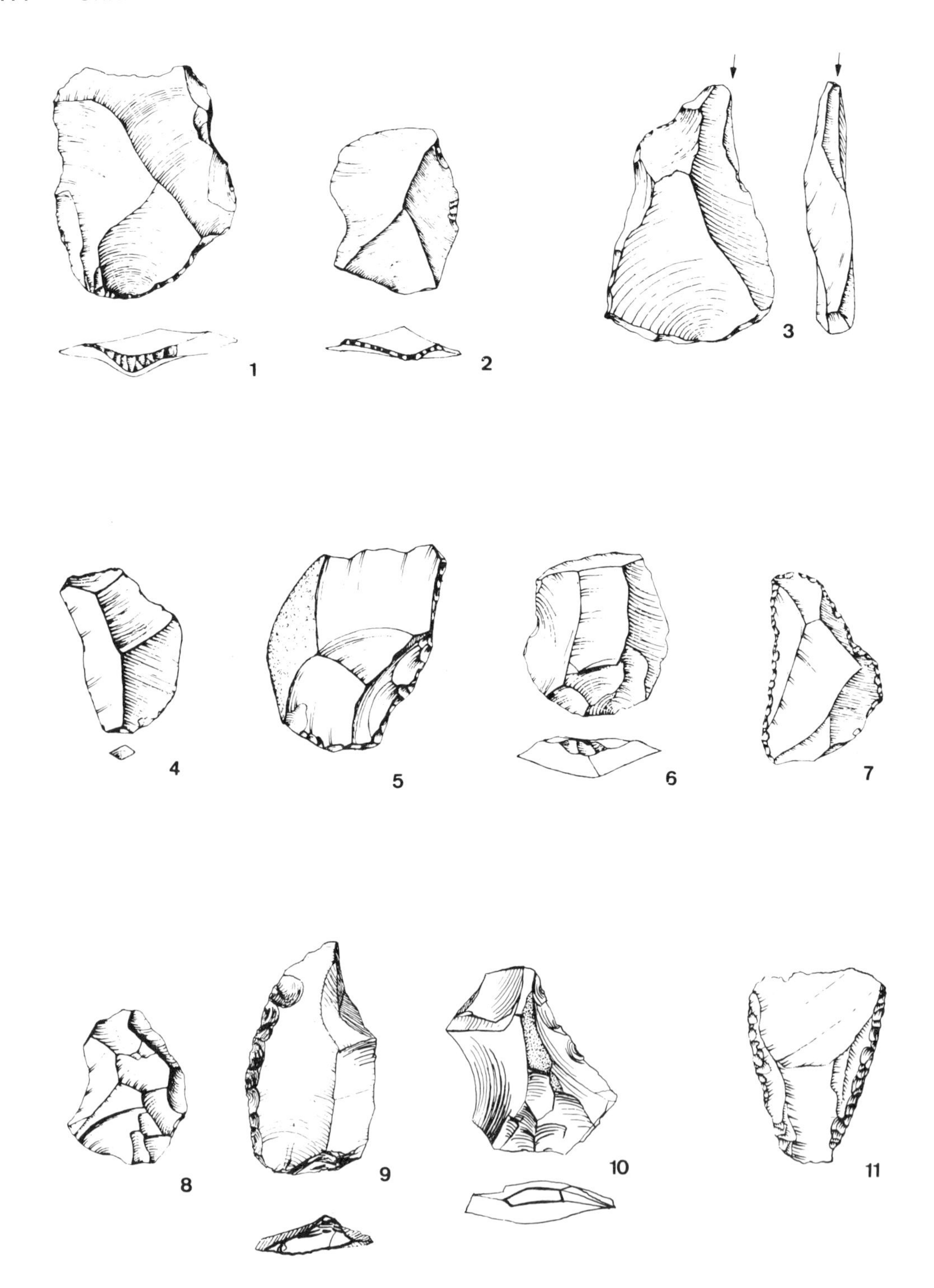

Figure 4.15. Examples of flakes and retouched tools attributed to centripetal core reduction. (1–8=Gr. Guattari; 9, 10=Gr. dei Moscerini; 11–13=Gr. Breuil; 14–18=Gr. di Sant'Agostino.)

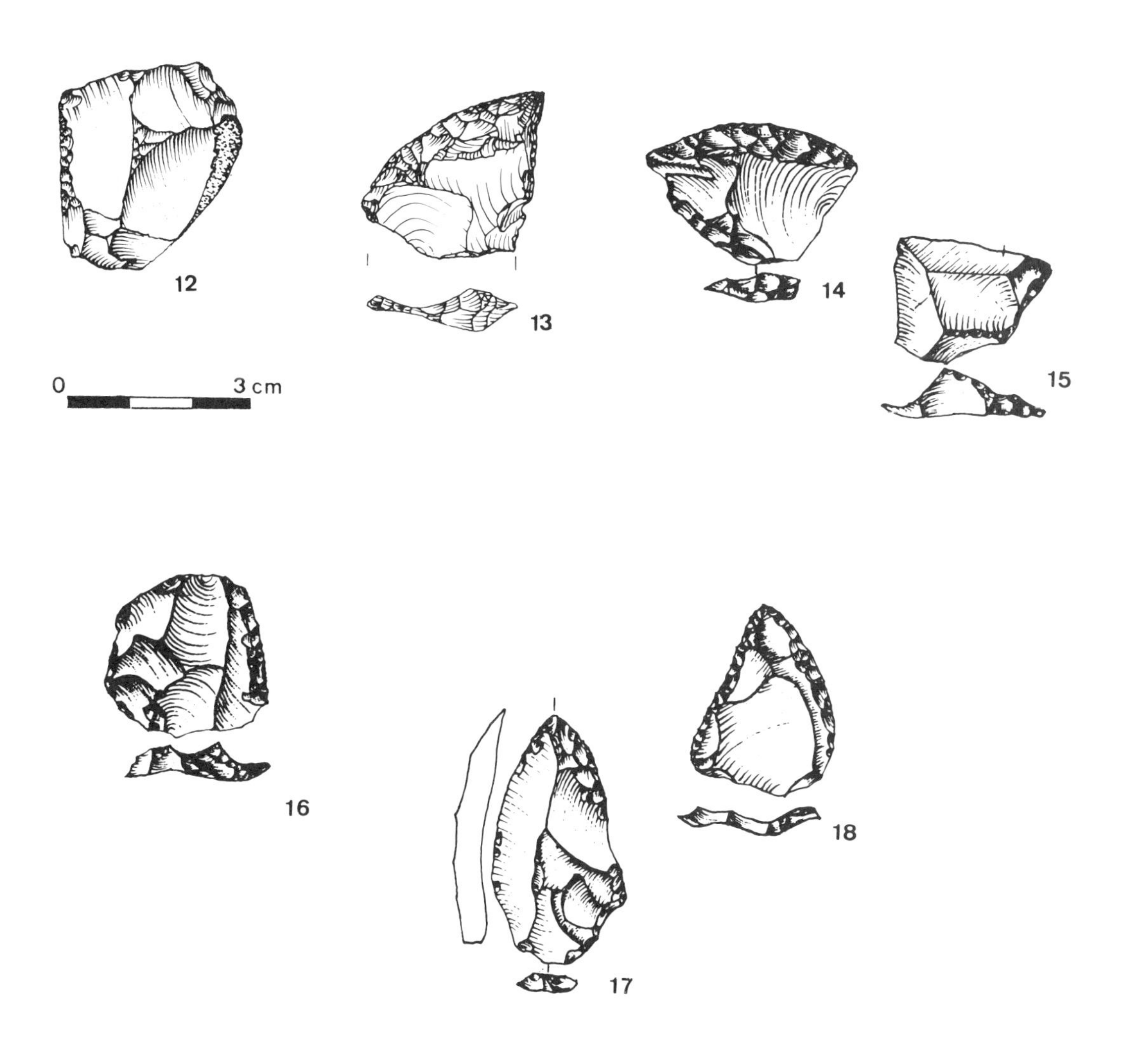

Figure 4.15. *cont.*

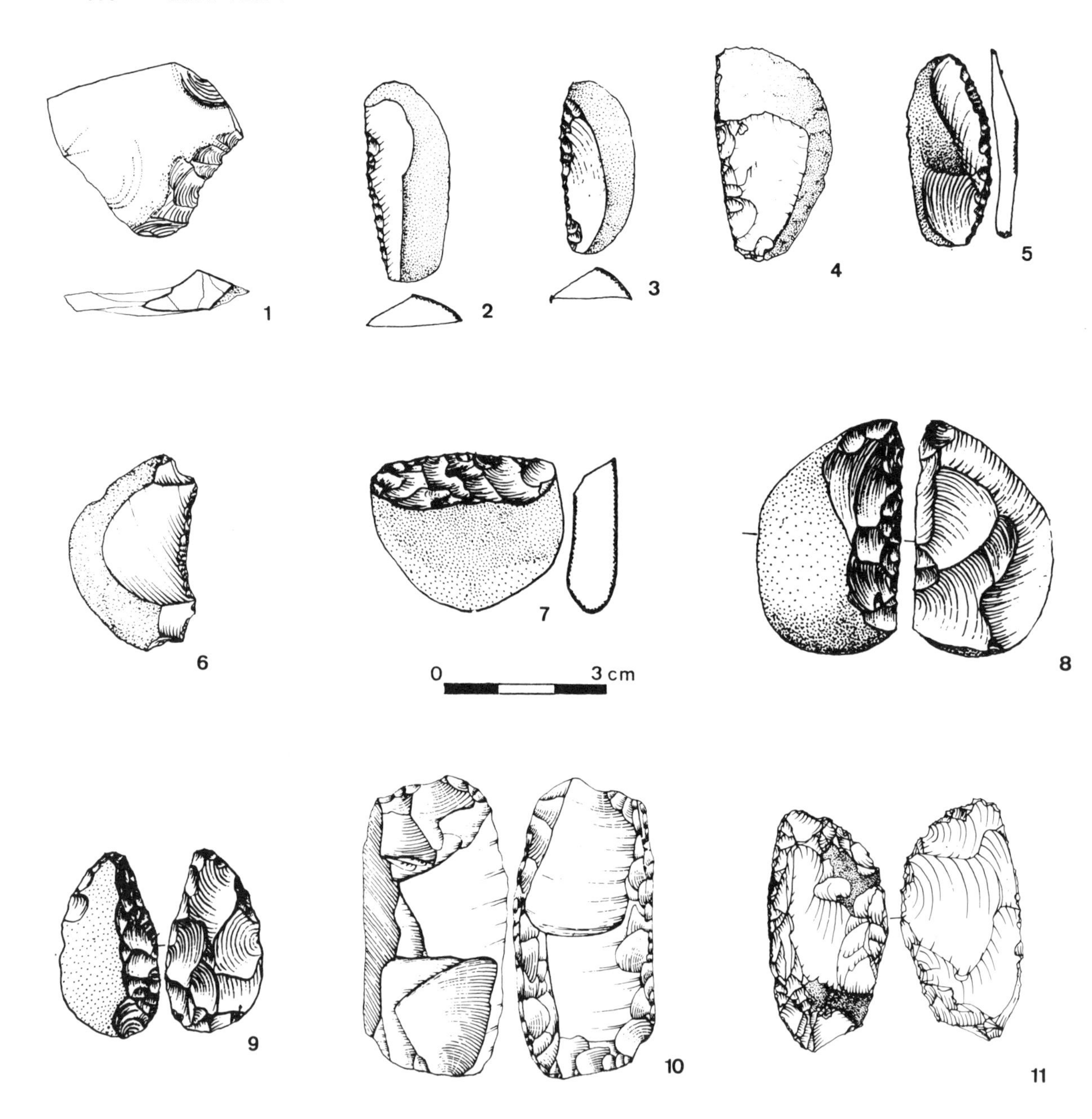

Figure 4.16. Examples of flakes and retouched tools considered "unattributable." (1=Gr. dei Moscerini; 2, 3, 5–9=Gr. di Sant'Agostino; 4, 11=Gr. Breuil; 10=Gr. Guattari.)

Table 4.13. Comparisons of maximum dimensions of unbroken flakes and retouched tools (only statistically significant results shown)

	Flake/blank category	*Maximum dimension (mm)*		*n*	*t-value*	*p*
		Mean	*sd*			
Grotta Guattari	Centripetal	32.8	10.4	90	–4.53	<.001
	Bipolar	28.7	5.7	215		
	Parallel	32.6	8.9	52	–3.93	<.001
	Bipolar	28.7	5.7	215		
Grotta dei Moscerini	Centripetal	30.8	9.1	96	–4.65	<.001
	Bipolar	26.1	5.2	131		
Grotta di Sant'Agostino	Centripetal	28.3	6.7	211	–3.65	<.001
	Bipolar	26.5	5.2	412		
	Parallel	27.9	8.4	172	–2.99	.003
	Bipolar	26.5	5.2	412		

Table 4.14. Length/width ratios for unbroken flakes

		Flake type		
Maximum dimension		*Parallel scar patterns*	*Radial scar patterns*	*Bipolar platform*
All	mean	1.51	1.28	1.10
	sd	0.53	0.46	0.48
	n	302	166	329
<20 mm	mean	1.20	1.14	1.07
	sd	0.31	0.26	0.24
	n	41	23	39
20–30 mm	mean	1.40	1.24	1.06
	sd	0.42	0.38	0.45
	n	189	108	231
30–40 mm	mean	1.93	1.38	1.27
	sd	0.54	0.54	0.64
	n	59	25	53
>40 mm	mean	2.25	1.82	1.24
	sd	0.72	0.72	0.76
	n	13	10	6

Table 4.15. Frequencies of retouched pieces, by size category (percent)

	Unretouched	*Retouched*	*n*
<20 mm	64.2	35.8	282
20–30 mm	46.3	53.7	1705
30–40 mm	24.7	75.3	794
>40 mm	23.0	77.0	196

Table 4.16. Median percentage of dorsal cortex cover

	Flake/blank type (percent)		
	Parallel scar patterns	*Centripetal scar patterns*	*Bipolar platform*
All sites	25	15	75
Grotta Guattari	25	10	70
Grotta dei Moscerini	15	10	70
Grotta di Sant'Agostino	25	15	75
Grotta Breuil	30	17.5	80
Agro Pontino	20	15	75

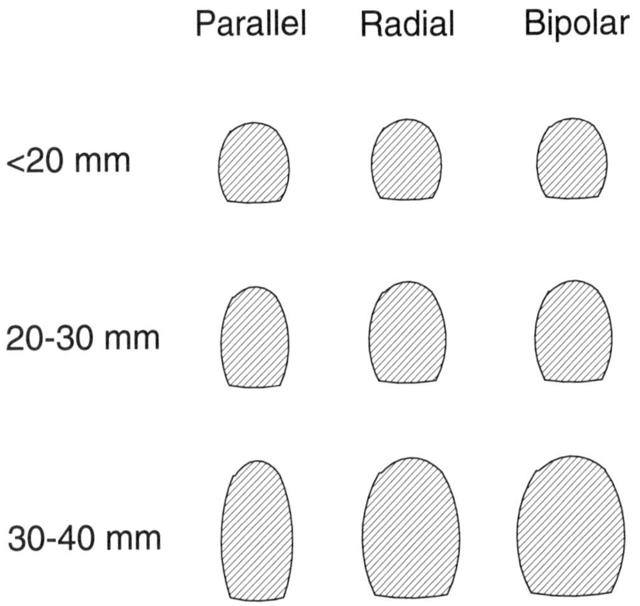

Figure 4.17. Schematic representation of average proportions of flakes in different size ranges (not to absolute scale).

radial or centripetal dorsal scar patterns normally preserve little or no dorsal cortex, and what cortex is present often forms an isolated patch near the center or distal end of the piece. As a consequence, the margins of such flakes tend to be relatively thin and sharp over most of the piece's circumference. In marked contrast, bipolar split pebbles often have complete or near-complete dorsal cortex cover, giving them thick obtuse or even rounded margins. A sharp working edge can be obtained from such blanks only by retouch. As mentioned previously, elongated flakes produced from prepared platform or pseudo-prismatic cores frequently have a strip of cortex along one margin and a sharp edge opposite: the classic naturally backed knife form.

Economy of production

In principle, economy of raw material should be quantified in terms of yield (i.e., number of usable flakes) per unit of raw material. Except where extensive refitting has been undertaken, it is impossible to reconstruct the size of the original piece of raw material. For the purposes of the present discussion it is sufficient to speak of the number of usable flakes obtained per core. It is reasonable to assume that a single pebble very rarely produced more than one core, so that flake-to-core ratios are good estimators of flakes obtained per unit of raw material by different methods.

Based on the global and site-specific figures,

Table 4.17. Core/blank ratios

	Parallel preparation			*Centripetal preparation*		
	Flakes and blanks	*Cores*	*Flakes + blanks per core*	*Flakes and blanks*	*Cores*	*Flakes + blanks per core*
All sites	550	155	*3.5*	582	334	*1.7*
Grotta Breuil	233	77	*3.0*	199	91	*2.2*
Grotta Guattari	71	19	*3.7*	128	133	*1.0*
Grotta dei Moscerini	81	9	*9.0*	129	53	*2.4*
Grotta di Sant'Agostino[a]	163	50	*3.3*	126	57	*2.2*

[a] Includes levels 2–4 only: samples from levels 0 and 1 are incomplete and nonuniform.

parallel core exploitation appears to have consistently resulted in the production of more identifiable flakes and tools per core than did methods of core reduction based on radial exploitation (Table 4.17). The ratios of flakes/blanks to cores attributed to different methods of reduction vary extensively from site to site for a variety of reasons (see below), but there are always fewer flakes and tools per centripetal core than per parallel core. Because of the large numbers of ambiguous, unattributable pieces, ratios of flakes and tools to cores in the table are certainly too low. If anything, however, the number of blanks is more severely *underestimated* for platform cores than for centripetal cores. Pseudo-prismatic and prepared platform cores would have yielded especially large numbers of naturally backed flakes with a single flake scar, all of which were arbitrarily classed as "unattributable." In this regard, it is worth noting that around 60% of flakes and blanks with recognizable dorsal scar patterns in the "unattributable" category have a single flake scar originating from the proximal end.

The notion that centripetal cores would be somewhat less productive than single or double platform cores with parallel preparation is consistent with several previous observations. First, flakes with radial dorsal scar patterns are relatively wide for their lengths compared to pieces from parallel cores, meaning that they contain a greater volume of raw material: one would consequently expect that fewer could be produced per core. Also, radial core preparation often requires more extensive preliminary shaping of the face of detachment, so more raw material would be expended in preparing the core. With parallel preparation and detachment of flakes, previous removals "set up" subsequent detachments. Even so, given the ambiguities inherent in determining what kind of core a particular flake came from, the numbers in Table 4.17 must be taken as estimates of *relative* rather than *absolute* productivity.

Hammer-on-anvil technology represents a special case, since this form of flake production rarely leaves a residual core. In experiments, a flat pebble could usually be split into two, and less often three, usable blanks by this method. Some of the resulting flakes broke spontaneously, however, while others were too thick for retouch and would have had to be further reduced by another means. An estimate of something less than two flakes per pebble seems justified as a rough average. By permitting one to actually split a pebble in half, hammer-on-anvil technique also maximizes the size of flakes relative to the original size of the object piece. Thus, it could be used to exploit pebbles too small to be reduced by other means. Furthermore, if successfully executed, bipolar technology converts virtually the entire piece of raw material into usable blanks.

It is worth noting that these calculations involve only flakes and tool blanks more than 15 mm in length, considered about the minimum size for retouched tools. Many of the by-products of core reduction consisted of small flakes and angular pieces, well below the minimum size that can be practically retouched. The largest flake scars remaining in evidence on cores were measured, but in fewer than 10% of the cases did these exceed the 15 mm limit. Unless all this work simply represents attempts to get one last large flake from a core, the fact that cores were normally reduced beyond the point at which they yielded flakes large enough to be retouched suggests that there was some use for the smaller products. The functionality of very small flakes and possible differences in the yield of tiny flakes for alternative methods of core reduction merit further investigation in another context.

Summary: The Organization of Flake Production in the Pontinian Mousterian

At least four alternative approaches to core reduction were employed to manufacture the range of flakes and tool blanks found in the Pontinian Mousterian assemblages. Three of these resulted in distinctive core forms, termed centripetal or radial, pseudo-prismatic, and prepared platform. The fourth method, bipolar hammer-on-anvil percussion, rarely yielded residual cores but left distinctive traces on platforms and ventral faces of the resulting flakes. Analyses of core and flake/blank sizes indicate that these four methods represent largely distinct trajectories, rather than a series of techniques applied sequentially during the reduction of a single core.

While it is possible to draw parallels between core technology in the subject assemblages and *chaînes opératoires* documented in other Middle Paleolithic contexts, the strong association between the forms of cores and the shapes of pebbles on which they were made reminds us that Pontinian Mousterian technology must be understood as a particular response to the raw materials used. It is most productive to view *chaînes opératoires* in the Mousterian of coastal Latium not in terms of a single core, but in terms of a hypothetical "pocket full of pebbles." The dynamic aspect of flake production can be found not in how tool makers dealt with changes in core form over the course of reduction, but in how they responded to and exploited the array of pebble shapes available to them. Cylindrical or chunky pebbles were normally worked into pseudo-prismatic cores or, less frequently, globular forms. Small lenticular pieces could be split by bipolar technique. Larger flat or lens-shaped pebbles were exploited by *either* centripetal or parallel (prepared platform) methods. Of all the variants, prepared platform and centripetal core technologies were probably practiced on the most similar arrays of pebble shapes, although there may be more subtle differences in the ranges of pebbles chosen.

Even taking into account the links between pebble shapes and core forms, Mousterian tool makers exhibited a certain degree of autonomy with regard to the pebbles they chose to use and the ways they chose to exploit them. Pebbles of similar shape could be, and were, worked by a number of different techniques (Tables 4.10a and b). Moreover, sites located close together do not tend to yield similar ranges of core forms, demonstrating that the choice of flake production techniques was not dictated by local variation in the shapes of clasts found within nearby beach deposits (Table 4.11; Fig. 4.12). Thus, although Mousterian populations were certainly constrained by their reliance on small pebbles, it appears that they could also take advantage of the opportunities offered by variation within the materials at hand, selecting stones because they were suitable for working in a particular manner, and not simply adopting different techniques according to the pebbles they happened to collect.

Although diverse core and flake forms are found in all assemblages, the degree of reliance on different modes of core reduction varies from site to site and from assemblage to assemblage. Overall, centripetal core technology and bipolar reduction are the most abundantly and consistently represented. The two technological variants involving parallel preparation and flake removal are less comprehensively distributed among strata. Significantly, the preference for alternative core reduction technologies appears to have changed over time in the study area. Radial cores and associated flakes and tool blanks are ubiquitous, but centripetal and bipolar reduction were far and away the dominant methods of flake production before 50,000 to 55,000 years ago. Parallel schemes of flake production became more common after this juncture, although they are predominant only in the most recent Mousterian assemblages (Figure 4.18). Pseudo-prismatic cores are present in relatively low numbers throughout the sample, but more common in the most recent levels from Grotta Breuil and Grotta di Sant'Agostino. Prepared platform cores are so scarce in as-

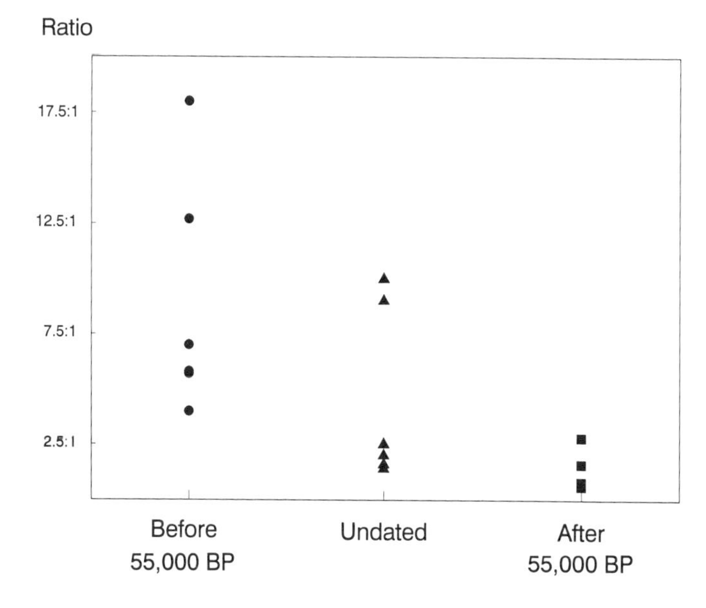

Figure 4.18. Ratios of centripetal cores to parallel cores in assemblages of different ages. Each symbol indicates a single assemblage.

semblages from Grotta Guattari and Grotta dei Moscerini (only 3 specimens) that it is tempting to view these as anomalies: in contrast, they are much more abundant in Breuil and Sant'Agostino (Table 4.18). It should be emphasized that the most important changes in core technology over time entail frequency shifts within a range of preexisting alternatives, and not the appearance of radically new and different methods for making flakes. Only with the most recent assemblage, from Grotta Breuil strata 3/4, is there evidence for the appearance of a novel *chaîne opératoire*, in which uni- and bidirectional platform cores were transformed into radial forms late in the reduction sequence. This assemblage may also mark the disappearance, or at least the partial abandonment, of bipolar technology, which accounts for only a very small proportion of its flakes and blanks (Table 4.18).

Given that Pontinian Mousterian tool makers exercised a certain degree of flexibility in the production of flakes and tool blanks, changes in technological "choices" over time may tell us something about the factors that influenced technological behavior. There is reason to believe that products of the various tactics of flake manufacture differed in form and number, such that a given core technology might have been more or less advantageous under different conditions. The contrasting economic and functional characteristics of the three main methods of flake production are summarized in Table 4.19. Parallel schemes of core preparation and exploitation (resulting in pseudo-prismatic and prepared platform cores) appear to have yielded the largest number of "usable" flakes and blanks (>15 mm) per pebble. The flakes produced, especially the larger specimens, tend to be quite narrow and elongated. Centripetal or radial approaches to core reduction appear to have resulted in the production of fewer flakes per pebble than parallel reduction. In compensation, flakes and blanks tend to be wider, with greater potential for resharpening or renewal, and a larger proportion of the margin is sharp and noncortical. The third major flake production method, bipolar (hammer-and-anvil) reduction, yielded relatively short, wide flakes and tool blanks. Bipolar pieces often preserve cortex over most of their dorsal surfaces, so that sharp edges could be obtained only through retouch. The single most advantageous property of bipolar technique may have been that it produced usable flakes from the smallest of pebbles, converting virtually the entire volume of raw material with little or no preparation.

In conclusion, economy of production and the sizes and shapes of flakes/blanks could well have influenced the choices between alternative reduc-

Table 4.18. Frequencies of cores and core products, by assemblage (percent)

	Centripetal cores[a]	*Centripetal flakes and blanks*[b]	*Parallel cores*[a]	*Parallel flakes and blanks*[b]	*Bipolar flakes and blanks*[b]
		AFTER 55,000 BP			
Breuil					
strata 3/4	28.6	35.2	49.0	55.8	10.0
Sant'Agostino					
level 1	65.0	21.3	29.0	24.5	54.2
level 2	42.5	27.3	52.5	37.0	35.7
level 3	54.6	25.8	31.8	26.7	47.5
		UNDATED			
Breuil					
1986–87 sample	53.6	36.2	34.3	30.1	33.7
old *sondage*	50.0	18.9	25.0	39.1	42.0
Sant'Agostino					
level 4	91.2	16.7	8.8	28.2	55.0
Guattari					
stratum 1	46.7	32.8	33.3	8.2	59.0
Moscerini					
strata group 5	100.0	28.6	0.0	7.6	24.5
strata group 6	62.5	23.7	25.0	25.4	50.1
		BEFORE 55,000 BP			
Guattari					
stratum 2	67.4	41.2	16.3	28.1	30.1
stratum 4	88.5	24.4	4.9	9.3	65.8
stratum 5	74.5	14.9	5.8	15.3	69.2
Moscerini					
strata group 2	81.0	33.1	14.3	14.2	52.8
strata group 3	70.0	44.6	10.0	31.7	23.8
strata group 4	74.2	37.2	12.9	29.5	33.3

[a] Percentage of formal cores: chopper/chopping tools and tested pebbles excluded.
[b] Percentage of identifiable specimens: unattributable pieces excluded.

Table 4.19. Characteristics of major flake production methods

	Flakes per core	Flake proportions	Flake morphology	Other features
Centripetal preparation	Lowest yield	Moderately elongated	Little cortex, sharp margins	
Parallel preparation	Highest yield	Elongated, laminar	Frequent cortical backing	Different techniques for different pebble forms (?)
Bipolar	Moderate yield	Short, wide	Mostly cortical, edges dull	Can be used on smallest pebbles

tion techniques by Middle Paleolithic tool makers in coastal Latium. Demonstrating that alternative variants of core technology have had the potential to fulfill different requirements for tools and tool blanks is not sufficient argument in itself, however. We cannot assume that the attributes identified by the analyst were the characteristics sought by Mousterian technologists. In order to mount a more comprehensive argument about the potential adaptive or evolutionary significance of this kind of technological variation, it is necessary to establish that conditions existed in the past that would have rendered any of the identified properties advantageous, and to investigate whether Mousterian tool makers actually exploited those properties. These topics are addressed in the sections that follow.

5

LITHIC RAW MATERIAL ECONOMY

A fundamental premise of this research is that the rate at which raw materials are consumed represents a pivotal limiting factor for lithic technologies. At the most basic level, strategies for exploiting stone are expected to vary in response to the "cost" of obtaining raw material and keeping it on hand. The cost of making raw material available to tool makers and users in turn varies as a function of three main factors: the natural distribution of stone sources, the movement of human groups relative to those sources, and the scheduling of labor investment in foraging. The second and third factors are obviously of considerably greater interest in this research, as patterns of land use and foraging organization represent an essential link between technology and the broad sphere of human subsistence adaptations. It is nonetheless necessary to account for the effects of raw material availability on stone tool manufacture and use, if only to identify and sort out those components of technological variation that are not related to the ways Mousterian populations made a living and moved about the landscape.

There are two dimensions to archaeological investigations of lithic raw material economies: the study of behavior related to the "management" of stone, and analyses of patterns of raw material procurement and transport. Although intimately related, the conservation and the movement of artifacts and raw materials are normally investigated using quite different domains of data.

Tactics for managing the flow of raw materials are often reflected in the extent to which artifacts have been reused, renewed, or consumed, as well as in strategies of artifact production. While relatively easy to work, crypto-crystalline stone is brittle, and tool edges wear out quickly. Tool makers have two options for producing fresh edges: they can rework worn margins, or they can make new implements. An assumption underlying most studies of lithic raw material economies is that fresh artifacts are more desirable or useful, whereas extensively consumed pieces would present some functional disadvantage: in other words, we presuppose that when time and raw materials are not limited, people would choose to make a new tool or core every time one was needed. Such an assumption certainly seems justifiable in the case of most Mousterian lithic artifacts. Large implements with fresh edges normally would be more effective than small, worn-down "slugs" with heavily reworked edges. The notion that newer or bigger is better probably also applies to cores, at least after they have been worked to the stage at which they began to yield usable blanks. The implication of this premise is that the degree to which artifacts were consumed, used up, or reduced is a direct reflection of the cost of replacing them. If artifacts are extensively consumed, we may conclude that it was relatively difficult to come by new ones. Conversely, if cores and tools were abandoned in relatively fresh condition, chances are it was relatively easy to obtain raw material and/or to maintain a ready supply.

In this study, cores and retouched tools are treated separately with respect to measuring the degree to which they have been consumed or reduced. In addition, several independent criteria are employed to measure the extent of reduction of different classes of artifact. There are two reasons for this. First, none of the tests is absolutely unambiguous, and it is simply more prudent to use more than one. Secondly, raw material economies are complex and internally differentiated. As discussed in Chapter 2 above, the treatment of diverse elements of technology may be influenced by a variety of independent factors (Kuhn 1991b). A more finely divided, multivariate perspective on artifact reduction affords a more detailed view of the economics of raw material use, and provides virtually the only chance of sorting out the effects of human land use patterns from those of raw material distributions.

Two basic kinds of measures are used to assess the extent to which pebble cores have been consumed. One is based on artifact size, either maximum length (employed in the previous section) or, in the case of cores worked on one or two parallel faces, thickness. The second is a more impressionistic assessment of the extent of core preparation, simply the ratio of "informal" core forms—tested pebbles and chopper/chopping tools—to more completely worked "formal" varieties. The extent of reliance on alternative methods of flake production, discussed in the previous chapter, provides a somewhat different perspective on the economics of raw material use. As different core technologies appear to produce varying numbers of flakes per core, it is possible that their use was conditioned in part by the ease of access to flint pebbles. However, core technology presents a somewhat more complex set of issues, because functional as well as economic factors must be taken into account. As we have seen, a "disadvantage" in the number of flakes produced may be offset by some particularly advantageous aspect of blank morphology.

Three different variables are employed to assess the extent to which available tool blanks have been used up. The first is simply a ratio of unmodified flakes to retouched tools, which provides a rough estimate of how many of the flakes produced were actually used. This is the main measure of "raw material use intensity" employed by many researchers (e.g., Gamble 1986; Rolland 1981; Rolland and Dibble 1990; Tavoso 1984). The frequency of retouch alone does not account for pieces used in their natural, unmodified states. As most of the collections employed in this study were not gathered and curated with an eye to protecting the edges, however, neither micro-wear analysis nor informal macroscopic assessments of utilization would be reliable.

A second measure of tool reduction is an index of resharpening developed by the author (Kuhn 1990b). This experimentally tested index estimates the amount of a blank removed by primary modification or resharpening. A typical sidescraper flake blank can be idealized as triangular in cross-section and thickest in the center. With progressive resharpening, the retouched edge(s) move closer to the central ridge of the flake. As this occurs the thickness of the blank at the point where retouch scars terminate (*t* in Figure 5.1) converges on the maximum thickness at the center of the blank (*T* in Fig. 5.1). The index is calculated as the ratio of *t*/*T*. Because the thickness of a blank at the termination of retouch scars may be difficult to measure directly, this value is estimated by multiplying the depth or extension of retouch scars (*D*) by the sine of the angle of retouch (*a*) (Figure 5.2). The index is then calculated using the following formula:

$$I = \frac{\sin a(D)}{T}$$

In principle, the index should increase with each additional resharpening to the point at which the

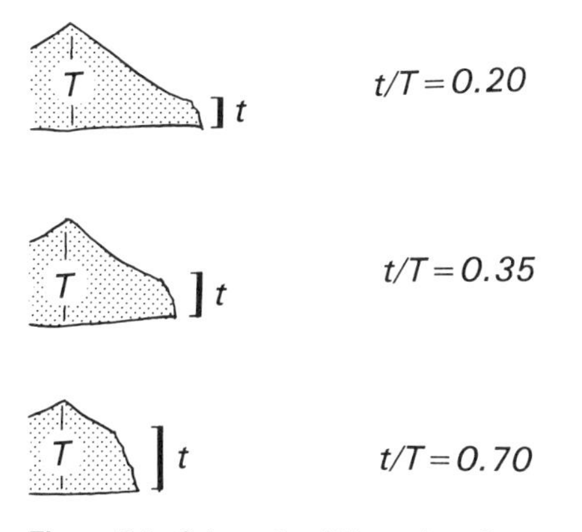

Figure 5.1. Schematized illustration of progressive changes in the index of scraper reduction.

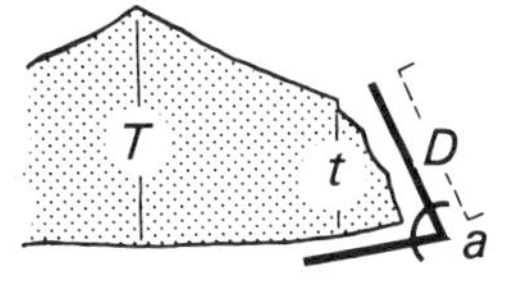

Reduction Index

$$t/T = \frac{\sin a(D)}{T}$$

Figure 5.2. Measurements used to calculate the index of scraper reduction.

retouched edge crosses the central crest of the flake, at which point the index value would be 1.0: it is unlikely that pieces were regularly reworked much past this point. In practice, values of the reduction index as great as 1.25 are obtained, probably as the result of combined error in the three measurements used in its calculation. (Aberrantly high values are converted to 1.0 for statistical analysis.) Experimental findings also show that the index increases in a non-linear fashion with resharpening events, in that the first or second resharpening event produces greater changes than the fifth or sixth (Kuhn 1990b). For these reasons, median-based ordinal statistics are preferred over parametric summary statistics for comparisons.

The third measure of tool consumption used here is a simple count of the number of retouched edges per tool, a criterion employed by other investigators (Barton 1990; Dibble 1988). Two distinct means of evaluating the state of retouched tools are necessary because different tactics for producing new edges were employed according to the shape of the tool blanks. Sharp-edged blanks from centripetal cores often had a second edge added at some point, while with other kinds of flakes it was more common to simply resharpen a single edge (Kuhn 1992a). Consequently, no single variable tells the whole story, and relying on just one could provide a confusing picture of artifact use patterns.

The second dimension of lithic raw material economies is defined by the sources of raw material exploited and patterns of movement of raw materials across the landscape. From one perspective, transport is nothing more than a means for making tools and toolmaking potential available at times and places where raw materials are scarce, nonexistent, or of poor quality. As such, information on the movement of artifactual materials can provide clues about the scheduling of artifact manufacture and maintenance, as well as on the qualities sought by tool makers and users. From another perspective, knowing where people went to obtain flint and how and how far they carried it provides direct indications about patterns of territorial exploitation. Given the limitations of local flint pebbles as raw material for toolmaking, one might expect that transport of raw material from more distant sources would have been an option that prehistoric populations in coastal Latium often chose.

Monitoring the movement of artifactual materials by prehistoric populations requires that one both identify and discriminate between potential sources of lithic raw materials. Coastal Latium presents daunting analytical challenges in this regard. There as in many other parts of the world, raw materials are relatively homogeneous over broad areas. The small flint pebbles constitute the principal source of chippable stone in a territory extending for hundreds of kilometers along the coast: "Pontinian-like" Mousterian industries made on heavily rolled marine pebbles are found as far away as Monte Argentario in Tuscany, over 200 km north of Monte Circeo (Segre 1959). The "pebble zone" is quite a bit narrower than it is long, but (as discussed above) the locations of the primary flint sources nearest the coast are poorly known. In addition, the pebbles in coastal gravel beds are in secondary or even tertiary position, so knowing the geological origin of a particular kind of flint does not tell one whether the stone was transported to the coast by geological forces or by human action.

Because of these problems, artifact transport must be dealt with in a somewhat circumspect fashion for the subject assemblages. In fact, the identification of flint sources and the mapping of primary and secondary deposits available to prehistoric populations are always somewhat probabilistic, although they are not often treated as such. In the case of the Pontinian Mousterian, it is not feasible to identify "exotic" artifacts with much reliability, much less to specify where they came from. However, one can provide a picture of the relative (not absolute) frequencies of non-local materials, and to rank assemblages in terms of the probable abundance of transported artifacts. While such information cannot help to reconstruct prehistoric territories, it provides valuable clues about the general scale of short-term movements by human groups, and about the relative extent of reliance on transported toolkits at different times and places.

The main criteria used to address artifact transport are the type of cortex present and the sizes of artifacts. Flint pebbles found in beach deposits throughout the coastal region bear a distinctive frosted and pitted cortex, quite different from that found on exotic nodular flints originating outside the coastal zone. This guideline is of limited utility, however, as it cannot help identify either noncortical pieces of non-local origin, or tools that might have been made on larger pebbles found to the north of Monte Circeo, around Anzio. In compensation, it is possible to make a probabilistic assessment of the number of non-local pieces in an assemblage on the basis of artifact sizes. As stated previously, pebbles found in deposits nearest to the coastal cave sites seem to be somewhat smaller than pebbles obtainable farther north and east. They would certainly have been much smaller than nodules or tabular pieces collected from primary sources. Assuming that at least a limited proportion of the artifacts present in the Pontinian sites originated at some distance from the places they were found, the most likely candidates are unusually large specimens. The number of specimens lying outside the expected normal size range can thus serve as an estimate of the *relative* frequency of non-local artifacts within a particular assemblage.

There is considerable variation across time and space within the study area in the economics of raw material utilization and the frequency of artifact transport. One component of this variation, the extent to which cores have been consumed, appears directly attributable to the "cost" of obtaining raw materials relative to particular locations (i.e., sites). On the whole, however, the availability of flint pebbles seems to have had relatively little influence over the ways retouched tools were produced, modified, and renewed. Patterns of retouched tool exploitation show much stronger relationships with aspects of the associated faunal assemblages, especially strategies for procuring game and patterns of anatomical-part transport. It would be difficult to sustain that these links were strictly functional ones, reflecting differences in the activities performed in association with the procurement and processing of game. Instead, as is discussed below, the main connection between evidence of game procurement and toolmaking may be found in patterns of foraging and territorial exploitation.

Raw Material Distribution and Quality

One of the principal goals of the analyses that follow is to sort out variation that reflects the local availability of flint pebbles from patterns that may reflect economic responses to the exigencies of land use and foraging organization. As discussed above, it is virtually impossible to identify the specific sources of flint pebbles that would have been closest to the four cave sites when Mousterian hominids occupied them. A general obstacle stems from the fact that "availability" is a relative term: the actual cost of obtaining raw materials depends in large part on where people would otherwise go to harvest other vital resources. Flint located near a commonly exploited source of food, for example, can be obtained with relatively little expense, while raw material that must be collected in a special excursion is inherently more costly, regardless of how far away it is located. An additional problem peculiar to the study area is the ephemerality of the beach deposits that yielded the flint pebbles. Those pebble beaches formed during colder parts of the Pleistocene are now covered by the sea. At the same time, although pebbles can be found today at a number of locations on the Agro Pontino, millennia of aeolian deposition insure that many beach deposits created during periods of high relative sea levels (marine "transgressions") have since been covered.

Despite these problems, it is possible to have some notion of broad *inter-site* differences in access to and quality of raw materials. As explained in Chapter 3, the results of systematic geological and archaeological surveys indicate that pebble flint has probably always been more abundant and ubiquitous in the northern part of the study area, on the Pontine Plain around Monte Circeo, than in the Fondi Basin to the south. Comparisons between the Fondi Basin sites (Grotta di Sant'Agostino and Grotta dei Moscerini) and the caves of Monte Circeo (Grotta Guattari and Grotta Breuil) can thus serve as one means for investigating general technological consequences of variation in raw material availability within the study sample. Smaller-scale variation in ease of access to flint can also be identified within the northern part of the study area. An outcropping of beach deposits, probably representing the oxygen-isotope stage 5e high sea stand, is situated directly adjacent to Grotta Guattari. None of the other cave sites is known to have a source of raw material adjacent to it, so Guattari would consequently have afforded more immediate access to flint than any of the other shelter sites, including Grotta Breuil, the other Monte Circeo cave.

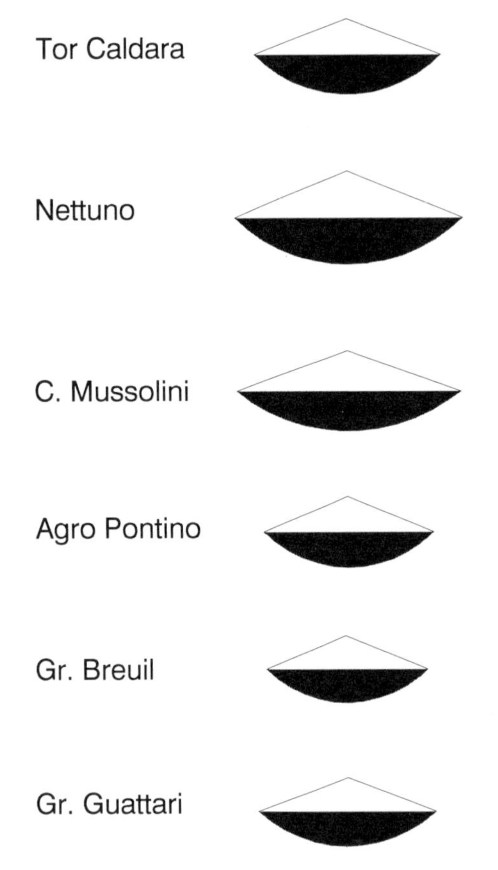

Figure 5.3. Average dimensions of cores from sites arranged top to bottom along a north–south axis.

Table 5.1. Dimensions of unbroken centripetal and prepared platform cores (mm)[a]

Locality		*Maximum dimension*	*Thickness*
Tor Caldara	mean	38.3	14.7
	sd	6.6	4.3
	n	7	7
Nettuno	mean	47.0	18.3
	sd	10.7	6.2
	n	104	104
Canale Mussolini	mean	46.2	15.8
	sd	10.1	9.1
	n	8	8
Agro Pontino (survey)	mean	34.6	13.1
	sd	7.6	4.7
	n	118	118
Grotta Breuil	mean	33.0	13.0
	sd	6.3	4.4
	n	111	111
Grotta Guattari	mean	36.6	13.8
	sd	7.7	5.2
	n	86	86
Grotta dei Moscerini	mean	29.3	10.0
	sd	6.8	2.4
	n	25	25
Grotta di Sant'Agostino	mean	28.1	10.5
	sd	4.2	3.2
	n	141	141

[a] Two categories combined.

Although no systematic assays of pebble deposits have been published to date, archaeologists working in the coastal region do share the impression that the pebbles found in the older beach ridges situated farther from the coast are both larger and of higher-quality raw materials than those found closest to the coast. It is also readily apparent that the flint pebbles found in the extreme northern part of coastal Latium, around Anzio and Nettuno, are considerably larger than those found in the vicinity of Monte Circeo and farther south. Figure 5.3 is a schematic illustration of the mean length and thickness of "flat" core forms (the radial and prepared platform types) found at various localities in the littoral zone, the Agro Pontino and Monte Circeo: summary data are presented in Table 5.1. The most northerly localities are at the top of the figure, the most southerly at the bottom. Cores from sites in the Anzio/Nettuno area, as well as artifacts collected from the deposits exposed in drainage canals in the Agro Pontino (the Canale Mussolini sample), are strikingly larger than those obtained from the two Circeo cave sites. I do not believe that the north–south gradient in core size can be entirely attributed to either collection strategies or differences between open-air and shelter sites. Instead, it probably reflects the reworking of flint pebbles by marine action as they were moved by littoral drift down the coast from their origins in the Tiber and rivers further north. Because all of the cave sites discussed here are located well to the south and west of the areas where the larger pebbles would have been found, these patterns illustrate the distances Paleolithic tool makers might have had to travel to obtain larger pebble flints, and not differences among the raw material sources located closest to the shelter sites.

The primary sources of *in situ* nodular or tabular

flints in Latium and the surrounding regions are poorly documented. Upper Paleolithic peoples in the study area, and quite probably Mousterian populations as well, did sometimes utilize non-pebble flints. Some of the principal sources, such as Monte Genzana in Abruzzo more than 100 km east of the study area, are well known. However, there is little systematic information about the possible existence of primary flint deposits or secondary alluvial sources in the mountains and valleys closer to the coast. As a consequence we cannot draw any reliable inferences about differential access to sources of non-pebble flints from different sites within the study area. We know only that, wherever they were found, raw materials other than small marine pebbles must have come from a considerable distance.

The Consequences of Raw Material Scarcity

One way to investigate the influence of access to flint on the economic behavior of Pontinian Mousterian tool makers is through a simple set of comparisons. If the distance between a site and a source of pebbles were the only determinant of the "cost" of procuring raw materials, then there should be greater evidence for "economizing" of raw material in those areas where flint pebbles were comparatively hard to come by. Where flint was more difficult to obtain, both cores and tools/blanks should be more extensively worked, more completely used up. Consequently, artifacts from the sites located around the Fondi Basin (Grotta dei Moscerini and Grotta di Sant'Agostino) should be more heavily worked than artifacts from the caves on Monte Circeo (Grotta Breuil, Grotta Guattari), where pebble beaches are more common. *Within* the Monte Circeo area, the intensity of core and tool exploitation should be greater at Grotta Breuil than at Grotta Guattari, since the latter site has a fossil pebble beach immediately adjacent to it.

The analyses below begin with cores and move on to address the differential exploitation of flakes and retouched tools. In conducting statistical tests, pairwise comparisons are made between sites, rather than between assemblages from individual strata. There are two reasons for this. First, raw material availability varies across subregions of the study area as the result of geomorphological factors, broad differences in topography, and sedimentary environments. In comparison with the tempos of change of either human activity or the depositional environments within caves, these geomorphological forces are relatively stable. If the overall accessibility of flint at a regional scale played an important role in structuring Mousterian lithic raw material economies, its consequences should be apparent throughout the histories of occupations of sites in different parts of the study area. The second reason for comparing sites rather than individual assemblages is simple economy of presentation. With 14 to 16 assemblages of sufficient size and completeness for statistical comparison, the number of possible pairs would be prohibitively large.

In each of the following tests, there are five possible site pairings of interest: four between the Circeo and Fondi Basin sites, and one between Grotta Guattari and Grotta Breuil on Monte Circeo. Contrasts between Grotta di Sant'Agostino and Grotta dei Moscerini are also presented, but they are not immediately pertinent to the argument at hand because there is no reason to believe that accessibility of flint varied between these two caves. Differences in the treatment of artifacts at Moscerini and Sant'Agostino are relevant to other issues, however, and will be addressed in ensuing chapters. The artifact sample from the Agro Pontino survey is also excluded from this analysis because it was selected using very different criteria from the excavated samples, and is thus subject to a different array of biases at the assemblage level.

Consumption of cores

If the high cost of flint induced tool makers to try and get as much as possible out of their supply of pebbles, we would expect cores to be more extensively and more completely modified. One general expectation would be that larger numbers of fully realized "formal" cores, and relatively fewer tested and partially worked pieces, should be found at sites where raw material was difficult to obtain. Table 5.2 contains appropriate data for the subject assemblages. In the first part of the table the combined frequencies of tested pebbles and chopper/chopping tools are presented as a measure of the number of incompletely worked cores, while the number of other "formal" core forms is used as an

Table 5.2. Percentages of informal and formal cores

	Tested pebbles + chopper/chopping tools	*Formal cores*
Grotta Guattari	45.2 (n = 144)	54.8 (n = 175)
Grotta Breuil	14.0 (n = 39)	86.0 (n = 240)
Grotta dei Moscerini	24.5 (n = 23)	75.5 (n = 71)
Grotta di Sant'Agostino	7.0 (n = 18)	93.0 (n = 238)

Chi-square results (df=1)

	Breuil	Moscerini	Sant'Agostino
Guattari	66.60 *p<<.001*	12.04 *p<.001*	100.06 *p<<.001*
Breuil	—	4.85 *.01<p<.05*	6.06 *.01<p<.05*
Moscerini	—	—	23.40 *p<.001*

estimate of more "completely worked" pieces. The second part of the table shows statistical (chi-square) comparisons of the core assemblages from various sites. The results of the tests are largely consistent with the expectations of a model in which access to raw material determines the treatment of cores. Grotta Guattari (located on a source of pebbles and in the more pebble-rich zone) has more "casually worked" cores than any other site. Grotta Breuil also yields more informal core types than Grotta di Sant'Agostino.

The only major inconsistency is that Grotta dei Moscerini contains fewer extensively modified formal cores than does Grotta Breuil, in spite of being located in the pebble-poor Fondi Basin. One explanation for the unexpectedly high frequency of choppers at Grotta dei Moscerini is that some methods of core reduction result in the production of greater numbers of choppers and chopping tools as intermediate stages. The core assemblage from Grotta dei Moscerini is dominated by centripetal forms (Table 4.1), which are likely to have passed through a chopping-tool stage (Fig. 4.11). Prepared platform cores, the only major type that probably never resembled choppers or chopping tools, are very rare at Moscerini but more common at Grotta Breuil. Thus, because of the general approaches to preparation and flake detachment used, fewer of the cores from Grotta Breuil would be likely to resemble choppers or chopping tools, even if they were reduced to a less advanced stage.

Extensively consumed cores should also be smaller than less reduced pieces. If access to raw material is a limiting factor in Pontinian Mousterian technology, cores from sites with poor access to raw materials should be smaller than those from sites where pebbles were more abundant. This possibility is addressed in Tables 5.3 and 5.4, which present summary statistics and *t*-tests comparing the maximum lengths and thicknesses of cores from the various sites. Data are presented only for "flat" cores—the radial and prepared platform groups—because they were made on similar ranges of pebble shapes and have comparable geometries. As above, the size differences between sites fit very well with a model of technological behavior constrained by raw material. Cores from the two Circeo caves are larger in both dimensions than cores from both Fondi Basin sites. The flat nuclei from Guattari are also longer, though not significantly thicker, than comparable specimens from Grotta Breuil.

Evidence introduced in the previous chapter indicated that methods of blank production involving parallel preparation and detachment yielded somewhat larger numbers of flakes per core than did methods employing centripetal or radial preparation. If the main impetus for employing alternative core technologies were the number of flakes that could be made from a single pebble, prepared platform and pseudo-prismatic cores, which employ parallel schemes of flake detachment, should be more abundant at localities where raw material was most scarce. Table 5.5 demonstrates that this prediction is not supported. Only two pairwise comparisons (Guattari–Breuil and Guattari–Sant'Agostino) are consistent with the model. In two cases (Guattari–Moscerini, Breuil–Sant'Agostino) there is no significant difference between the core assemblages, even though one would be expected based on the distribution of flint pebbles. The differences between core assemblages from the remaining site pair (Breuil–Moscerini) are contrary to the predictions of the model, in that the less "economical"

Table 5.3. Maximum lengths of unbroken centripetal and prepared platform cores

	Mean (mm)	*sd*	*n*	*t-value*	*p*
G. Guattari	36.6	7.7	86	−6.516	*<.001*
G. Breuil	33.6	6.3	111		
G. Guattari	36.6	7.7	86	4.309	*<.001*
G. dei Moscerini	29.3	6.8	25		
G. Guattari	36.6	7.7	86	10.747	*<.001*
G. di Sant'Agostino	28.1	4.2	141		
G. Breuil				2.653	*.009*
G. dei Moscerini					
G. Breuil				7.381	*<.001*
G. di Sant'Agostino					

Table 5.4. Thicknesses of unbroken centripetal and prepared platform cores

	Mean (mm)	*sd*	*n*	*t-value*	*p*
G. Guattari	13.8	5.2	86	−1.158	.248
G. Breuil	13.0	4.4	111		
G. Guattari	13.8	5.2	86	3.485	*.001*
G. dei Moscerini	10.4	2.4	25		
G. Guattari	13.8	7.7	86	5.851	*<.001*
G. di Sant'Agostino	10.5	3.2	141		
G. Breuil				3.201	*.002*
G. dei Moscerini					
G. Breuil				5.100	*<.001*
G. di Sant'Agostino					

Table 5.5. Frequencies of centripetal and parallel cores

	Centripetal cores	*Pseudo-prismatic + prepared platform cores*
	FREQUENCIES	
Grotta Guattari	114	21
Grotta Breuil	77	66
Grotta dei Moscerini	53	9
Grotta di Sant'Agostino	138	86

CHI-SQUARE RESULTS (df=1)

	Breuil	Moscerini	Sant'Agostino
Guattari	28.83	0.001	19.92
	p<.001	p>>.05	*p<.001*
Breuil	—	17.30	1.86
		p<.001	p>.05
Moscerini	—	—	11.43
			p<.001

centripetal core technology is more abundant in the site (Moscerini) from which raw materials would have been less accessible. These results clearly imply that simple numbers of flakes produced cannot alone explain variation in the use of alternative core technologies, and that other factors, such as functional attributes of different flake forms, must be investigated.

Consumption of flakes and tools

The abundance of, or proximity to, sources of flint pebbles clearly had some influence over the extent to which cores were modified and reduced in the Pontinian Mousterian assemblages. If the same factors regulated the treatment of retouched tools, we would expect to see greater intensity of tool modification and more extensive consumption of artifacts from the two Fondi Basin sites, compared with materials from the Monte Circeo sites; tools and blanks from Grotta Breuil should also be more extensively worked than artifacts from Grotta Guattari. These expectations are evaluated using data in Tables 5.6–5.9 below.

Table 5.6 presents frequencies of unmodified flakes and retouched tools for the various sites and assemblages. Although archaeologists did sieve sediments in excavations at all of the sites discussed here, there is some question about the sizes of screens employed in the earlier projects. Retouched tools are generally larger than unretouched flakes and are more likely to be picked out during initial sieving, so that use of coarse screens could result in a bias towards retouched tools. In order to counteract potential biases introduced by collection procedures, only specimens greater than 2.0 cm in length are reported in the table. The two incompletely sampled assemblages from Grotta di Sant'Agostino (from level 1 and the disturbed level "0") have been excluded from the analysis because the study sample is biased towards retouched tools.

Frequencies of retouched pieces are comparatively high throughout the sample, undoubtedly a function of the overall scarcity of flint in coastal Latium, as well as of the difficulty of obtaining flakes of retouchable size. Nonetheless, constraints imposed by the local pebble raw materials have not totally depressed variation among sites. The results of chi-square comparisons between tool and flake assemblages are strikingly inconsistent with the notion that proximity to raw materials was the primary influence on the frequency of retouch. Grotta Guattari, the site closest to a pebble beach, has a frequency of retouch as high as or higher than at each of the three other sites. Grotta di Sant'Agostino, in the raw-material-poor Fondi Basin, actually has the lowest frequency of retouched pieces of any site taken as a whole (a fact that also argues against the supposition that recovery in the early excavations was biased towards retouched pieces). Only the contrast between Breuil and Moscerini fits the simple economic model. The implication is clear: the frequency with which flakes were converted into retouched tools appears to have been largely unaffected by ease of access to raw material, at least at the scale monitored here.

Table 5.7 presents an inter-site comparison of statistics for the index of scraper reduction described in Figs. 5.1 and 5.2 (indices for individual assemblages are found in Table 5.8). The values shown represent group medians, combining values for single-edged pieces with the indices for the "principal" (most completely modified) margin of pieces with multiple edges. The small size and the scarcity of raw materials in the study area certainly attenuates the range of variation in the scraper reduction index. Nonetheless, there are significant differences between sites and assemblages. As in the case of retouch frequencies, however, there appears to be no consistent relationship between ease of access to raw material and the extent to which scrapers were resharpened. Most notably, the index of scraper reduction for Grotta Guattari is as high as or higher than for any of the other three sites. As might be expected, Grotta Breuil does show a scraper reduction index that is significantly lower than Grotta dei Moscerini, but contrary to expectations, its scrapers are no less heavily reduced than the implements from Grotta di Sant'Agostino.

Frequencies of tools with multiple retouched edges are also at odds with the idea that the availability of stone was the primary driving force behind Pontinian Mousterian raw material economies (Table 5.9). The numbers of tools with two or more retouched edges differ significantly between only one pair of sites, Grotta Guattari and Grotta di

Table 5.6. Percentages of retouched tools and unretouched flakes (—2.0 cm)

	Unretouched	*Retouched*
Grotta Guattari	32.2	67.8
	(n = 157)	(n = 330)
Grotta Breuil	54.6	45.4
	(n = 328)	(n = 273)
Grotta dei Moscerini	28.1	71.9
	(n = 107)	(n = 275)
Grotta di Sant'Agostino[a]	65.9	34.1
	(n = 323)	(n = 167)

CHI-SQUARE RESULTS (df=1)

	Breuil	Moscerini	Sant'Agostino
Guattari	53.43	1.63	111.05
	$p<<.001$	$p>.05$	$p<<.001$
Breuil	—	65.74	14.63
		$p<<.001$	$.001<p<.01$
Moscerini	—	—	123.42
			$p<<.001$

[a] Assemblages from levels S2, S3, and S4 only.

Table 5.7. Comparison of indices of scraper reduction

	Median	*n*	*Kruskal-Wallis chi-square*	*p*
G. Guattari	0.64	436	18.13	*<.001*
G. Breuil	0.55	422		
G. Guattari	0.64	436	0.98	.989
G. dei Moscerini	0.62	390		
G. Guattari	0.64	436	14.38	*<.001*
G. di Sant'Agostino	0.57	838		
G. Breuil			10.11	*.001*
G. dei Moscerini				
G. Breuil			1.34	*.215*
G. di Sant'Agostino				

Table 5.8. Summary statistics, index of scraper reduction

	Mean	*sd*	*n*
Agro Pontino	0.65	0.20	148
Grotta Breuil			
1986–87 sample	0.56	0.22	193
old *sondage*	0.57	0.27	33
strata 3/4	0.45	0.22	57
Grotta Guattari			
stratum 1	0.62	0.22	52
stratum 2	0.59	0.23	105
stratum 4	0.66	0.22	189
stratum 5	0.65	0.23	92
Grotta dei Moscerini			
strata group 2	0.65	0.24	115
strata group 3	0.60	0.23	123
strata group 4	0.61	0.25	81
strata group 5	0.62	0.21	39
strata group 6	0.61	0.22	74
Grotta di Sant'Agostino			
level 1	0.58	0.22	448
level 2	0.58	0.24	122
level 3	0.56	0.24	69
level 4	0.57	0.19	36

Table 5.9. Percentages of artifacts with single and multiple retouched edges

	One retouched edge	*More than one retouched edge*
Grotta Guattari	77.3	22.7
	(n = 357)	(n = 105)
Grotta Breuil	78.1	21.9
	(n = 282)	(n = 79)
Grotta dei Moscerini	76.9	23.1
	(n = 316)	(n = 95)
Grotta di Sant'Agostino	84.2	15.8
	(n = 208)	(n = 39)

CHI-SQUARE RESULTS (df=1)

	Breuil	Moscerini	Sant'Agostino
Guattari	0.04	0.01	4.37
	$p>>.05$	$p>>.05$	$.01<p<.05$
Breuil	—	0.10	3.10
		$p>>.05$	$.05<p<.10$
Moscerini	—	—	4.66
			$.01<p<.05$

Sant'Agostino. Moreover, in this case the site with better access to flint (Guattari) contained a higher frequency of pieces with multiple retouched edges. Once again, there appears to be no direct relationship between the accessibility of raw materials and the extent to which retouched tools were modified or consumed.

Summary: Effects of proximity to raw material

The sheer number of comparisons presented above makes it difficult to grasp the total picture of local "raw material effects" on Pontinian Mousterian technology. Table 5.10 summarizes the outcomes of all pairwise comparisons between sites with differential access to lithic raw material. For each test or criterion, there are five significant pairings: the pairing of Grotta dei Moscerini and Grotta di Sant'Agostino is irrelevant here, as there is no known difference in their proximities to pebble beds. The results of each pairing fall into one of three categories: a statistically significant difference, *consistent with* the expectation that artifacts from the site with easier access to raw materials will be less completely consumed or reduced; no statistical difference for the measure in question; or a statistically significant difference contrary to expectations. Of these, only results in the first category support the idea that raw material availability had a dominant influence on the behavior of Mousterian tool makers in the cases at hand. Pairwise comparisons falling into either of the last two categories fail to confirm the model.

When viewed in this distilled form, the results presented thus far are more easily interpreted. Findings on the intensity of core exploitation are quite clearly consistent with a model positing access to raw material as the primary force in the economics of lithic raw material use. Out of 15 possible tests comparing the extent to which cores were used up (5 pairwise comparisons on three criteria), 13 are consistent with the expectations of the model. In other words, in 87% of the comparisons, cores from sites with less ready access to pebbles were more extensively worked or consumed than were cores from sites with a source of pebbles closer at hand. However, blank production technology does not fit expectations. The frequency of more productive methods—involving parallel preparation and detachment of flakes—seems largely independent of variation in the availability of flint pebbles.

Analyses of tool and tool blank use intensity furnish quite different results. The immediate availability of flint pebbles appears to have very little influence on the frequency or intensity of retouch. Examining retouch frequency and the extent of scraper modification/reduction, only 2 of 15 cases (13%) are consistent with the expectations of the model. In 7 cases there is no significant difference

Table 5.10. Summary of tests relating artifact reduction to raw material access

	Model confirmed	*Model disconfirmed*	
	Significant diff. consistent	*No difference*	*Significant diff. contrary*
CORES			
Formal/informal cores	++++		+
Core maximum length	+++++		
Core thickness	++++	+	
CORE TECHNOLOGY			
Centripetal/parallel cores	++	+	++
TOOLS			
Tool/flake ratio	+	+	+++
Reduction index	+	++	++
Number of retouched edges		++++	+

between sites with differential access to raw material, while in the remaining 6 cases inter-site contrasts are contrary to expectations, in that sites which presumably afforded better access to raw material yield artifacts that are actually more extensively retouched or reworked.

We can conclude from these findings that the economy of raw material use varied among different classes of artifacts and different stages in the tool "life histories." Raw material availability strongly influenced the measures that Mousterian tool makers took to extend the lives of cores: where pebbles were more difficult to come by, cores were exploited more completely. In contrast, the proximity of pebble beds had few if any consequences for "choices" about either flake production technology or the treatment of the resulting blanks and retouched tools. In the Mousterian cases discussed here, the intensity of core exploitation appears to have been closely linked to the characteristics of particular locations. More changeable variables, such as the nature of occupations and patterns of land use, may also affect the relative cost of raw materials, but they apparently did not have sufficient weight to counteract the influence of the short-term costs of procuring flint.

That the reduction of cores, but not of tools, is so closely tied to the intrinsic geological or geographical characteristics of places (i.e., sites) probably reflects the energy costs and benefits of transporting different artifact forms. Every core contains a relatively large amount of potentially wasted material, stone that cannot be transformed into usable implements. Wastage includes material removed in reshaping the striking platform and face of detachment, mis-strikes, and flakes that break during manufacture, as well as the residual "slug" from which it is impossible to manufacture additional usable flakes. Cores made on small pebbles are especially inefficient in this regard because only a few flakes can be removed before the core is exhausted, and the unusable part is relatively large compared with the original size of the piece. The inevitable wastage involved in transforming raw material into working edges means that a substantial energy price may be exacted for carrying lots of unworked chunks of raw material around the landscape. Consequently, decisions about how and when to remove material from cores and how extensively to use them should be highly contingent on transport costs (e.g., Metcalfe and Barlow 1992), and we would expect the distance between a site and the raw material source to have comparatively strong effects on the measures people took to extend the lives of cores.

Core reduction technology is a different matter. The *économie du débitage*, the method of flake production, can be quite independent of the management of raw materials (Perlès 1991), and the Pontinian Mousterian is a case in point. While proximity to the source of raw materials strongly impacted *how extensively* Mousterian cores were worked, it apparently had little influence on the methods for producing flakes and blanks. Every mode of core reduction involves a trade-off between maximizing the number of flakes and controlling their functional properties. "Choices" of which technique to use may be governed by either the economy of production or functional properties of the flakes produced. In the subject assemblages, it would be difficult to argue that a need to maximize the number of flakes produced when pebbles are scarce explains inter-assemblage variation in the frequency of different core technologies. We can only conclude that some other properties of the alternative methods or their products stand behind variability in the ways Mousterian tool makers chose to work the pebbles available to them.

The observation that the treatment of tools and flake blanks in the subject assemblage varies independently of a site's proximity to raw material is also likely to reflect the efficiency of transporting different kinds of artifacts. Finished tools and tool blanks afford a relatively efficient means of delivering artifacts to different points on the landscape, providing the maximum number of usable edges per unit mass (Kuhn 1994). Because tools and flakes are less costly to move over either short or long distances, the effort needed to procure workable stone from a specific place should have fewer direct consequences for their treatment. Certainly, raw material availability can, and has often been shown to, influence variation in the intensity of tool exploitation (e.g., Bamforth 1986, 1991; Marks and Friedel 1977), especially at the regional level. The point is that, in the case of the Pontinian

Mousterian, known differences in access to raw material do *not* seem to have significant consequences for the treatment of tools and tool blanks. Apparently, intra-regional variation in the cost of procuring flint was simply not sufficiently pronounced to override other influences.

Evidence for Artifact Transport

Evidence for the movement of artifacts and raw materials can help to further define the variation in hominid land use patterns that may stand behind the "unexplained" aspects of stone tool manufacture and use in the subject assemblages. The presence of lithic artifacts in sites like Grotta di Sant'Agostino and Grotta Breuil, where flint is not found on site, attests to the transport of artifacts over at least a few kilometers. Of greater interest is the movement of artifacts over longer distances, transport in the context of provisioning sites or individuals at a scale that might explain the differential reduction or modification of artifacts in the various assemblages. Although the data must be approached from a critical stance, there is reason to believe that the frequency of "exotic" artifacts varied across time and space within the study sample.

The frequencies of different types of cortex provide the most direct evidence for long-distance transport and the exploitation of raw material sources lying outside the coastal raw material zone. The heavily rolled pebbles found throughout the "Pontinian heartland" of coastal Latium bear a characteristic frosted cortex with many tiny pits marking impact with other sedimentary particles. A very different kind of chalky white cortex is characteristic of at least some sources of nodular flint located farther inland. In some cases, remnant nodular cortex found on artifacts appears somewhat eroded and smoothed, suggesting that sources of derived materials (e.g., alluvial or colluvial deposits) near the primary sources were exploited. This "rolled nodule" cortex is quite easy to distinguish from the more extensively altered surfaces of the pebbles from marine beaches and gravel layers.

Table 5.11 shows frequencies of artifacts with pebble cortex, "non-pebble" (fresh or rolled nodular) cortex, and no cortex at all, in Pontinian Mousterian assemblages from the four coastal cave sites. Data are also presented for several Upper Paleolithic sites in this region. In order to insure comparability between samples, only retouched artifacts are considered in the tabulations. Grotta Jolanda and Riparo Salvini yielded late Epigravettian assemblages. The materials from Riparo del Sambucco, one of the Cavernette Falische (Fig. 3.1, no. 1), are thought to date from the Gravettian (Mussi and Zampetti 1985). The assemblage from stratum 21 at Grotta del Fossellone is "typical" Aurignacian. Le Grottacce, an open-air locality, has yielded exclusively Upper Paleolithic materials, but the assemblage is not assignable to a specific period or "culture."

Several things are immediately apparent from an inspection of the data in Table 5.11. First, the frequency of non-pebble cortex in the Upper Paleolithic samples confirms the supposition that sources of nodular raw materials are in fact located within reach of the coastal zone. It is interesting that Grotta Jolanda, situated about 35 km north of Monte Circeo, contains only a slightly larger proportion of artifacts with non-pebble cortex than the Fossellone Aurignacian assemblage, and a much lower proportion than Riparo Salvini. This suggests that wherever the primary sources are situated, their distances from the coast alone cannot explain variation in their exploitation.

Perhaps the most striking feature of Table 5.11 is the negligible frequency of non-pebble cortex among the Mousterian assemblages. Not apparent from the table is the fact that the few Mousterian specimens listed as having "non-pebble" cortex in fact bear *probable* "rolled nodule" cortex. No specimens with fresh, unrolled nodular cortex are known from Middle Paleolithic layers excavated to date. Both the Mousterian and Upper Paleolithic samples discussed here represent long spans of time, and patterns of raw material exploitation undoubtedly varied considerably over those intervals. Nonetheless, one can only conclude that, as a group, Mousterian populations were less likely to utilize and transport flints from primary inland sources than later Gravettian and Epigravettian groups. These data do not imply that Mousterian populations never left the coastal zone, or that they never moved artifacts from place to place. Instead,

Table 5.11. Frequencies of different types of cortex, Mousterian and Upper Paleolithic assemblages (unbroken retouched tools only) (percent)

	No cortex	*Pebble cortex*	*Non-pebble cortex*	*Percentage of all cortex non-pebble*	*n*
Gr. del Fossellone 21	32.4	55.8	11.8	17.4	713
Gr. Iolanda	46.4	40.6	13.0	24.3	69
Le Grottacce	46.7	40.0	13.3	25.0	30
R. Salvini	66.5	22.9	10.6	31.6	624
R. del Sambucco	68.0	18.7	13.3	41.2	75
Gr. Breuil					
1986–87 sample	16.5	82.5	1.0?	1.2?	206
strata 3/4	20.0	76.7	3.3	4.1	60
Gr. Guattari					
stratum 1	24.6	75.4	0.0	0.0	57
stratum 2	31.8	67.3	0.9?	1.3?	110
stratum 4	14.4	85.1	0.5	0.6	195
stratum 5	19.0	81.0	0.0	0.0	100
Gr. dei Moscerini					
str. group 2	33.0	65.0	1.6?	2.4?	123
str. group 3	42.8	55.0	2.2?	4.0?	131
str. group 4	35.5	64.5	0.0	0.0	93
str. group 5	33.3	64.1	2.5?	3.9?	39
str. group 6	17.1	82.9	0.0	0.0	75
Gr. di Sant'Agostino					
level 1	10.4	89.0	0.6?	0.7?	492
level 2	18.1	81.9	0.0	0.0	133
level 3	12.3	87.7	0.0	0.0	73
level 4	4.9	92.7	2.4?	2.6?	41

as we shall see, they are more likely to attest to fundamental changes over time in patterns of movement and place use (Kuhn 1992b).

Evidence from regions where raw material sources are more easily distinguished indicates that Mousterian groups regularly transported artifacts, especially retouched tools and Levallois flakes, over distances of 50 km or more (e.g., Geneste 1989; Rensink et al. 1991; Roebroeks et al. 1988). It seems reasonable to assume that artifacts were moved around by the makers of the Pontinian Mousterian as well, particularly in light of the functional advantages afforded by sources of larger raw materials (pebbles or nodules) located some distance from the coastal caves. In the absence of much direct (cortical) evidence for the use of nodular or tabular flints from outside the pebble raw material zone, however, it is necessary to turn to more indirect sources of evidence in order to get some assessment of the variation among assemblages.

It is readily apparent from Table 5.11 that the frequencies of *noncortical* artifacts vary considerably among the various assemblages. Some of the strata groups at Grotta dei Moscerini yielded numbers of noncortical artifacts that approach the figures for Upper Paleolithic assemblages. Noncortical specimens are highly ambiguous, since they could be from either local pebbles or more distant primary sources. As the coastal gravels are derived geologically from *in situ* flint deposits, it is virtually impossible to distinguish between local and exotic artifacts on the basis of the character of the stone itself. In any event, the frequency of artifacts that are *not necessarily* local does not give a very satisfying estimate of the number of transported pieces.

More interesting and compelling results are obtained when artifact size is considered. All assemblages contain at least a few artifacts that simply appear too large to have been manufactured from the normal range of pebbles known to be available near the coastal sites. Such "oversized" pieces might have come either from gravel deposits in the area of Anzio, north of Monte Circeo, or from primary sources even farther afield. However, the appearance of being too large is a difficult measure to quantify or replicate, and a better estimate is needed. The mean sizes of unbroken flakes and blanks can provide a baseline for identifying pieces likely to have been made on exotic raw materials. For the purpose of this study, a specimen is defined as "oversized," and is very likely to have been made on something other than the typical pebble raw materials, if its length or width exceeds the global average for whole flakes and blanks by two standard deviations or more. Appropriate threshold values appear in Table 5.12. Assuming that the sizes of pebbles ultimately limit the sizes of artifacts, a length or width that exceeds the mean by more than two standard deviations implies there is a very good chance that the artifact in question was made on a piece of raw material falling outside the normal range, either a very large marine pebble or possibly (if no cortex is present) some type of nonpebble flint.

It is difficult to calculate the exact probability associated with the criteria used to define "oversized" artifacts. For a large sample, only 2.5% of items within a normal distribution will fall two standard deviations above the mean (Snedecor and Cochran 1980:55). In the present case, however, the sample distribution has been artificially truncated by selecting only artifacts larger than 1.5 cm in maximum dimension, thereby raising the mean but lowering the standard deviation. Even so, it is likely that 95% or more of the sample of artifacts should have dimensions *less than* the sample average plus two standard deviations.

Of course, one still expects a small proportion of artifacts to be more than two standard deviations beyond the mean, even if drawn from a uniform population. In the case of the subject assemblages, slightly fewer than 8% of all whole artifacts are classed as "unusually large" (Table 5.13). More interesting is variation in the frequencies of the largest pieces. First, specimens in the largest size class are almost twice as likely to be free of dorsal cortex as are "normal-sized" artifacts (Table 5.13). Since flakes without cortex are produced later in the reduction chain, we might normally expect them to be slightly smaller than cortical flakes on average, but here the noncortical specimens tend more often to be oversized. As the sample of specimens illustrated in Figure 5.4 shows, many of these very large artifacts are classical "typological" Levallois or Levallois-like pieces, produced after extensive core preparation. This strongly implies that at least some artifacts without any cortex were made on unusual raw materials, either very large pebbles or nodular flints.

The frequencies of oversized artifacts also vary strikingly among assemblages (Table 5.14; Figures 5.5a and b). The two bar charts show the proportions of oversized artifacts among all whole flakes and blanks (Fig. 5.5a) and among noncortical pieces only (Fig. 5.5b). Dotted lines mark sample-wide proportions. Interestingly, like core technology and artifact reduction intensity, the frequencies of oversized artifacts separate by site (and by chronology). All assemblages from Grotta di Sant'Agostino, along with the Grotta Breuil 1986–87 sample (B86–87), contain fewer uncommonly large pieces than the population average. In contrast, assemblages from Grotta Guattari and Grotta dei Moscerini exceed the population average markedly in one or both charts. The collection from Breuil strata 3/4 also contains unusually high numbers of oversized artifacts. The abundance of oversized specimens and, by inference, the frequencies of transported artifacts do not seem to be strictly tied to the local abundance of flint: assemblages from the two Fondi Basin sites (Sant'Agostino and Moscerini) cover both the high and low extremes of the spectrum.

I emphasize that neither unusual size nor the presence or absence of pebble cortex unequivocally proves that an artifact was carried a great distance by Mousterian hominids: by the same token, we cannot exclude that smaller artifacts with pebble cortex were transported from place to place within

Table 5.12. Population summary statistics, lengths and widths of unbroken flakes and retouched tools (mm)

	Mean	*sd*	*Mean + (sd × 2)*
Length	26.3	8.0	42.3
Width	21.2	6.3	33.8

Table 5.13. Percentages of "oversized" artifacts among whole flakes and tool blanks (all assemblages)

	"Normal-sized"[a]	*"Oversized"*[b]
Cortex absent	83.3 (n = 415)	16.7 (n = 83)
Cortex present	91.9 (n = 2326)	8.1 (n = 205)

Chi-square = 34.51, $p << .001$ (df=1)

[a] "Normal-sized" includes all pieces less than two standard deviations larger than population mean length or width.

[b] "Oversized" includes all pieces more than two standard deviations larger than population mean length or width.

Table 5.14. Frequencies of "oversized" artifacts (by assemblage)

	Oversized noncortical (n)	*Oversized cortical (n)*	*Oversized as percentage of unbroken*	*Oversized as percentage of unbroken noncortical*
Grotta Breuil				
1986–87 sample	2	14	5.1	5.9
strata 3/4	5	21	14.4	19.2
Grotta Guattari				
stratum 1	2	6	11.6	10.5
stratum 2	13	18	20.4	26.0
stratum 4	6	23	14.7	25.0
stratum 5	6	10	14.7	37.5
Grotta dei Moscerini				
strata group 2	6	1	5.0	14.6
strata group 3	10	8	14.9	25.0
strata group 4	5	14	23.8	20.8
strata group 5	4	3	15.2	36.4
strata group 6	5	4	10.2	23.8
Grotta di Sant'Agostino				
level 1	7	22	5.4	12.7
level 2	3	5	2.5	7.5
level 3	0	3	2.1	0.0
level 4	0	5	5.6	0.0

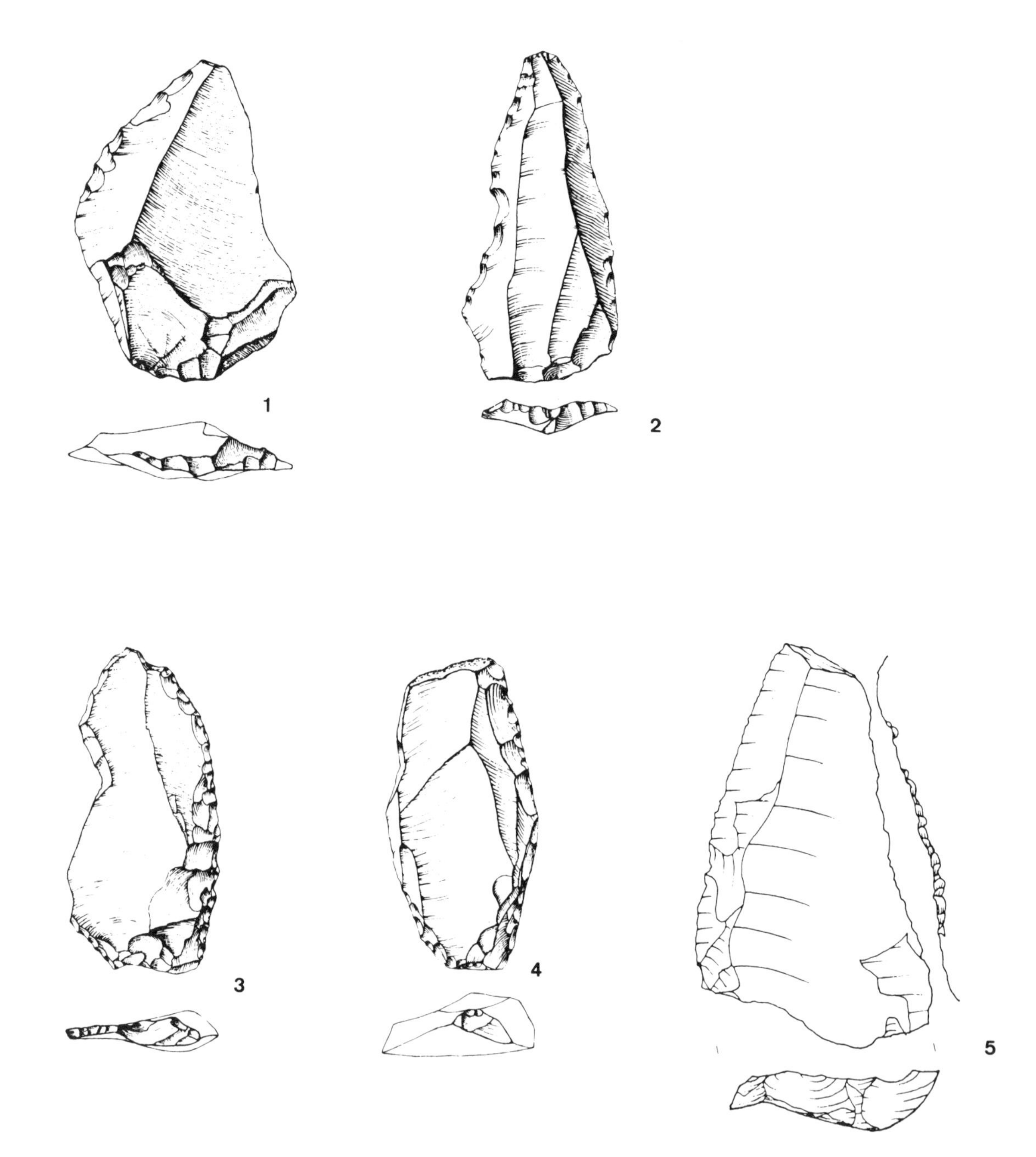

Figure 5.4. Examples of "oversized" artifacts from Pontinian Mousterian sites. (1–4=Gr. Guattari; 5–6=Gr. Breuil; 7, 8=Gr. di Sant'Agostino.)

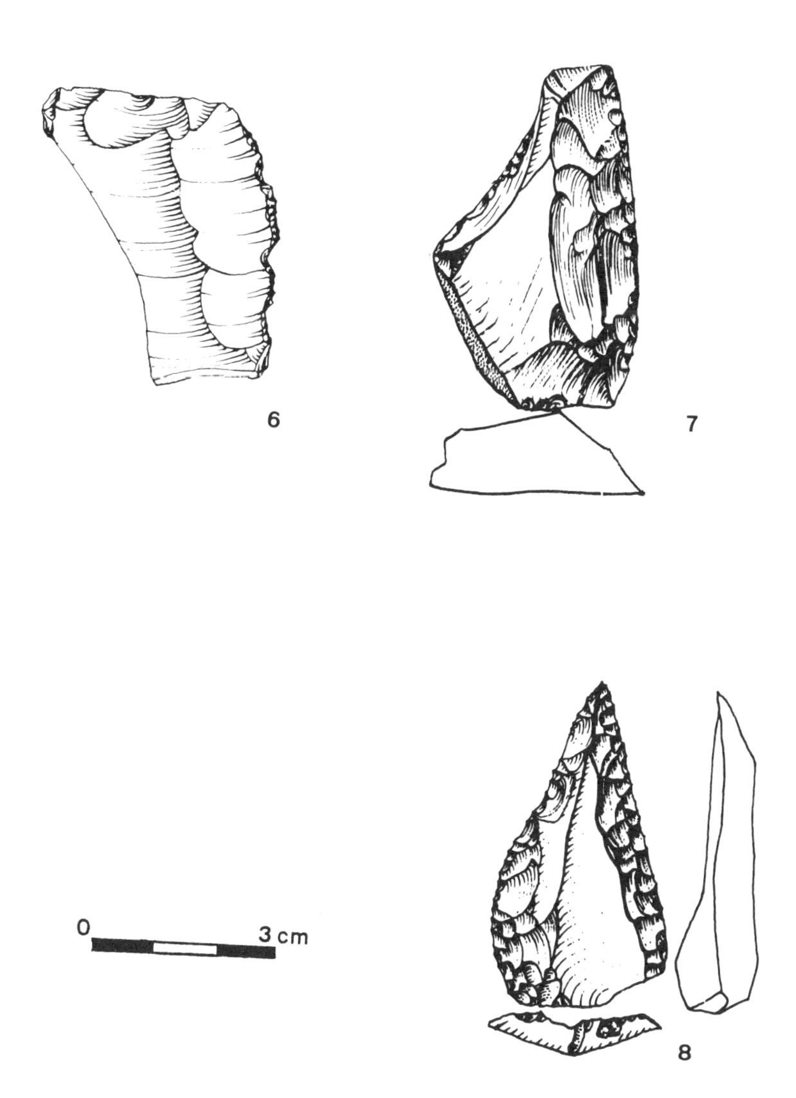

Figure 5.4. *cont.*

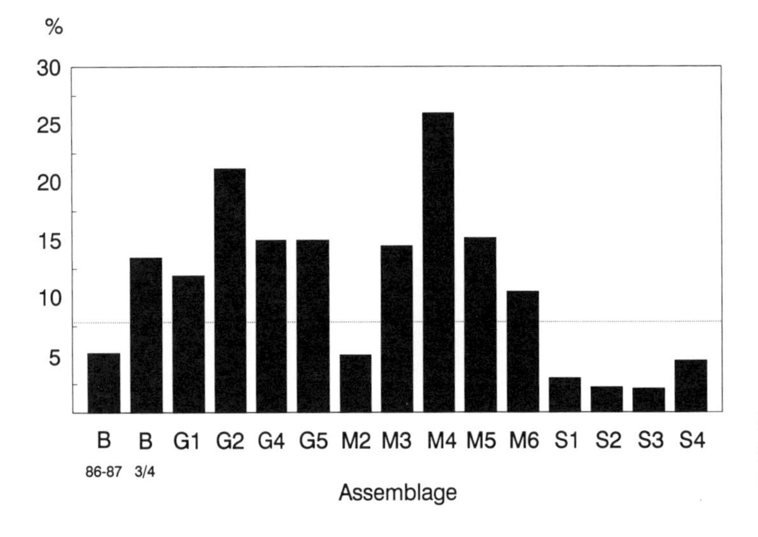

Figure 5.5a. Percentages of oversized artifacts among all whole flakes and retouched tools.

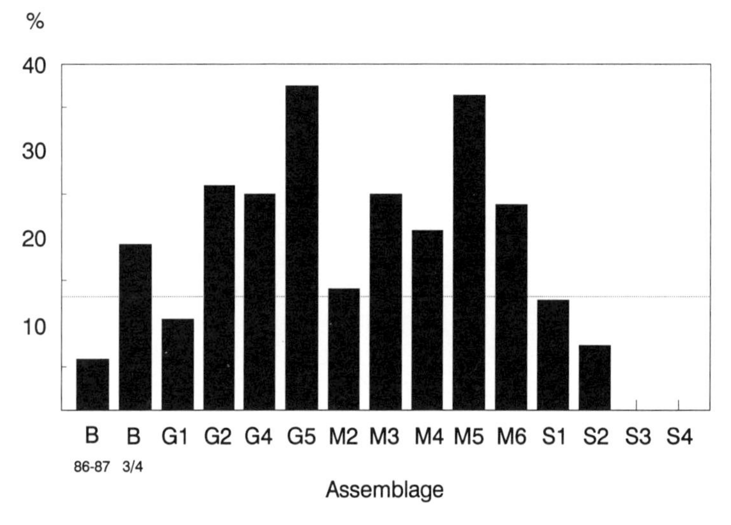

Figure 5.5b. Percentages of oversized pieces among whole, noncortical flakes and retouched tools.

the coastal raw material zone. Information about artifact size and cortex cover serves as a "substitute measure" for the number of transported artifacts in the various sites and assemblages, a means of addressing this important phenomenon in an area where it cannot be monitored more directly. Because they are framed in terms of general probabilities, rather than true frequencies of artifacts derived from specific known sources, it is difficult to compare these findings with information on raw material transport from other regions. The significance of this line of evidence emerges from its relationship with the other technological phenomena discussed above.

The Other Side of the Economy: Technology and Fauna

In the preceding sections, I have mapped the ways Mousterian stone tool manufacture and use varied within and among sites in coastal west-central Italy. In spite of the general constraints imposed by small and scarce pebble raw materials, the intensity of artifact exploitation, methods of flake production, and movement of tools or raw materials all vary among the subject assemblages. Of all these different phenomena, only the extent to which cores were consumed appears to be attributable to localized variation in the abundance of flint. Tactics of flake manufacture, the frequency of transported or exotic artifacts, and the intensity of retouch and resharpening seem to vary independently of the immediate accessibility of raw materials. So, while there is clearly an important economic dimension to variation in raw material exploitation, it remains unclear how it was organized, and what factors stand behind the changing strategies of stone tool manufacture and use.

As described in Chapter 2, ease of access is only one element in the equation that determines the "costs" of raw materials, and human responses to them. Just how the supply of flint is managed also depends on a whole range of exigencies arising from the general structure of subsistence and land use. Most notably, the frequency of residential mobility determines the extent to which people must rely on what they carry with them, and how much transport costs will govern artifact size and maintenance or renewal. The duration and predictability of occupational events, independent of overall levels of mobility, also influence the practicality of creating artificial local abundances of raw material by provisioning places. Because they are intimately tied to how people move around the landscape, foraging patterns may have important indirect consequences for when it is practical to repair and replace tools, and how long they must last. Variation in Mousterian raw material economy not explained by the simple availability of pebbles may be related to one or more of these factors.

Exploring the potential relationships between lithic technology and mobility requires some form of independent control over patterns of foraging and land use. The archaeofaunas associated with the lithic assemblages are an essential source of information about the general realm of Mousterian subsistence adaptations in west-central Italy. The faunal remains tell something about how game was procured, and about the nature and quantity of (bony) elements introduced into the shelter sites. Knowing how animal foods were obtained in turn provides clues as to the foraging patterns of Mousterian hominids. The quantity and variety of bones carried to caves have implications for the nature and duration of occupations. Large-animal bones obviously do not tell the whole story, as ungulates were not necessarily the only—or even the most important—components of Middle Paleolithic diets in the study area; however, along with some fragmentary evidence for shellfish exploitation, they are the best we have.

Archaeofaunas from the four Mousterian cave sites were described in Chapter 3, so it is only necessary to reiterate some salient aspects here. Contrasts among faunas are striking, more so than among lithic assemblages: although the variation can be described as continuous, the faunal assemblages do appear to fall into three rather dissimilar groups. All hominid-generated faunas from Grotta Breuil and Grotta di Sant'Agostino (with the exception of level S4) are characterized by the nearly complete anatomical representation and numerical predominance of prime-age individuals. They have been interpreted as representing the products of some form of ambush hunting. Apparently, virtually entire carcasses of red deer and other medium-sized animals were transported to cave sites, heavily processed there, and presumably consumed there. Levels containing this type of bone assemblage constitute what will be referred to as the "hunting-associated" group.

The second body of faunal assemblages includes materials from strata 2, 4, and 5 at Grotta Guattari and strata groups 2, 3, 4, and 6 from Grotta dei Moscerini. These faunas consist almost entirely of cranial elements. Even when teeth are excluded and only bony fragments counted, head parts are found in frequencies well out of proportion with postcrania. Compared with any reasonable expectation for a living population there are also unexpectedly large numbers of old individuals, past their reproductive prime. The head-dominated faunas appear to represent primarily the results of scavenging activities. Head parts would have been preferentially selected because they preserve more of their fat content in nutritionally stressed animals, and they were probably carried to the caves for further processing and consumption. This group of strata and assemblages is termed "scavenging-associated."

The final group of archaeofaunas includes materials from stratum 1 at Grotta Guattari, level 4 at Grotta di Sant'Agostino, and strata group 5 at Grotta dei Moscerini. All three appear to have been accumulated primarily or exclusively by nonhuman carnivores: spotted hyaenas were the culprits in the case of G1 and M5, while wolves appear to have done the damage in the case of S4. The "carnivore-dominated" bone collections yield little direct information about human behavior but do have possible implications with regard to coexistence and/or competition between hominids and other predators.

As discussed previously, the variation among faunas associated with the Pontinian Mousterian assemblages has a strong temporal component. Head-dominated, scavenged faunas date to between 110,000 and 55,000 years before present. Dated assemblages with more complete anatomical representation and prime-dominated mortality patterns indicative of ambush hunting occur after 55,000 BP. The chronological separation between different modes of game procurement demonstrates that the two main groups of hominid-generated faunal assemblages cannot represent synchronic, perhaps seasonal, components of the same foraging system. Instead, they attest to two quite distinct patterns of resource exploitation that persisted over long periods of time, at least in the vicinity of the coastal cave sites. The carnivore-dominated faunas do not form a chronologically distinct group. The assemblages from Guattari stratum 1 and level 4 at Sant'Agostino probably date to right around the 55,000 BP juncture. Though not directly correlated with the external series, strata group 5 from the internal sequence at Grotta dei Moscerini is probably somewhat older than either of these.

Statistical links between faunal exploitation and stone tool production and use

Patterns of stone tool manufacture and use covary in a remarkably consistent manner with the contents of associated archaeofaunas. Contrasts among faunal assemblages appear to "account for," in a statistical sense, much of the variation not attributable to raw material availability. Relationships between the first two groups of hominid-generated faunas and various indicators of artifact consumption and transport are especially strong. Not surprisingly, there are no distinctive technological tendencies linked with the carnivore-dominated archaeofaunas, which tend to overlap almost completely with the other two groups. The statistical associations between the contents of faunal and lithic assemblages are described below. For the sake of economy of presentation, co-variation between lithics and faunal materials is approached by contrasting the three distinct classes of faunas (scavenged, hunted, and carnivore-produced). More detailed, stratum-by-stratum analyses of associations between individual technological and faunal variables have been presented previously and are not reprised here (Kuhn 1990a:480–500).

All three of the indicators of stone tool consumption or reduction exhibit identical correlations with faunal evidence, although the relationships vary somewhat in strength. Generally speaking, artifacts in lithic assemblages associated with head-dominated, scavenged faunas are more extensively modified and more completely used up than those associated with hunted faunas. Two measures of the intensity of stone tool utilization, the frequencies of retouched tools and artifacts with more than one modified edge, are addressed in Figures 5.6 and 5.7 respectively. The bars in the graphs represent composite values for each group of assemblages, while the solid dots indicate values for individual assemblages. Fig. 5.6 portrays the strongest relationship. There is absolutely no overlap between the frequencies of retouched pieces associated with hunted and scavenged faunas, and a chi-square comparison of retouch frequency between the two groups produces highly significant results (chi-square=209.08, $p<<0.0001$, df=1). Proportions of retouched tools with multiple edges in the two principal faunal groups separate much less cleanly, but the group frequencies do differ significantly (chi-square=17.40, $p<0.001$, df=1). In both graphs, the lithic materials associated with carnivore-produced faunas show intermediate group values.

Figure 5.8 shows a tendency analogous to that expressed in the two previous graphs. The figure is a box plot of the index of scraper reduction for lithic assemblages associated with different faunal groups. Each plot shows the median (central line), the upper and lower quartiles (edges of the box), and the range ("whiskers") of the distribution of values for artifacts within each assemblage group: median values for individual assemblages are denoted by circles. The parentheses within each box

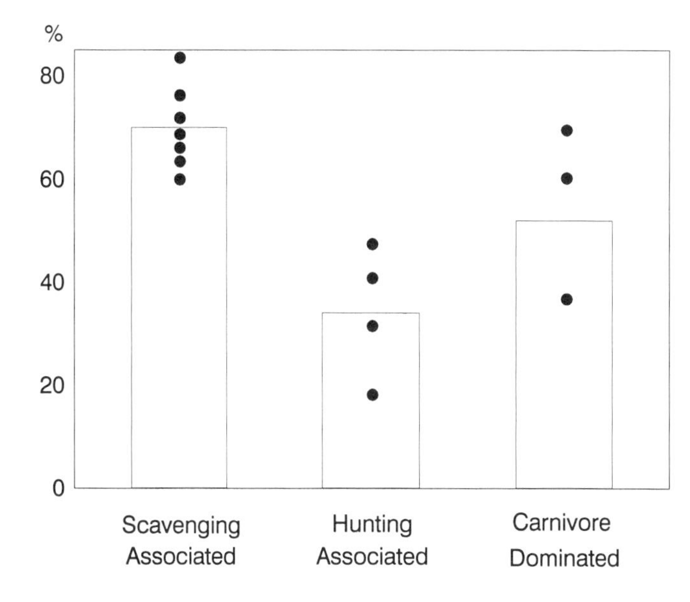

Figure 5.6. Percentages of flakes over 20 mm that have been retouched. Assemblages are grouped according to faunal association. Bars indicate group means; circles indicate percentages in individual assemblages.

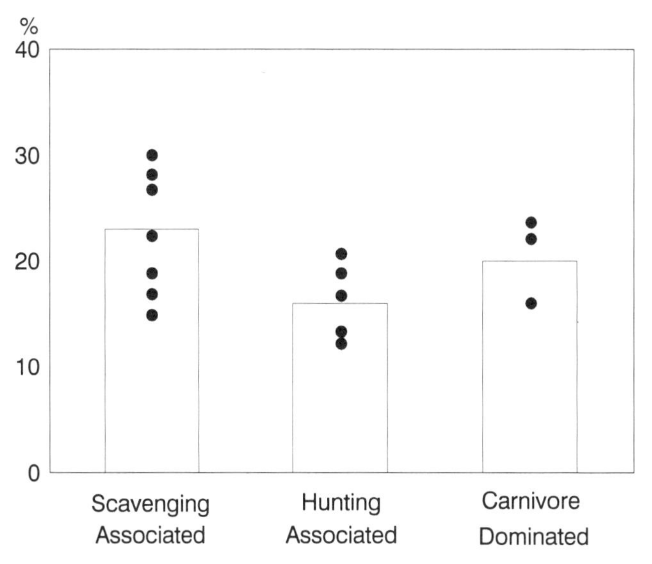

Figure 5.7. Percentages of tools with more than one retouched edge. Assemblages are grouped according to faunal association. Bars indicate group means; circles indicate percentages in individual assemblages.

indicate the location of "notches," 95% confidence intervals about the group medians (McGill et al. 1978). If the areas bounded by parentheses within two "boxes" do not overlap, there is about a 95% chance that the sample medians truly are different. Because all values for individual artifacts within each group are used, there is extensive overlap among the major assemblage groupings. Even so, the confidence intervals about the group medians do not overlap, indicating that the median index of scraper reduction for lithic assemblages associated with hunted faunas (0.55) is significantly lower than that associated with the scavenged faunas (0.63). Simply stated, although all collections contain both lightly and heavily reduced pieces, flake scrapers from assemblages affiliated with faunas obtained by ambush hunting tended to be less extensively resharpened than implements linked with evidence of scavenging. As in the previous cases, the materials associated with faunas collected mainly by non-human predators exhibit values overlapping those of the other two groups.

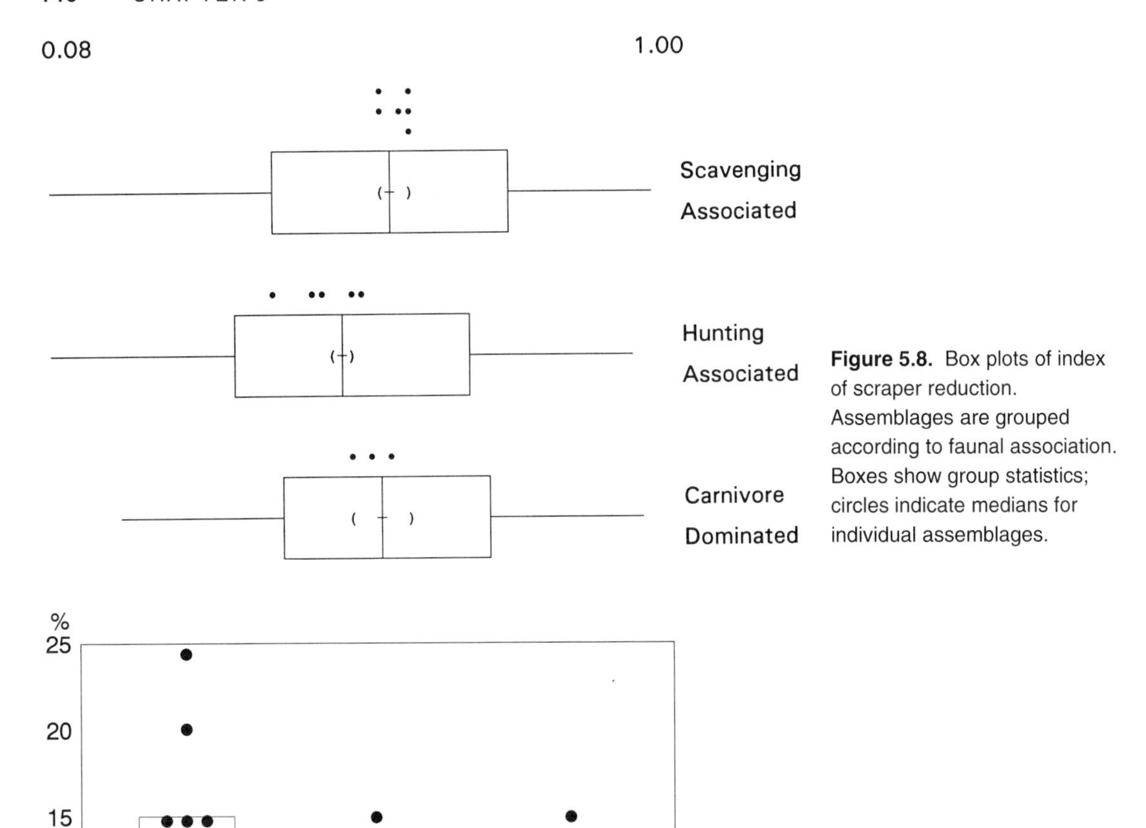

Figure 5.8. Box plots of index of scraper reduction. Assemblages are grouped according to faunal association. Boxes show group statistics; circles indicate medians for individual assemblages.

Figure 5.9. Percentages of "oversized" artifacts. Assemblages are grouped according to faunal association. Bars indicate group means; circles indicate percentages in individual assemblages.

Indirect evidence for the transport of artifacts also shows notable covariance with the contents of archaeofaunas. Figure 5.9 graphs the proportions of "oversized" specimens in the various assemblages. Values for lithic assemblages linked to the two different types of hominid-procured faunas overlap only slightly. On the whole, the lithic assemblages found in context with head-dominated, scavenged faunas contain more oversized artifacts than those associated with more complete, hunted faunas. One of the cases within the hunting-associated group does have an anomalously high value that puts it well within the range of scavenging-associated assemblages. This outlier, the very recent Middle Paleolithic assemblage from strata 3/4 at Grotta Breuil, is distinctive in other respects as well (e.g., core technology). Even with the strongly deviant value of this one case, the frequency of oversized pieces is significantly higher for the group of assemblages associated with scavenged faunas (chi-square=48.74, $p<<0.001$, df=1). The implication is that artifacts derived from exogenous sources of larger pebble or non-pebble raw materials are more frequently encountered in levels formed when un-

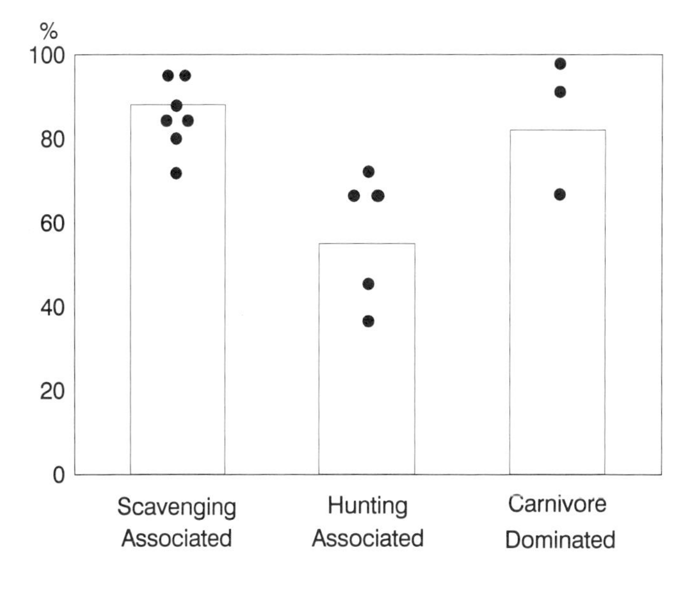

Figure 5.10a. Percentages of centripetal cores (only centripetal and platform cores considered). Assemblages are grouped according to faunal association. Bars indicate group means; circles indicate percentages in individual assemblages.

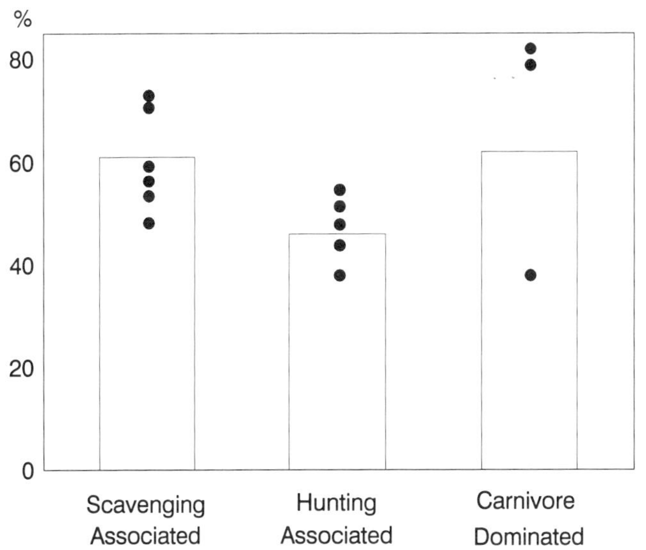

Figure 5.10b. Percentages of centripetal core products. Assemblages are grouped according to faunal association. Bars indicate group means; circles indicate percentages in individual assemblages.

gulate resources were obtained mostly by scavenging than they are in strata deposited when hunting was the dominant mode of game procurement.

Even the reliance on different methods of flake production varies in association with foraging and subsistence, as monitored in the faunal remains. Links between core technology and faunal assemblages are summarized in Figures 5.10a and b. In order to avoid "noise" introduced by the presence of several core and blank categories, only cores and flakes/tool blanks attributed to centripetal and parallel core technologies are considered: thus, the percentages shown refer only to these two modes of flake manufacture. Centripetal cores clearly dominate assemblages associated with head-dominated faunas indicative of scavenging large ungulates. Pseudo-prismatic and prepared platform cores are much more abundant in assemblages found with more complete, hunted faunas (Fig. 5.10a): a chi-square comparison of group frequencies is highly significant (chi-square=62.08, $p<<0.001$, df=1). An analogous pattern can be seen in the abundances of flakes and retouched tools attributed to different core technologies (Fig. 5.10b), although overlap

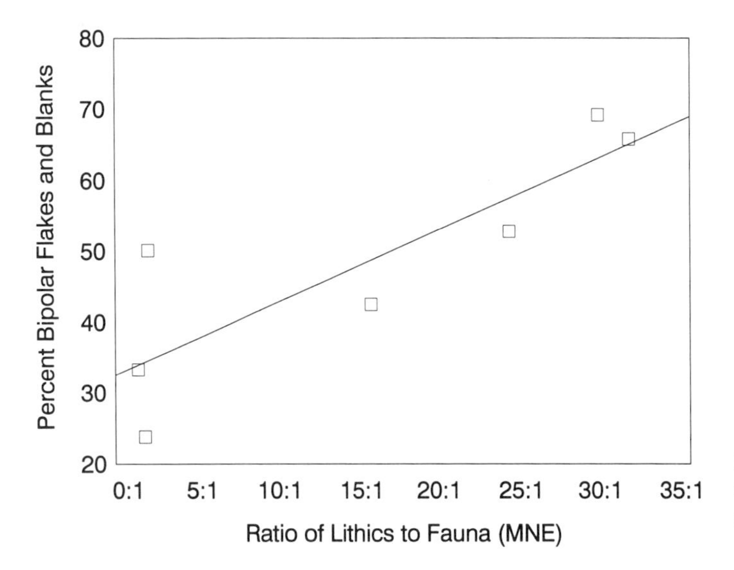

Figure 5.11. Percentage of bipolar flakes and tool blanks plotted against ratios of lithic artifacts to faunal remains.

between the two groups of lithic assemblages is somewhat more extensive. The fact that cores separate the two groups more cleanly than flakes and blanks is probably the result of ambiguities in attributing the latter to a particular mode of production. Nonetheless, the frequencies of centripetal and parallel blanks do differ quite significantly between the two groups of assemblages (chi-square= 22.85, $p<0.001$, df=1). Once again, the three lithic assemblages associated with carnivore-produced faunas exhibit a range of values spanning the first two groups.

The third important mode of flake production, bipolar or hammer-on-anvil technology, does not show any patterned co-variation with the mode of ungulate procurement or the range of anatomical elements present in archaeological deposits. However, the frequency of bipolar technology *is* closely linked to the relative abundance of animal remains, at least among those assemblages associated with scavenged faunas. This relationship is summarized in Figure 5.11. The figure plots the frequency of bipolar flakes and blanks, as a percentage of "attributable" specimens, against the ratio of lithics to bone, calculated as the ratio of tools, flakes, and cores to bone specimens identifiable as to anatomical element (Minimum Number of identifiable faunal Elements, or MNE): all ungulate species are represented. There is a significant positive correlation between the proportion of bipolar flakes and tools, and the ratio of lithics to bones ($r=0.851$, $0.01<p<0.05$, df=5). In fact, levels dominated by evidence of hammer-on-anvil technology yielded exceedingly small numbers of faunal remains. In the case of strata 4 and 5 at Grotta Guattari, lithic artifacts outnumbered identifiable bones by around thirty to one! The few elements preserved are not excessively weathered or broken up, so it is not a question of differential presentation; instead, there are simply fewer bones where bipolar technology was most heavily emphasized. This relationship holds true, however, *only* for the seven levels or strata yielding head-dominated, scavenged faunas. There is no relationship at all between bipolar technology and bone abundance, either in the hunted faunas from Grotta Breuil and Grotta di Sant'Agostino (Kuhn 1990a:493–500) or among the carnivore-dominated assemblages.

Covariant patterning in the lithic and faunal data sets can be summed up as follows. Lithic assemblages from deposits yielding apparently scavenged, head-dominated faunas tend to show evidence of extensive modification and reduction, with very high frequencies of retouch, elevated indices of scraper reduction, and higher than average numbers of double-edged tools. Unusually large artifacts are also common in these collections, implying relatively high frequencies of transported or non-local artifacts. Centripetal or radial core reduction (which yields comparatively few large, wide flakes per core) and bipolar technique are the predominant modes of flake production. Within this group, the degree of reliance on bipolar technique (versus centripetal core technology) is negatively correlated with the abundance of faunal remains. Strata containing more anatomically complete, prime-dominated faunas apparently obtained by ambush hunting are characterized by somewhat different technological indicators. Overall, tools and flakes are much less completely "used up" than in assemblages associated with scavenged faunas. Frequencies of retouch, indices of scraper reduction, and the numbers of scrapers with more than one retouched edge are all low. With the sole exception of the assemblage from the uppermost strata at Grotta Breuil, frequencies of unusually large and possibly "exotic" artifacts are also low. Although centripetal and bipolar core technologies remain quite common, other methods, typified by parallel schemes of preparation and flake removal, are more abundantly represented. These "parallel" methods of blank production appear to yield somewhat larger numbers of flakes than do centripetal or bipolar methods, although the resulting blanks tend to be narrow and elongated, and thus not suitable for heavy retouch or repeated resharpening.

Foraging, Land Use, and Technological Provisioning in Coastal Latium

The production and treatment of stone tools in the Pontinian Mousterian co-varied closely with the procurement of ungulate resources. Although demonstrating a statistical association reveals that there is some relationship between technology and modes of ungulate procurement, it does not explain the co-variation. The links between lithics and faunal exploitation are not direct; bones are not an inevitable by-product of toolmaking, and vice versa. It also seems unlikely that the associations between faunal exploitation and technology described here are purely functional, in the sense that different activities were conducted or different raw materials were worked in the context of procuring ungulates by hunting versus scavenging. Processing the carcasses of hunted animals could have favored lightly resharpened or unretouched edges, thus contributing to the low frequency and intensity of retouch in hunting-associated lithic assemblages. Still, this scenario would not account for differences in core technology, nor does it clarify the association between scavenged ungulate faunas, more extensive transport, heavy reduction of tools, and centripetal core reduction techniques.

A more satisfactory and comprehensive explanation for the co-variation between faunal and lithic assemblages derives from foraging patterns implied by different modes of ungulate procurement, and their implications for mobility and the duration of occupational events. The ways people search for food and the amount of time they spend in particular locations are directly linked to the kinds of resources they target and the returns those resources provide. Patterns of food search, mobility, and site use in turn have important consequences for what demands they place on their tools and the technological provisioning strategies they use to keep themselves supplied with artifacts and raw materials.

Contrasts between the general technological profiles of lithic assemblages associated with two main groups of faunas are strongly reminiscent of the alternative strategies of technological provisioning outlined in Chapter 2 above. The *provisioning of individuals* should be marked by extensive maintenance and renewal of artifacts, heavy dependence on transported toolkits, and the manufacture of tools with good potential for renewal. This corresponds with the distinguishing characteristics of the artifact assemblages found with *scavenged faunas*, namely heavy retouch and reduction, comparatively large numbers of potentially exotic artifacts, and the use of flake production technologies (centripetal and bipolar) that produce broad blanks suitable for repeated resharpening. In contrast, the *provisioning of places* entails less extensive dependence on what people are able to carry with them from place to place, relaxing the need to maximize utility per unit weight. The strategy of keeping places rather than people supplied with artifacts and raw materials should be distinguished by greater emphasis on producing new tools when fresh edges are needed, and concomitantly less extensive reduction and modification of tools. If there is a need to economize, it should be seen in the treatment of cores or raw materials, assuming that these are the units of transport and provisioning. The technological patterns associated with *hunted faunas*—limited retouch and resharpening, comparatively low frequencies of "exotic" material, and the use of core reduction technologies that produce many laminar blanks with little potential for renewal—fit much more closely with this second strategy. I emphasize that we are dealing not with two qualitatively distinct classes of lithic assemblage, but with a range of quantitative characteristics that tend towards one or another polarity within a continuous range of variation. In light of the kind of variation expected within any cultural system, and especially considering the fact that each assemblage represents the material remains of hundreds of years of hominid activity, it would be unrealistic to expect any single assemblage to be "strategically pure." Instead, the lithic assemblages associated with scavenged faunas suggest a greater dependence on provisioning individuals, while those found in levels yielding hunted ungulates imply more of a focus on provisioning of places.

The main factors argued to determine the suitability of alternative strategies of technological provisioning are residential mobility and the duration of occupations. The more people move around, the greater is their dependence on keeping mobile individuals supplied with tools and toolmaking potential. Provisioning of places becomes practical only when occupational events are relatively prolonged, or when particular locations are reused repeatedly and predictably. The contrasting modes of ungulate procurement seen in the Pontinian Mousterian sites fit quite well with these expectations about mobility and land use.

Scavenging as a tactic of animal procurement is entirely consistent with a highly mobile lifestyle and short-term, even ephemeral, occupations. By their very nature, scavengeable carcasses are dispersed resources. Animals already or almost dead, be they natural casualties or the leavings of other predators, are thinly distributed in most contexts. Temporary, localized gluts of carrion do occur in nature, but unusual events of this type cannot be behind the Mousterian pattern, which persists over many thousands of years in at least two locations quite distant from one another. Targeting highly scattered food "patches" in turn implies spatially extensive, highly diffuse food search patterns. Moreover, individual scavenged carcasses are likely to have provided relatively small nutritional yields, at least compared with live animals. If the kinds of anatomical parts found in archaeological sites are a good indicator, only heads and some lower limb bones were commonly worth carrying back to shelters for processing and consumption. Thus, along with wide-ranging foraging patterns, evidence of much dependence on scavenging in some Mousterian levels implies relatively little provisioning of occupations with animal foods, consistent with frequent mobility and residential occupations of limited duration.

Even when it was the primary mode of procuring meat, scavenging of ungulate carcasses was probably not the mainstay of Mousterian subsistence. After all, we cannot assume that hunting and scavenging were the only ways Mousterian hominids could have obtained food (cf. Price 1993:242): animal bones are simply the most abundant and best preserved indicators of subsistence available. It is more realistic to think that cranial tissue and marrow scavenged from the carcasses of ungulates served as a fat- and protein-rich (Stiner 1991b, 1993a) supplement to a more varied diet including marine invertebrates, reptiles, vegetable resources, and even small terrestrial game. Only one site, Grotta dei Moscerini, provides any direct evidence for exploitation of other kinds of resources, the sparse remains of shellfish and turtles collected and processed by hominids. However, there is other, indirect evidence that strongly suggests that scavenged game was not the principal source of nutrition, even when it was the dominant means of procuring medium and large ungulates. Within that group of levels with head-dominated, scavenged faunas, there is virtually no correlation between the quantity of lithic artifacts and the number of bones present (Figure 5.12) ($r=-.290$, $p>>.05$, df=5) (see also Stiner and Kuhn 1992). Assuming that the amounts of both bone and flint discarded should have increased with the duration or number of occupational events sampled by a particular excavation trench, we would expect some correlation

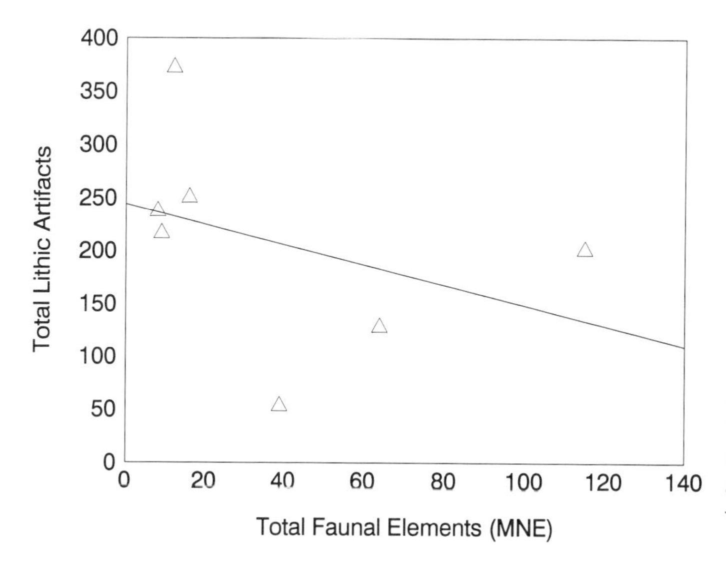

Figure 5.12. MNE (Minimum Number of identifiable faunal Elements) plotted against lithic artifact counts. Only assemblages associated with scavenged faunas are shown.

almost as a matter of course. The absence of any clear relationship implies that the hominid presence in these levels was not exclusively dependent on how much food could be gotten by scavenging.

In all likelihood, the apparent dependence on scavenging in some Pontinian Mousterian assemblages is more a symptom than a cause of highly diffuse food search patterns and spatially extensive foraging. Mousterian foragers probably included scavenging of game as part of a more generalized strategy targeting other food sources also widely scattered about the territory. Marine invertebrates, turtles, and vegetable foods are also generally dispersed, low-yield food sources. Shellfish were intensively exploited by late Pleistocene and Holocene coastal foragers, but Mousterian populations in coastal Latium seem to have made only casual use of them: marine mollusk remains are found in significant numbers only at Grotta dei Moscerini, and are not especially abundant even there. Similarly, although some wild plant foods can certainly support intensive utilization, the Pontinian collections, like other Mousterian artifact assemblages, lack any evidence for the kinds of implements—large grinding or pounding tools—that are generally associated with the harvesting and *intensive* processing of vegetable resources. The exploitation of small game as part of a spatially diffuse, dispersed foraging regime is problematic. Although bones of rabbits and rodents are found in some of the Mousterian sites, taphonomic evidence clearly points to non-human predators, especially canids, as the primary accumulators (Stiner 1993a, 1994).

Ambush hunting, the main tactic of game acquisition behind the second group of Mousterian faunas, fits well with a heightened emphasis on provisioning of more prolonged and substantial occupations with tools or raw materials. Compared with scavenging or the use of other small package-size resources, ambush hunting offers relatively high returns per procurement event. A live red deer, or a small group of them, is a much more productive food "patch" than the carcass of a dead animal. The anatomical representation demonstrates that Mousterian hunters transported virtually entire carcasses of red deer, roe deer, and similar-sized game to cave sites, where they were extensively processed and the bones broken to extract the marrow. The transport of entire carcasses to a shelter for consumption represents a substantial provisioning of occupations with food. We have no independent information about the sizes of consumer groups, or about the number of kills that occurred in the course of the average episode of shelter occupation: it is possible that Mousterian groups stayed in one place no longer than it took to process and consume a single kill. Nonetheless, the fact that whole or nearly whole carcasses of medium-sized ungulates were brought to shelter sites implies a duration of occupations surpassing that suggested by evidence from the levels with scavenged faunas. Not surprisingly, the abundance of lithics associated with the hunted faunas does increase in accordance with the amount of bone present (Stiner and Kuhn 1992: Fig. 12), implying that the amount of technological activity and, by inference, the human presence did depend to some extent on the amount of meat brought to the site.

Acquisition of game by hunting would have enabled Mousterian foragers to remain in one place for a relatively prolonged period, making it both economical and advantageous to collect artifacts and raw materials at those locations. Still, raw materials were neither ubiquitous nor plentiful. However, in a context where the cost of continuous transport was not significant, the focus of "economizing behavior" shifted from retouched tools to cores, as reflected in the use of parallel-core preparation methods yielding larger numbers of flakes per core. Making and using artifacts *in situ* also relaxes the need to repeatedly resharpen tools, and stockpiling flint at a residential base makes it practical to produce fresh edges from stocks on hand as needs arise. In fact, it appears to have counteracted the general scarcity of flint in some locales, to the extent that retouched tools from Grotta di Sant'Agostino and Grotta Breuil are less extensively reduced and less frequently retouched than implements from Grotta Guattari, the only site affording immediate access to raw materials. Flakes with fresh, unmodified or lightly modified edges are sharper and should be more effective for many tasks. They would be particularly useful in dividing and processing the carcasses of game animals.

The fact that "oversized" artifacts are generally quite scarce in levels yielding evidence for hunting (with the exception of Breuil strata 3/4) implies that raw materials brought in to provision places originated largely, if not exclusively, in the coastal deposits yielding relatively small pebbles. As discussed in Chapter 2, the distances from which things are *brought to* places should increase with the duration or predictability of an occupation. Mousterian foragers apparently did not carry quantities of larger pebbles or nodular flints from more distant sources to the "substantial" camps supplied by hunting, even though implements of such materials would have provided distinct functional advantages. Instead, coastal pebbles might well have been collected in the course of short-range foraging forays within the immediate area. Apparently, occupational events were not so prolonged that people traveled out long distances and returned, nor was it the case that groups anticipated and prepared for these stays well in advance. The implications of this observation are discussed in greater detail in the final chapter.

The negative correlation between bipolar technology and the abundance of faunal remains in the context of scavenging is less easy to interpret in terms of provisioning strategies and the logistics of stone tool manufacture and use. Hammer-on-anvil technique maximizes the flake size relative to pebble size, and for this reason it is particularly suitable for working the comparatively small pebbles found in the gravels nearest the cave sites. It is conceivable that tool makers turned to the tactic of bipolar reduction most when they were forced to rely on whatever pebbles could be found closest at hand. Because of the inverse correlation between bone frequencies and the importance of bipolar technique, it would follow that the need to get the most out of immediately available raw materials was independent of, or even in conflict with, dependence on scavenged resources. Bipolar technology might have been especially favored in contexts in which foods other than hunted or scavenged game, the other small package-size resources discussed previously, were the main focus of subsistence. Whether there might have been some functional or strategic connection between this kind of blank and the exploitation of non-game resources remains obscure, but it certainly merits future investigation.

Artifact consumption and recycling

Some researchers (Dibble 1991; Rolland 1981; Rolland and Dibble 1990) have argued that the reuse and recycling of artifacts previously abandoned at archaeological sites is a major, even predominant, factor behind intra-site variation in artifact consumption or reduction. In attempting to account for variation among *facies* of the southwestern French Mousterian, N. Rolland argues that recycling increases as a function of the duration of occupations, so that the longer people stay in one place, the more extensively they use up whatever artifacts and raw material they have previously brought to the spot and abandoned. In short, Rolland's "recycling hypothesis" has opposite implications from the model of technological provisioning, in that it equates extensive consumption of artifacts with prolonged, stable occupations.

One of the patterns in the French Mousterian that led Rolland to equate prolonged occupations with heavy reduction was contrasts between open-air and cave or rockshelter sites. It is often observed that Quina Mousterian assemblages, with high implement frequencies and presumably advanced artifact reduction, are found only within natural shelters. Open-air sites in southwest France much more frequently yield Denticulate or Levallois-dominated Mousterian assemblages, thought to be lightly modified. The assumption is that caves were occupied more often and for longer periods than open sites, with recycling leading to the production of the heavily reworked, scraper-dominated Quina Mousterian. In the Pontinian case, contrasts in faunal remains do not appear to be consistent with the notion that advanced reduction derives from prolonged occupations. The common supposition that caves and rockshelters would have been occupied more intensively or for longer periods can also be questioned. Since these kinds of localities provide instant—if somewhat cold and damp—shelter, we might expect people to use them most often when they were "just passing through" and lacked the time or materials to construct artificial housing. Nonetheless, the empirical pattern persists, and it is of interest to know whether the contrasts between Middle Paleolithic open-air and shelter sites in France also obtain in the study area.

Differences in collection strategies make it impossible to compare tool/flake ratios between cave sites and open-air locations. Two other measures of stone tool consumption are applicable, however. Figure 5.13 summarizes data on mean indices of scraper reduction for both sheltered and non-sheltered localities. Obviously, scrapers for the open-air assemblages are not consistently less reduced than pieces from cave sites: rather, values of the reduction index fall across much of the range for the shelter assemblages. The same is true for the frequencies of double-edged scrapers, another measure of the extent of artifact modification (Figure 5.14). The parity between assemblages from different classes of sites is particularly noteworthy because two of the open-air sites (Tor Caldara and Nettuno) are located on, or quite near to, deposits of large flint gravels, so that raw material would have been especially easy to come by. The presence of some heavily reduced tools in such locations may reflect "retooling," replacement of well-

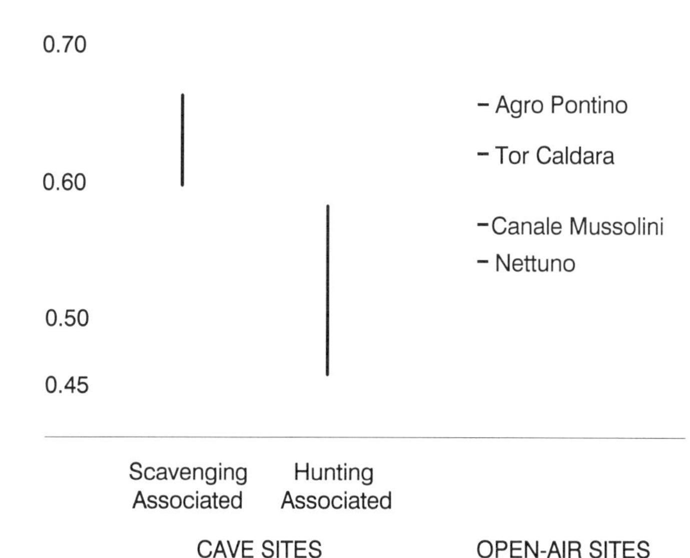

Figure 5.13. Ranges of mean index of scraper reduction in assemblages from open-air localities (dashes) and cave sites (vertical bars).

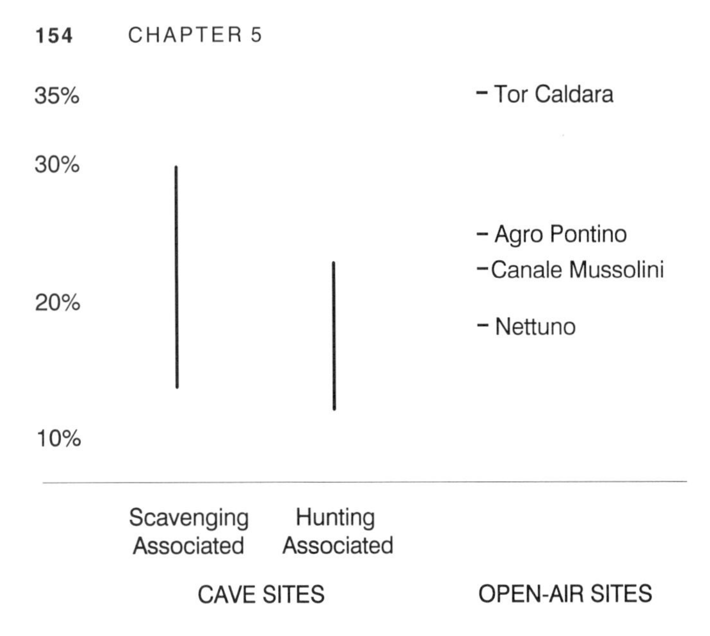

Figure 5.14. Frequencies of multiple-edged scrapers in assemblages from open-air localities (dashes) and cave sites (vertical bars).

worn implements with new ones made on the spot (e.g., Gramly 1980). These results also reinforce the observation, made previously, that the treatment of tools in the subject assemblages is less indicative of nature of particular locations than of "system-wide" factors such as land use patterns.

Reuse of artifacts abandoned by past inhabitants of a place has been documented ethnographically (e.g., Fowler 1992:106–9), and there is no doubt that artifact reuse and recycling did occur in the past. Moreover, the opportunistic reuse of things found in place could well have profound implications for the frequency and extent of artifact modification. However, the findings reported here show that it is important to separate the phenomenon of tool "scavenging" *per se* from assumptions about the context in which it occurs. I would argue that recycling of *in situ* material does not necessarily occur as a result of more prolonged occupations. In fact, holding availability constant, if recycling of "found objects" occurs in response to the difficulty of obtaining raw materials, then it ought to be more frequent in the context of relatively brief residential stays, rather than extended occupations. A highly mobile forager group would arrive at a location carrying a limited array of implements. In order to keep from using up their "curated" gear, they might well take advantage of any serviceable artifacts or raw materials they encountered on the spot. In addition, very brief occupations would provide few opportunities to collect stone from the surrounding landscape, making artifacts abandoned during earlier visits more attractive. In contrast, collection of raw materials as part of normal foraging forays should be less costly and more frequent in the context of extended stays. Given the functional disadvantages of extensively used and reworked implements, it is difficult to see the payoff in recycling previously discarded tools in situations where it is relatively cheap and easy to collect and store raw materials.

Summary: Lithic Raw Material Economy in the Pontinian Mousterian

Mousterian tool makers occupying the coastal caves in western Latium during the Upper Pleistocene had to cope with a variety of forces in maintaining a regular supply of finished implements and raw materials. There is good reason to believe that their approaches to making, modifying, and extending the lives of artifacts varied in response to both the availability of raw material and the exigencies of food procurement and mobility. Patterns of stone tool manufacture, retouch, and resharpening point to the existence of multiple strategies of technological provisioning, which are associated in turn with different modes of ungulate procurement. Technology and food-getting are linked by two factors, the duration of site occupations and patterns of movement across the landscape. Heavy modification and reduction of stone tools, relatively high frequencies of possible "exotics," and the use of core technologies yielding large, durable blanks—all suggestive of a strategy of provisioning mobile individuals with tools—characterize assemblages associated with scavenged faunas. Artifact assemblages found with faunas obtained through ambush hunting are distinguished by less extensive modification and reduction of tools and blanks, fewer apparently transported artifacts, and a shift to core technologies yielding larger numbers of blanks less suitable for prolonged use: these criteria are more consistent with a strategy of provisioning places rather than individuals with technology. It is inferred from this that the scavenged faunas and intensively exploited artifact assemblages reflect relatively ephemeral occupations and wide-ranging, diffuse foraging, focusing on a variety of resources. In contrast, hunted faunas and less heavily used artifact assemblages would have occurred as the result of more prolonged occupations provisioned with substantial quantities of meat.

It is important to point out that we can assess only the *relative* duration or intensity of the hypothetical occupational episodes associated with differing tactics of stone tool manufacture and use. The patterns identified do not represent fully sedentary versus highly residentially mobile lifeways, "foraging" versus "collecting" modes of hunter-gatherer subsistence (*sensu* Binford 1980), or "radiating" versus "circulating" patterns of territorial exploitation (e.g., Lieberman 1993; Marks 1988a). On the basis of "provisioning" with both meat and flint, it is reasonable to infer that some occupations were more prolonged or involved larger, less mobile residential groups than others. Beyond the likelihood that the "more substantial" occupations lasted sufficiently long for a Mousterian group to process and consume a red deer carcass, it is not possible to reliably assess their duration.

It should also be kept in mind that the assemblages discussed here come from thick natural or arbitrary levels. They are not single occupational episodes, but accumulations of material from many such episodes occurring over hundreds or even thousands of years. As such, no fauna or lithic assemblage is an absolutely "pure" sample of hunting or scavenging, provisioning of individuals or provisioning of places. The two groups of collections that structure the analyses above are, in all probability, mixtures of different kinds of occupational episodes, different types of game procurement patterns. Contrasts between them should thus be taken to reflect the frequency of the different kinds of events or episodes that make up any long-term accumulation of debris. In analyzing variation among the individual assemblages or groups of assemblages we are dealing the results of *more or less* hunting, *greater or lesser reliance* on transported toolkits, and not pure examples of categorically different phenomena.

Despite the composite nature of the assemblages, the contrasts among Mousterian faunas from coastal Latium are surprisingly sharp. With the technology we see greater overlap among cases and groups of cases, more subtle variation that is quite clearly quantitative rather than qualitative. The contrasting patterns of variation are a natural consequence of the indirect connections between game exploitation and technology. Even radical differences in how game was procured would not be expected to involve complete and total reorganization of hominid land use and settlement patterns. Some occupations associated with scavenging of game undoubtedly lasted longer than others, even if, on the whole, they were more brief than stays associated with the consumption of hunted game. Moreover, while the materials present in a geological deposit represent the fallout of events occurring at that place, the *conditions* of the stone tools found there reflect past actions as well as anticipated events. Evidence pertaining to the manufacture and subsequent treatment of stone tools is likely to reflect long-term "strategies" as much as the events occurring at the location of discard. Even where hominids were spending longer periods in the coastal cave sites consuming hunted animals, they must still have moved about, probably quite frequently, and they must still have depended to some extent on transported toolkits. In other words, no matter how dissimilar the faunas associated with them are, there are no "pure" lithic assemblages containing only one type of core technology or a single provisioning strategy.

The results presented here show that Pontinian Mousterian technology was not monolithic or robotic, at least when viewed in terms of the economics of tool manufacture and use. At the very least, Middle Paleolithic tool makers adjusted their behavior to local contingencies, and possibly to more long-range needs as well. Patterns of co-variation between technology and game use also provide clues about hominid foraging and mobility that could not be obtained from studying either class of data in isolation. Understanding how the manufacture and maintenance of artifacts varied in response to external and internal factors in this one case provides an initial view of Middle Paleolithic technology as an integral part of the larger suite of hominid adaptive behavior. These findings cannot tell us precisely how other Mousterian technologies in other regions were organized, but they do point to where it might be most productive to focus research. In the specific case at hand, variation in both the faunal and lithic data also reveals a directional component, a chronological trend, inasmuch as scavenged faunas and lithic assemblages suggesting high mobility and ephemeral use of places all date to before about 55,000 BP, whereas hunted faunas associated with evidence for more provisioning of places with technological material are no older than 55,000 BP—some considerably younger. The implications of these and other observations are explored in the final chapter.

6

IMPLICATIONS AND CONCLUSIONS

In the previous two chapters, I have described how Mousterian groups produced and modified stone tools, and how their tactics of artifact manufacture and maintenance varied according to the availability of raw materials and the way food resources were procured. Changing patterns of land use link lithic technology and subsistence in these cases. Variation in core reduction technology and the economics of lithic raw material exploitation can be explained as responses to the exigencies of keeping people supplied with artifacts under differing mobility regimes. This study is primarily exploratory in nature, an attempt to systematically examine relationships among two domains of Middle Paleolithic behavior—toolmaking and subsistence—that have been treated as largely independent in the past. However, exploration for exploration's own sake is of limited value. The findings reported here do have relevance beyond simply illuminating some aspects of lithic technological behavior within a distinctive and little-discussed set of Mousterian assemblages. They also bear on a number of broader issues of concern to researchers interested in the Mousterian and the Paleolithic in general. In this chapter I discuss some of the general implications of the research described in the preceding chapters. First, I examine points of correspondence and divergence between findings on the Pontinian Mousterian reported here, and results of research in other Mousterian regions. A second issue concerns evidence for change over time in Middle Paleolithic foraging and technology: is it real, and what does it imply about the so-called Middle to Upper Paleolithic "transition" in Latium? Finally, I consider what these findings may contribute to the debate over whether or not Mousterian populations behaved in a fundamentally different manner from anatomically modern populations, particularly in the arena of "planning."

Diversity or Discrepancy? The Pontinian in Its Mousterian Context

The Pontinian Mousterian assemblages are the work of hominids living in a biotic and lithic raw material environment quite different from those experienced by most other Middle Paleolithic populations. There is no reason to expect patterns and correlations observed among the sites in coastal Latium to hold true across the vast range of the Middle Paleolithic. Rather, they present a kind of experimental situation, in which we have an opportunity to study Mousterian behavior under somewhat unusual, indeed unique, circumstances. It is therefore instructive to examine some of the commonalities and the contradictions between the findings of this study and the results of comparable research in other parts of the world.

In the subject assemblages there is an apparent link between centripetal core technology (reminiscent of radial, "recurrent" Levallois method), the heavy reduction of tools, and the high mobility/brief occupations associated with scavenging. The general connection between (radial) Levallois technology, mobility, and transported toolkits appears to be widespread, at least within the European and North African Middle Paleolithic. Where appropriate data exist, it is frequently observed that Levallois flakes and tool blanks are, as a class, the artifacts most likely to be made of "exotic" raw materials (Bar-Yosef et al. 1992:511; Geneste 1988, 1989; Otte 1991; Wengler 1990, 1991). As argued above, there may be a simple explanation in energy terms for this general tendency: Levallois flakes from radial cores and similar artifacts are especially suitable as supports for transported artifacts (Kuhn 1994). Centripetal Levallois technology in particular tends to produce large, broad flakes that are comparatively thin for their size. Such flakes provide a maximum of edge per unit weight. Moreover, the breadth of flakes permits them to be reworked or resharpened several times—an important attribute for people that must rely on a restricted number of transported implements.

Associations between raw material economy and mobility provide a more complex and challenging set of global patterns. I have argued that much of the inter-assemblage variation in retouch frequency and stone tool reduction is not attributable to the availability of pebbles, but instead reflects the effects of foraging and site use patterns. Very frequent mobility in the older assemblages (associated with scavenged faunas) would have made Mousterian groups highly dependent on provisioning individuals with mobile toolkits, a strategy generally characterized (in the modern world) by intensive and prolonged maintenance and renewal of tools. Moreover, with very brief occupations, the sparse traces of such "curated" implements would not have been swamped by the debris from *in situ* manufacture. The associational link between transport and retouch/reduction intensity is strengthened by the observation that "oversized" artifacts, the specimens most likely to be exotic to the region around the four cave sites, are more frequently retouched and more often have multiple edges than "normal-sized" blanks bearing typical pebble cortex (Table 6.1).

Table 6.1. Modification of "oversized" and local pebble artifacts

Artifact class	*Unretouched*	*Retouched*
"Oversized"	28.1 (n = 207)	71.9 (n = 81)
"Normal-sized," pebble cortex	42.7 (n = 1337)	57.3 (n = 1795)

Chi-square=22.45, p<.001 (df=1)

Artifact class	*One edge*	*Multiple edges*
"Oversized"	76.3 (n = 158)	23.7 (n = 49)
"Normal-sized," pebble cortex	84.6 (n = 1519)	15.4 (n = 276)

Chi-square=8.79, p<.001 (df=1)

Here again, the *general* association between transport and artifact reduction seems to hold true across a variety of Mousterian cases. Although data permitting direct inferences about land use patterns are few, it is almost always the case that the artifacts made of non-local raw materials are more frequently retouched and more extensively worked than those made on local stone (e.g., Caspar 1984; Geneste 1988; Geneste and Rigaud 1989; Meignen 1988; Rensink et al. 1991; Roebroeks 1988; Roebroeks et al. 1988; Otte 1991; Wengler 1991). This suggests that they arrived in sites as part of regularly transported toolkits (Kuhn 1992b), and that their abundance should therefore be correlated with dependence on different provisioning strategies. However, when taken as a unit the Quina Mousterian of southwest France represents a major exception to this general pattern. Quina assemblages appear to be the most heavily reduced and extensively exploited of all Middle Paleolithic occurrences in western Europe, and yet they tend to be made on raw materials available close at hand, provided that they are of comparatively high quality (Turq 1989, 1992). While exotic artifacts *within* Quina assemblages may be even more extensively modified than other implements (e.g., Meignen 1988), the Quina group as a whole presents an interesting anomaly.

Noting that assemblages of the Quina type were more often associated with cold climatic conditions, N. Rolland (1981, 1986; Rolland and Dibble 1990) contends that the heavy resharpening and reduction of tools in those assemblages reflect two main factors: decreased access to flint because of deeply frozen ground and river beds, and decreased mobility linked to exploitation of presumably aggregated game herds. The assumption is that during cold periods populations were more tethered to natural shelters; in spending more time in these places, they tended to recycle previously deposited artifacts. Thus, in Rolland's scheme a *lack* of mobility is associated with heavy modification and reuse of tools inferred for the Quina Mousterian.

While Rolland's thesis has much to recommend it, especially the focus on artifact use histories and on raw material economy as a major force in assemblage structure, it relies on explanations which cannot be, or have not yet been, documented independently. It is reasonable to expect that any decrease in access to raw material would cause tool makers to take measures to conserve raw material, including perhaps resharpening their tools more often. However, there is no reliable means of independently ascertaining the accessibility of flint in the past under differing climatic conditions. Nor is there, at least at present, solid independent evidence of decreasing mobility and more prolonged occupations. Although it might be reasonable to expect hominids to have moved less during cold periods, it cannot be assumed. Modern Arctic and subarctic foragers employ a great variety of seasonal mobility strategies, from winter sedentism to frequently shifting residences. Among northern foraging populations, seasonal or year-round sedentism seems to have been predicated either on storing large quantities of meat, fish, or vegetable resources or on exploiting marine mammals during the cold months. We know that French Mousterian hominids were not typically hunters of pinnipeds or cetaceans, nor were they heavy users of anadromous fish. By that same token, it is by no means self-evident that the postulated large aggregations of game would have been present in southwest France (cf. Rolland and Dibble 1990).

How can we reconcile these conflicting data and contradictory explanations? The model of mobility-linked technological provisioning outlined in Chapter 2 seems consistent with much patterning in the Middle Paleolithic, as well as later time periods, but it cannot account for the "Quina phenomenon." Conversely, Rolland's model might accommodate the Quina Mousterian cases, provided they could be shown to represent especially prolonged occupations, but it cannot explain the apparent correlation between intensity of tool reduction and mobility in other contexts. One possible answer is that we may be dealing with a peculiar and distinctive combination of tactics in the Quina Mousterian. Variation in core technology and in the forms of flakes and blanks provides some clues. Flat, thin radial Levallois flakes provide a lot of edge per unit weight, making them eminently suitable for mobile toolkits. In contrast, the thick, short blanks characteristic of Quina assemblages (Bordes 1961a, 1968; Turq 1992) seem ill-adapted to such use. Moreover, typical Quina core technology, if such a thing can be said to exist, involves little or no preparation (Turq 1992), suggesting a preference for producing flakes rapidly and with comparatively little effort. Turq suggests that the French Quina Mousterian represents a functionally or strategically specialized phenomenon, oriented towards rapid production of tools for some specialized application (1992:80). The heavy retouch typical of these materials may therefore not be a function of *either* long-term maintenance *or* recycling of artifacts, but rather of the particular functional context in which they were applied. Parenthetically, it would be of great interest to have information on raw material and faunal exploitation for Ferrassie Mousterian assemblages, which are characterized by both high retouched tool frequencies and heavy use of Levallois technology, to see which they fit better—the Quina pattern or the predictions of the technological provisioning model.

The French Quina Mousterian, with its unexpectedly heavy modification of locally made artifacts, provides an opportunity to evaluate some of the premises underlying the model of technological provisioning outlined in Chapter 2. In an effort to keep things simple, it was assumed that the frequency of artifact retouch and resharpening would directly reflect the cost of keeping raw materials on hand. This cost in turn was expected to vary according to two factors: whether workable stone was generally available, and whether or not the entire inventory of tools and raw materials was continuously transported. The Quina case points to a third factor that might also strongly influence the frequency and intensity of retouch: the nature and/or duration of the activities in which tools are employed. Similarly, the negative correlation between the abundance of fauna and the use of bipolar technique in the older group of Pontinian Mousterian assemblages provides some indication that the nature of functional applications may also have influenced blank form and core technology in west-central Italy (Kuhn 1990a:537–38). Although it would be difficult to come up with global predictions about tool function and its consequences for either retouch or blank morphology, this eventuality certainly can and should be investigated directly, through use-wear analysis.

The distinctive and anomalous nature of the classic French Quina Mousterian also has implications for the identity of Pontinian as "facies." Because of the high frequencies of sidescrapers, the Pontinian was formerly classified as a variant of the southern Charentian/Quina group (Bordes 1984; Taschini 1979; Bietti 1980). While they do share other features with the corresponding French assemblages, including the high frequency of retouched tools and the scarcity of true Levallois flakes (Turq 1992:76), the west-central Italian materials are quite distinct. Most notably, the *chaînes opératoires* of the Pontinian, which has its own characteristic ways of working small pebbles, are quite different from those described for the French Quina assemblages. The Quina Mousterian may represent a homogeneous and logistically or functionally "specialized" phenomenon, whereas the Pontinian is an internally variable but regionally distinct unit. The simple lesson is that not all scraper-dominated assemblages are the same, even if they are poor in Levallois technology *sensu stricto*.

Some specific findings from this research arise from the unique aspects of the ecology and geology of coastal Latium, and are *not* expected to be easily generalized to other Mousterian cases. A case in point is the variety of core reduction technology. Researchers have questioned whether groups of prehistoric tool makers employed more than one "equivalent" *chaîne opératoire* at a time. The simultaneous existence of several independent methods of flake production in the Pontinian assemblages is probably a direct consequence of working with very small raw materials. Because the original forms of the pebbles had a great deal of influence on the geometry of the cores, tool makers had to employ different types of preparation to obtain products of usable size and morphology from pebbles of differing shapes. With larger raw materials, initial preparation can be used to create cores with a more uniform geometry, and multiple independent technological pathways may be redundant. This is not to say that other Middle Paleolithic populations could not have employed a variety of different methods of flake production: only that the case at hand represents the product of special conditions.

Another pattern not expected to extend beyond the boundaries of the study area involves the specific associations between faunal exploitation and lithic technology. There is no direct mechanical or functional link between radial core technology and the activity of scavenging, or between parallel core preparation and hunting. The common threads between technology and faunal exploitation are foraging and patterns of land use. While the residential mobility of Mousterian groups must certainly have varied in other contexts, there is no reason to expect that frequent movement would always be affiliated with scavenging, and more prolonged occupations tied to hunting. In colder continental climates, scavenging is likely to have played a somewhat different role in subsistence, and seasonal variation in the aggregation and movement of game species might well have had more profound consequences for land use patterns and the duration of occupations. The tactical "signatures" of alternative technological provisioning strategies will also inevitably change according to the constraints imposed by raw materials. For instance, I have argued that Pontinian Mousterian tool makers adopted special methods to make larger, broader, more "long-lasting" blanks for transported scrapers. In places where hominids exploited much larger raw materials, artifacts produced for expedient use could be quite massive, and the tools people chose to carry around with them might be relatively small (Ebert 1979; Kuhn 1994). The general approach developed in this study provides a framework for investigating the relationship between lithic technology and subsistence. It cannot predict how the relationship may vary across time and space within the Mousterian. However, I do know that this is exactly the kind of variation we need to appreciate in order to better understand adaptive and evolutionary processes in the Upper Pleistocene.

Change before the Transition?

Variation among both the lithic and faunal assemblages from the four cave sites that form the backbone of this study has a marked temporal component. In the previous section, technological and faunal variation was portrayed in an essentially categorical fashion, contrasting quantitative attributes of assemblages from layers yielding very different kinds of faunas. These same patterns can also be examined on an ordinal chronological scale (see Stiner and Kuhn 1992). Figures 6.1a, b, and c depict chronological variation in three technological indicators, the use of "parallel" schemes of core reduction and flake production (as indicated by the frequencies of pseudo-prismatic and prepared platform cores), retouch frequency, and the abundance of "oversized" artifacts. Assemblages are arranged in rank order based on ESR and U/Th dates. Estimated age declines from left to right: the black arrow below the *x* axis of each graph marks the position of the 55,000 BP juncture.

Only assemblages associated with hominid-produced faunas are included in the graphs. Three assemblages that are not directly dated—M2, M6, and B86–87—have been inserted into the ranking based on their estimated stratigraphic positions. If we extend the sedimentation rates for the dated strata at Grotta dei Moscerini, it is likely that strata group 2 is younger *on average* than strata 4 and 5 at Guattari, but older than Guattari stratum 2. As the internal layers forming strata group 6 at Moscerini are thought to correlate with the lower part of strata group 1 or the top of strata group 2 (see Chapter 3 above), they are given an adjacent ranking. The 1986–87 sample from Grotta Breuil (B86–87) is of uncertain age, but almost certainly belongs with the post-55,000 BP group. Somewhat arbitrarily, the assemblage has been ranked before Sant'Agostino level 3 and after Guattari stratum 2.

For core technology and retouch frequency at least, variation over time appears to be fairly continuous. The increased use of parallel core preparation (platform cores), and the decline in the frequency of retouched pieces, may have begun *before* 55,000 BP (Figs. 6.1a and b), although one could argue rates of change accelerated after this interval. Changes in faunal variables such as carcass completeness also appear to start well before 55,000 years ago (Stiner and Kuhn 1992:320). The frequencies of oversized pieces most likely to be exotic to the coastal zone (Fig. 6.1c) exhibit a much less regular decline over time, although one could also interpret the trend towards fewer unusually large pieces as beginning prior to 55,000 years ago. It is also worth emphasizing, as noted above, that the most recent assemblage (B3/4, from strata 3 and 4 at Grotta Breuil) provides evidence for a resurgence in the abundance of probable transported artifacts.

The technological trends are statistically robust as well as being visually obvious. Table 6.2 contains two alternative orderings of the assemblages, one including all 12 assemblages associated with hominid-produced faunas, and the other using only assemblages from directly dated strata. Table 6.3 shows results of Spearman's rank-order correlations of various technological indicators against the age ranking. Because error terms for the dates are often rather large, time is treated as ordinal rather than as a continuous variable. Two sets of statistics are shown, one for all 12 assemblages, the other for the 9 directly dated levels only. Rank-order correlations between the time variable and all technological indicators are statistically significant at or beyond the 0.05 probability level, regardless of the ranking used. Platform cores and their products become more abundant over time at the expense of cores and products from radial preparation. The frequency of retouch and the intensity of reduction or resharpening decline in more recent assemblages. "Oversized" artifacts likely to have come from some distance away from the coastal caves also decline in abundance over time. Trends in faunal variables such as carcass completeness and age structure are also statistically significant when subject to rank-order analysis (Stiner and Kuhn 1992:325).

Other researchers, most notably P. Mellars (1969, 1988), have argued for the existence of time trends in the European and Near Eastern Mousterian (see also Bar-Yosef 1989; Jelinek 1981; Rolland 1988; Van Peer 1991:140). However, in addition to being distinctly directional, trends in technology and faunal exploitation among the Italian Mousterian assemblages also appear somewhat "progressive." Although not especially profound or revolutionary, the shift from scavenging and radial preparation of flake cores to hunting and more "laminar" blanks from single and double platform cores could be seen to foreshadow the Upper

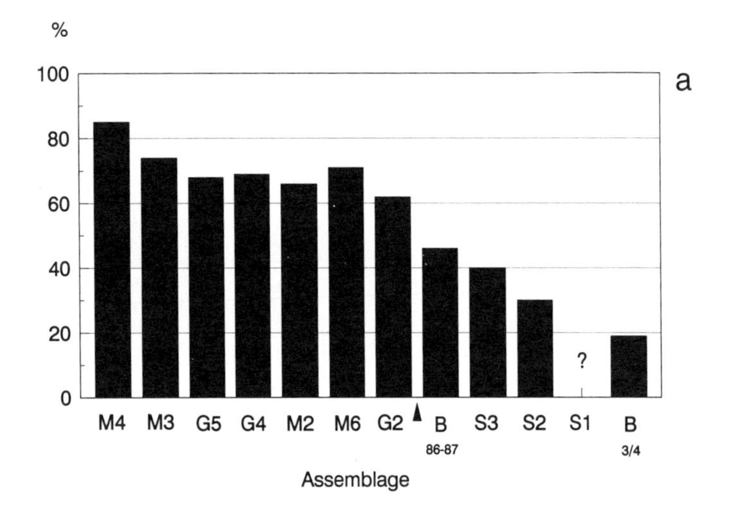

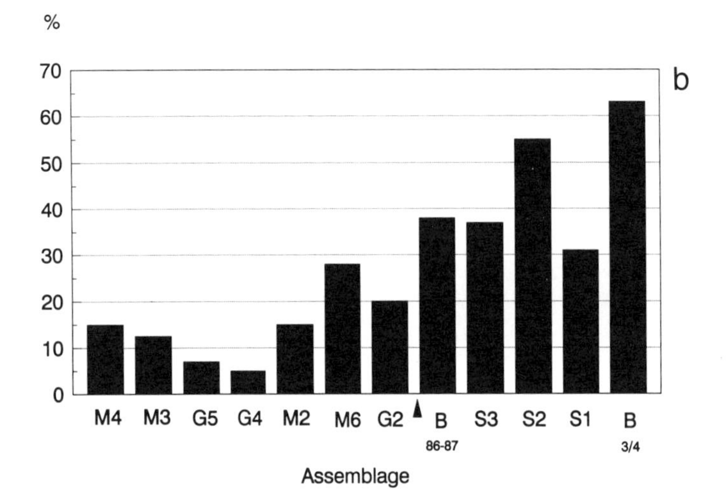

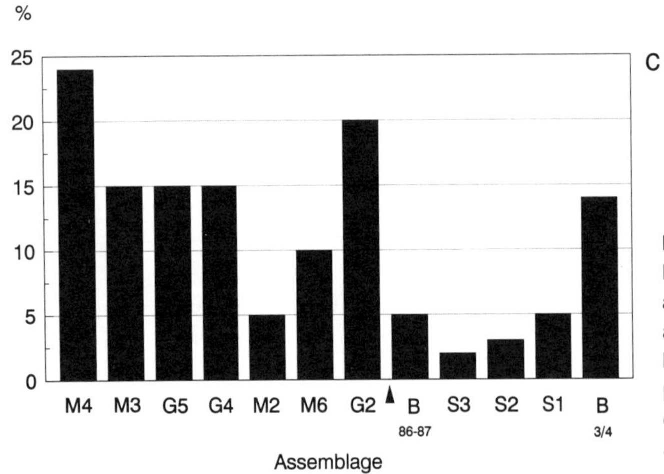

Figure 6.1. Three time trends among Pontinian Mousterian assemblages (ages of assemblages decline from left to right). a. Retouched pieces as a percentage of all blanks >1.5 cm. b. Platform cores as a percentage of all platform + radial cores. c. "Oversized" specimens as a percentage of all whole flakes and tools >1.5 cm.

Table 6.2. Alternative chronological rankings of assemblages[a]

		All assemblages	*Directly dated assemblages*
OLDEST			
↑	M4	1	1
	M3	2	2
	G5	3	3
	G4	4	4
	M2	5	—
	M6	6	—
	G2	7	5
	B86–87	8	—
	S3	9	6
	S2	10	7
	S1	11	8
↓	B3/4	12	9
YOUNGEST			

[a] Only assemblages associated with hominid-collected faunas included.

Table 6.3. Rank-order correlations of technological variables against chronology

Variable	*Spearman's r*	*df*	*probability*
ALL ASSEMBLAGES			
Mean index of scraper reduction	–.778	10	$p<.01$
Frequency of retouch	–.936	9	$p<<.01$
Frequency of "oversized" artifacts	–.613	10	$.02<p<.05$
Percentage of platform cores	.874	10	$p<.01$
Percentage of blanks with parallel preparation	.746	10	$p<.01$
DIRECTLY DATED ASSEMBLAGES ONLY			
Variable	*Spearman's r*	*df*	*probability*
Mean index of scraper reduction	–.795	7	$.01<p<.02$
Frequency of retouch	–.976	6	$p<.01$
Frequency of "oversized" artifacts	–.728	7	$.02<p<.05$
Percentage of platform cores	.800	7	$.01<p<.02$
Percentage of blanks with parallel preparation	.783	7	$p\approx.02$

Paleolithic (Aurignacian and Gravettian) assemblages that follow the Mousterian in central Italy. The possible implications of these findings for the archaeological continuity between Middle and Upper Paleolithic are obvious. However, in order to better appreciate the significance of the temporal trends among the Pontinian Mousterian assemblages, it is first necessary to consider two things. One is their geographic scale: an evolutionary trend that is confined to a distinctive, bounded region like the coastal plains of Latium is one thing, a more widespread or generalized tendency quite another. Secondly, it is necessary to question whether the changes in evidence represent a fundamental alteration in how hominids responded to their environment, or just a simple case of adjustment to changing local conditions, drawing on a more stable range of options.

Change over time: local or global?

An obvious question concerns whether the technological trends observed in this study have parallels in the sites of coastal Latium, as well as within Italy more generally. Two sites on Monte Circeo—Grotta del Fossellone and Grotta Barbara—have yielded Mousterian assemblages that, by virtue of their stratigraphic positions, appear to be quite recent. Layer 27 at Grotta del Fossellone, under study by S. Vitagliano and M. Piperno of the Pigorini Museum, was stratified near the top of more than 3 m of Mousterian deposits, and about 30 cm beneath a stratum yielding a "Typical" Aurignacian assemblage (Blanc and Segre 1953; Vitagliano and Piperno 1990–91). Grotta Barbara, a small cave excavated during the 1980s by M. Mussi and D. Zampetti (University of Rome), has a complex and somewhat altered stratigraphy. Nonetheless, it appears that the Mousterian layer there also was directly overlain by a stratum containing Aurignacian material (Mussi and Zampetti 1990–91). As no major stratigraphic hiatus is reported at either site, it is likely that the late Mousterian assemblages from both Fossellone layer 27 and Grotta Barbara predate the Aurignacian at those sites by relatively short intervals, which would surely place them well after 55,000 BP.

Unfortunately, few technological data are currently available for either the Fossellone or the Grotta Barbara assemblages. However, it is worth noting that naturally backed knives, a typical product of pseudo-prismatic and prepared platform cores, are reported as being especially abundant at Grotta Barbara (Mussi and Zampetti 1990–91). The *débitage* at Grotta Barbara is not reported as being notably laminar (Mussi and Zampetti 1990–91), but true blades are also rare in the other comparatively recent Mousterian assemblages described here. Unfortunately, neither descriptions nor illustrations of the cores themselves have as yet been published for either site.

Looking more widely, systematic presentations of data on retouch frequencies, tool reduction, and transport are seldom included in the typologically oriented studies that predominate in the Italian Paleolithic literature, so it is difficult to evaluate these variables. However, for the Italian peninsula as a whole, there does *not* appear to be any tendency for parallel schemes of flake production to replace centripetal or radial core technology over time in Mousterian assemblages. Flakes and blanks in some apparently late Mousterian assemblages, such as that from the open-air site of San Francesco di Sanremo in Liguria, are very laminar, but these materials may actually represent a northern Italian variant of the Castelperronian (Tavoso 1988). Other sequences from Italy actually seem to run contrary to a gradual shift towards more parallel preparation. Speaking with the benefit of comparatively few absolute dates, A. Palma di Cesnola suggests that there may actually be a general trend towards *decreasing* laminarity over time within the Mousterian of the peninsula as a whole (Palma di Cesnola 1986:167).

Riparo Mochi, one of the so-called "Grimaldi" sites in Liguria, is one sequence that shows a technological trend opposite to that found in Latium. Riparo Mochi contains a series of Mousterian layers more than 3 m thick. Although there are no absolute dates, a stratum yielding a classic early Aurignacian type of assemblage with many *lamelles Dufour* is stratified above the Middle Paleolithic levels, separated from them by a band of semi-sterile sediments 30–50 cm thick (Kuhn and Stiner

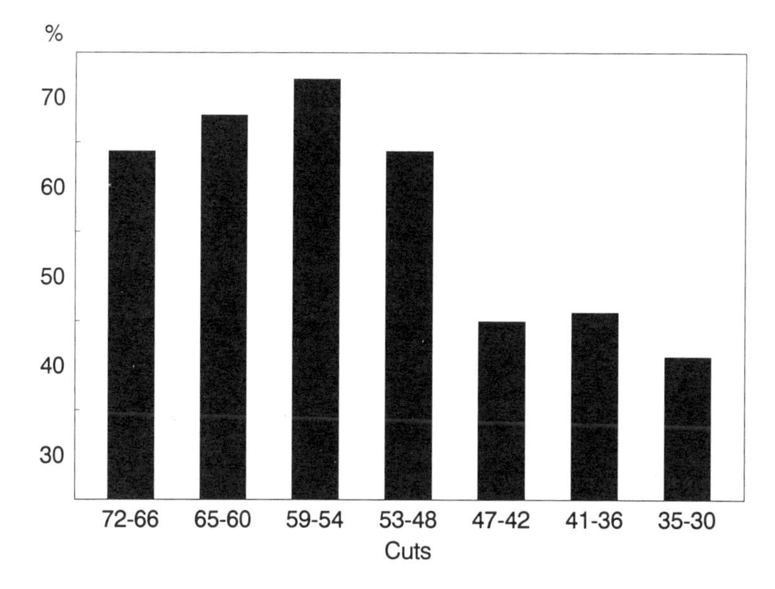

Figure 6.2. Percentages of flakes and retouched pieces with parallel dorsal scars in Mousterian assemblages from Riparo Mochi (ages of strata decline from left to right).

1992; Laplace 1977). As there is no evidence of a major stratigraphic discontinuity, it is likely that the uppermost Mousterian layers at Riparo Mochi are relatively recent. On the basis of geological indicators, de Lumley (1969:90) attributes the upper part of the Mousterian sequence at Riparo Mochi (roughly cuts 45–30) to the latter part of the "Würm II," conventionally dated between about 50,000 and 38,000 BP (Miskovsky 1974:294): this would make it approximately coeval with the assemblages from Grotta Breuil and Grotta di Sant'Agostino. Figure 6.2 shows that at Riparo Mochi *radial* core preparation increases through time at the expense of parallel flake detachment schemes (Kuhn and Stiner 1992). The levels at Mochi contemporaneous with comparatively recent Mousterian assemblages from Latium show the greatest use of radial or centripetal flake production technologies, while the oldest Mousterian layers, potentially comparable in age to Grotta Guattari or Moscerini, exhibit more evidence for the use of single or double platform cores with parallel preparation.

There is one demonstrably widespread temporal trend within the Italian Mousterian, involving changes in the typological composition of assemblages. "Terminal" Mousterian materials from throughout the Italian peninsula tend to be quite rich in denticulates, whereas earlier Middle Paleolithic assemblages are more often scraper-dominated (Barker 1981:41; Mussi 1990; Palma di Cesnola 1986). This trend seems to be echoed in at least some of the sites of coastal Latium. Layers 26 and 27 at Grotta del Fossellone yielded what was called a "denticulate micro-Mousterian," originally described as being quite different from the typical Pontinian materials from the other strata at the site (Blanc and Segre 1953; Bordes 1984:206; Grifoni 1960). Vitagliano and Piperno (1990–91) confirm that denticulates and notched pieces are more abundant than in many other local assemblages (34.8% of the retouched tool count), but still less common than sidescrapers (>37.5% of retouched tools). Curiously, the allegedly "microscopic" nature of the artifacts does not seem to hold up to metrical comparison (Vitagliano, personal communication, 1992). The limited assemblage (128 tools and flakes) from Grotta Barbara is even more distinctive (Mussi and Zampetti 1990–91). Compared with other collections from the sites of Monte Circeo, it has a relatively high Levallois index (ILty=13.8), and denticulates are nearly twice as abundant as sidescrapers (18.9%, versus less than 10% of the real count) (Mussi and Zampetti 1990–91:284). Curiously, however, there is no evidence of change over time in tool type frequencies among materials from the four cave sites used in this study

(Table 4.2b). Denticulates and notches are *slightly* more abundant in the most recent assemblage, strata 3/4 at Grotta Breuil, where they make up about 20% of the retouched tool inventory; even here, however, they are much less numerous than sidescrapers. A rank-order correlation of the percentage of denticulates and notches against time for all assemblages produces highly non-significant results. For all 12 hominid-dominated assemblages shown in Table 6.2, the Spearman's correlation between chronological rank and abundance of notches and denticulates is 0.175 ($p>>.05$): using only the 9 dated assemblages, the correlation declines to 0.067 ($p>>.05$).

There are several possible explanations for why the four sites discussed here do not follow the widespread tendency for very late Italian Mousterian assemblages to be comparatively rich in denticulates. First, increasing frequencies of denticulates over time may represent a statistical tendency but not a universal rule. Secondly, the trend may involve only the *most* recent Middle Paleolithic assemblages, the so-called "terminal" Mousterian (Palma di Cesnola 1986:167). In terms of absolute dates, only strata 3/4 at Grotta Breuil could be argued to date to this time range. Thirdly, and perhaps most important, there is always a certain amount of variation in how different investigators define artifact classes. The distinction between flakes with sporadic, irregular retouch and true denticulates or notches is a fuzzy one indeed: what one analyst calls a denticulate another may call a retouched flake, or types 45–49 in the Bordes (1961a) typology. The typological data presented here (Tables 4.2a and b) were produced by several different investigators, and it is not unlikely that the individuals employed somewhat different criteria in defining artifact classes. In this regard, it is interesting that the most recent Pontinian Mousterian assemblages do contain comparatively high percentages of "retouched flakes," pieces with partial or discontinuous retouch (Table 3.1).

In the final analysis, it would appear that the technological changes revealed in this study represent a fairly localized development, perhaps even one confined to coastal Latium. Shifts in core technology, transport, and tool resharpening are most likely to represent adjustments or responses to local conditions, whether environmental, demographic, or social. Because the apparently "progressive" technological trends observed here seem to be limited to one small region, they cannot easily be attributed to changes in the fundamental capacities or basic propensities of Mousterian tool makers as a biological group. Typological trends may be more universal, but they do not necessarily appear to foreshadow later developments and do not concern the subject assemblages.

Foraging, climate, and past landscapes

I believe that local changes in Mousterian foraging patterns and technology within coastal Latium may best be explained in terms of changes in sea level, and the consequent alterations of the landscapes immediately surrounding the four cave sites. The driving forces were the foraging opportunities encountered in the vicinity of the caves. All of the sites are situated within a narrow strip of (presently) coastal real estate. Depending on global climate and its influence on sea levels, the distances between the caves and the sea would have varied from a few meters to ten kilometers or more. The proximity of the littoral zone in turn had profound implications for the kinds of resources and foraging opportunities that Mousterian populations would have encountered.

As is well known, studies of changing sea levels during the Upper Pleistocene record a general trend of decreasing sea level after about 80,000 BP (e.g., Bloom et al. 1974; Shackleton and Opdyke 1973). Changes in the caves' topographic and environmental situations with declining sea level are illustrated schematically in Figure 6.3, an idealized portrayal of the situation at one site, Grotta Breuil. Assuming that the caves were even accessible during the warmer parts of the Upper Pleistocene, their surroundings probably resembled the conditions one encounters today. Under "interglacial" conditions, the caves would have opened directly on the sea: if they were even accessible, they would have been of limited use to terrestrial foragers (Fig. 6.3a). With slightly lower seas (0–20 m below current levels), there would have been a cordon of sand dunes, perhaps with brackish water lagoons behind them, between the caves and the sea

Figure 6.3. Illustration of possible influence of changing sea levels on topographic and environmental contexts of cave sites.

(Fig. 6.3b). Beaches, coastal dunes, and brackish water swamps would have provided "small-package" resources—including shellfish, tortoises, and vegetable products—in some abundance. Significantly, evidence for use of such resources is confined to layers (at Grotta dei Moscerini) dating to well before 55,000 to 60,000 years ago. It is also intriguing that surface finds of Mousterian artifacts on the Borgo Ermada beach level (age unknown, but probably late in isotope stage 5) are concentrated around lagoon sediments (Kamermans 1991: 28).

Beaches may also have been an especially rich source of scavengeable carcasses, a focus of Mousterian subsistence in the older faunas from Latium. Although the seashore is not normally a prime environment for ungulate species like red deer and aurochs, there are some conditions under which these creatures are actually attracted there. Animals starved for vegetation and minerals after a long winter have been observed to migrate to the coast in the spring in order to feed on seaweed washed ashore (Clutton-Brock et al. 1982). Not only is seaweed one of the few types of vegetation available in early spring, but it also contains salt and other valuable minerals depleted over the winter. Nutritionally stressed animals are in turn most likely to fall victim to predation and natural death, and thus to become available to scavengers. And, as stated above, the greatest concentrations of nutritive tissues in starved animals are in the neuro-cranium and mandible, exactly those parts preferentially selected by Mousterian foragers early on in the Mousterian record from these sites.

Even colder conditions and lower sea levels would have placed all four caves some distance from the sea. A further lowering of seas by 20 m or more would have placed all four caves a kilometer or more from the shore. With a 50-m drop the sites would have been between 5 and 10 km inland (see Table 3.2). Thus, when the global climate was especially cold the four Mousterian caves would have been situated well within the coastal plains, probably in the midst of some type of broken parkland vegetation (Fig. 6.3c). This is clearly a much better kind of territory for hunting healthy animals, the mode of ungulate procurement indicated in the faunal records from the more recent Mousterian strata.

Although sea levels certainly declined over the period in question, the decline was by no means monotonic. There were a number of relatively colder and warmer intervals between the end of oxygen-isotope stage 5 and the date of the most recent Mousterian assemblages (ca. 37,000 BP). Within this long span of time, it is unclear which specific climatic episodes or stages might have been associated with the developments in Mousterian subsistence, mobility, and technology in coastal Latium. Vagaries in absolute dates from archaeological strata make it difficult to correlate the archaeological assemblages with specific climatic intervals. Moreover, the estimates of sea levels are not consistent on a global basis (e.g., Hopkins 1982:12; Smart and Richards 1992) and do not always coincide well with other climatic indicators. For example, the 55,000 BP date used as a convenient point for dividing the archaeological cases into two groups corresponds quite well with the beginning of a particularly cold interval in both the Lake Vico and Valle di Castiglione cores (Table 3.1, above; Follieri et al. 1988; Frank 1969). Miskovsky (1974) also reports geological evidence for the onset of particularly severe conditions in the central Mediterranean at the beginning of the "Würm II," conventionally dated at about 55,000 BP. However, there is no corresponding cold phase in either of the widely used sea-level curves. In Shackleton and Opdyke's (1973) oxygen-isotope curve, a broad localized "peak" (stage 3)—albeit with the sea still about 50 m below current levels—is thought to occur at about this juncture. According to the estimates of Bloom et al. (1974), there was a minor drop in sea level between about 60,000 and 50,000 years ago, but it was no more extreme than several events occurring earlier. As a consequence, the way the localized developments in Pontinian Mousterian technology, game procurement, and land use might fit into global climatic sequences remains an issue to be resolved.

The meaning of change

If changes in Pontinian Mousterian foraging and technology did in fact occur as responses to changing sea levels and cave site contexts, then they were probably confined to the narrow coastal zone: after all, the ecological and topographic consequences of changing coastlines would have been much less drastic a few dozen kilometers inland. Unfortunately, the only dated Mousterian cave sites with preserved organic remains in the region are situated close to the present coastline, so we have only a vague notion of what the Mousterian hominids who occupied the coastal cave sites might have been doing in the plains, foothills, or inland mountains. In addition, the temporal trends discussed here are almost entirely quantitative in nature. For the most part, changes in technology appear to involve shifts in emphasis within a larger array of techniques and tactics, rather than the introduction of entirely new technological options. Several alternative methods of flake production are represented in most of the assemblages, albeit in varying quantities; it is only their frequencies that change. Similarly, there is evidence for both artifact transport and tool resharpening throughout the sequence: both simply decline over time. The same is likely to be the case with game procurement tactics and other aspects of foraging (Stiner 1994; Stiner and Kuhn 1992:331). In other words, we cannot assume that Mousterian hominids never hunted or used parallel core preparation before 55,000 BP, or that they never scavenged, ate shellfish, or worked their cores in a radial fashion after that juncture. All we know is that they relied to a varying degree on different modes of game procurement and technological options at various times while using the four cave sites discussed here.

If linked changes in technology and subsistence in the Pontinian Mousterian represent adjustment or adaptation to changing local conditions, then the variation appears *directional* only because the key underlying environmental variable—sea levels—itself changed in a somewhat directional fashion between 120,000 and 35,000 years ago. However, after a long period of shifting mixes of strategies there may be some indication of *qualitative* or structural change, as is manifest in Breuil strata 3/4. In many ways, the materials from the uppermost layers at Grotta Breuil appear to embody a somewhat extreme extension of trends observed in other assemblages. This assemblage is less extensively retouched and less heavily reduced by far than any other collection, and it is more strongly dominated by parallel core reduction schemes than any of the other comparatively "recent" Mousterian materials. The faunal assemblages also seem to represent an extreme extension of the time trends (Stiner 1990–91), in that they are more heavily skewed towards prime individuals, and contain a greater abundance of meat-bearing limb parts, than the other comparatively late Mousterian cases. But there are some qualitative differences as well. Most important, core technology may have undergone a fundamental reorganization. Rather than representing two distinct pathways, there is evidence for the appearance of a new *chaîne opératoire* in which parallel and centripetal preparation were applied successively to the same cores. In essence, it appears that radial core technology was abandoned, except perhaps as a means of getting the last usable flakes out of a core worked by other methods. This is apparent in the shapes of flakes and tool blanks: tools and flakes from Breuil strata 3/4 are significantly more elongated or blade-like on average than artifacts from the other comparatively recent Mousterian assemblages (Figure 6.4; Table 6.4). A somewhat different "mix" of economic tactics may also be in evidence. Whereas the Breuil strata 3/4 assemblage exhibits very light retouch and little evidence for resharpening—both logical extensions of the time trends—the frequency of "oversized," potentially exotic artifacts is relatively high, contradicting the general tendency (Fig. 6.1c; Table 5.14).

Table 6.4. Length/width ratios, artifacts from recent Mousterian assemblages

		Untretouched flakes	*Retouched tools*
Gr. Breuil strata 3/4	mean	1.53	1.64
	sd	0.60	0.45
Gr. Breuil 1986–87 sample	mean	1.30	1.26
	sd	0.55	0.50
Gr. di Sant'Agostino level 1	mean	1.32	1.35
	sd	0.46	0.54
Gr. di Sant'Agostino level 2	mean	1.28	1.42
	sd	0.45	0.56
Gr. di Sant'Agostino level 3	mean	1.44	1.26
	sd	0.52	0.50

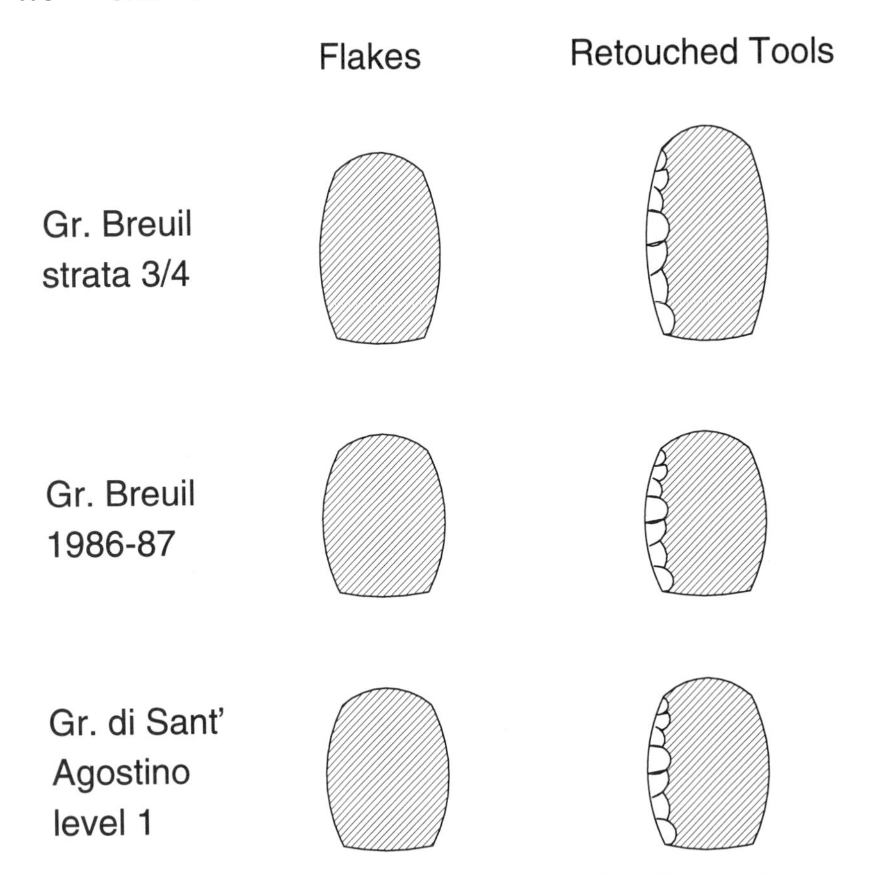

Figure 6.4. Schematic representation of flake and retouched tool dimensions in three recent Mousterian assemblages (not to absolute scale).

The general pattern of behavioral change in Latium may be analogous to the different scales or mechanisms of physiological adaptation (Bateson 1963). When first faced with novel or extreme conditions, living organisms respond through phenotypic adjustment. For example, when they are exposed to extreme cold, animals may begin to shiver, and may eventually undergo even more profound metabolic responses. If unusually cold conditions persist for generations, however, they may produce a genetic "response," favoring individuals with a predisposition towards thicker fur, increased size, or more rapid accumulation of fat. Similarly, Mousterian tool makers in Latium responded to varying mobility and different ways of using the coastal caves (brought on by sea-level fluctuations) by shifting emphasis within an array of technological strategies. Only after a prolonged and extreme alteration of conditions, and perhaps a particularly severe cold interval, is there evidence for a possible change in the range of technological options, as represented by the distinctive *chaîne opératoire* and atypical mix of tactics found in the very recent Mousterian assemblage from Grotta Breuil.

If changes in foraging and land use were in fact responses to alterations of local landscapes brought on by declining sea levels, the Mousterian of Latium does not necessarily support an argument for behavioral (or biological) continuity between the Middle and Upper Paleolithic, even though shifts in flake production technology and game procurement appear to anticipate Upper Paleolithic patterns somewhat. It is clear that the behavior of Mousterian hominids changed over time, but mutability alone does not imply cultural and biological continuity. After all, the technological developments at least seem to be confined to one small region. Unless one wishes to argue that the Upper Paleolithic and/or anatomically modern populations evolved around Monte Circeo but not in the rest of the Italian peninsula—and I do not—then this must be treated as an endemic phenomenon.

If there is a general lesson to be drawn from the results discussed here, it is that Middle Paleolithic hominids could adjust or adapt without necessarily turning into modern humans: that the Mousterian could change without becoming the Upper Paleolithic. This neither precludes nor guarantees the possibility that somewhere, sometime, this evolutionary transition did occur. Rather, it highlights the fact that the view of the Mousterian as a monolithic, static thing stemmed largely from the hegemony of a very limited descriptive and analytic perspective. The conventional systematics and terminology fall far short of accounting for all significant variation: even in the subject assemblages, where many things changed over time, the typological indicators are remarkably constant. Regardless of one's position on the origin of anatomically modern humans and the "transition" from Middle to Upper Paleolithic, it is obvious that the description and explanation of behavioral *variation* in evolutionarily relevant terms is a leading priority. The word "Mousterian" is nothing more than a blanket term referring to a group of assemblages made by two hominid subspecies over a limited range of time and space. It clearly does not refer to a homogeneous and fixed entity. Knowing how hominids responded to variation in their natural, social, and demographic environments *within* the boundaries of what we know as the Middle Paleolithic is a necessary prerequisite to understanding its eventual disappearance. We are just starting to appreciate behavioral diversity within the Middle Paleolithic, and have barely begun to understand what it may mean.

Mousterians and Moderns

The range of potential differences or similarities between Neandertals, or other archaic hominids, and fully modern humans has long been a topic of fascination for paleoanthropologists. Images of Neandertals have varied from subhuman brutes to enlightened, if craggy, flower children. By and large, changes in the image of archaic *Homo sapiens* reflect changes in how we would like to see our predecessors as much as, if not more than, the actual archaeological and osteological evidence (Trinkaus and Shipman 1993). Attempts to either modernize or "archaicize" Neandertals are also often polarized according to views about the nature of the biological transition and the origin of anatomically modern humans. Researchers favoring biological continuity tend to play down differences between modern humans and Neandertals, whereas those supporting population replacement emphasize the contrasts.

There is no necessary logical connection between the question of whether Neandertals acted just like modern humans and the question of whether they could have been human ancestors. Neandertals as a group might have been quite unlike modern humans as a group yet still have been ancestral to them. Evolution does happen, after all; it may even proceed quite rapidly in the case of biologically unconstrained behavior. Obversely, demonstrating differences neither proves nor explains replacement. There is a tendency to assume that different means primitive or ineffective. Yet as a group, archaic *Homo sapiens* was an extremely successful species, equaling or surpassing other predators in its geographic range. Searching for contrasts between hominid subspecies simply highlights the kinds of evolutionary developments that need to be explained, but it cannot explain evolution.

"Economic" behavior and economic models

In the preceding chapters I have argued that Mousterian tool makers adapted their procedures for making and maintaining stone tools to the requirements imposed by residential mobility. The kinds of technological "strategies" represented in the Pontinian Mousterian assemblages correspond well with the model of technological provisioning described in Chapter 2. If the interpretations of covariation between technological and faunal data are correct, Middle Paleolithic hominids responded to changing land use in much the same way they might be expected to, based on the model of technological provisioning. Some researchers have taken evidence of just this kind of "economizing behavior" as proof that Mousterian hominids were quite similar to fully modern humans (e.g., Marks 1988b; Munday 1976). Do the results presented here mean that Neandertals (the presumed makers of the Pontinian Mousterian) behaved in manner essentially identical to modern humans? Or is this even a valid question to ask of the findings?

The notion of technological provisioning strategies is basically an abstract, heuristic model. It assumes that transport costs and raw material are limiting variables in the technologies of mobile foragers, and it tries to predict how tool makers will respond to changes in the frequency and scale of residential mobility and the length of time they stay in a given location. It is a safe bet that workable stone was hard for Mousterian foragers in coastal Latium to come by, and that they were limited in the amount of *matériel* they could carry around with them. In other words, conditions were such that one might expect some sort of behavioral response on the part of Middle Paleolithic tool makers. However, all that the application of abstract economic models really tells us is that early hominids behaved *as if* they were following economic strategies. A variety of insects, reptiles, birds, and rodents have also been shown to behave in a "forward-looking" manner consistent with economic principles (see Stephens and Krebs 1986; Vander Wall 1990), and yet no one would argue that these organisms too were acting like modern human beings. Certainly, the Middle Paleolithic case involved the manufacture and use of stone tools, a domain of behavior in which few other animals are even engaged. The basic principle remains the same, however.

Demonstrating that an organism acts or acted in accordance with expectations created by a particular model reveals something about the basic constraints or influences on its behavior. The findings discussed here provide a perspective on the ways technology was tied to the larger domain of subsistence adaptations, how long-term variation in technological behavior might reflect responses to foraging and land use. They cannot tell us how (or if) choices among behavioral alternatives were made, or how the responses were induced. Optimal foraging research conducted by biologists provides ample demonstration that many animals incapable of complex reasoning do sometimes perform just as if they were assimilating and acting on vast bodies of economic information. However, there are several possible channels for passing information from one individual to another—ranging from genetic transmission to symbolic communication—and a vast number of mechanisms for regulating behavioral responses to that information (e.g., Bonner 1980). Ultimately, then, the significance of "economizing behavior" among Paleolithic hominids as a possible sign of their "modernity" hinges on whether the sets of integrated procedures we call strategies actually existed as mental constructs in the minds of tool makers.

A specific example can help to illustrate the distinction between the abstract and specific definitions of strategies. Under conditions of very frequent residential mobility and ephemeral site use, Mousterian tool makers appear to have preferentially employed radial core technology that yielded especially broad flakes. This observation alone provides no clues about how hominids actually came to "decisions" about which tactics of core reduction to employ. On one hand, individual tool makers might well have been aware of the pros and cons of alternative approaches to making tool blanks, choosing among them according to immediate and anticipated requirements. On the other hand, living tool makers may have had little clear knowledge of the advantages of alternative technologies as they have been articulated here, simply using whatever techniques they had learned by imitating others. In such a case, the accumulated experience and the experiments (or accidents) of individuals would many times result in the appearance and adoption of new methods suitable for novel circumstances—a process of innovation observed among living non-human primates (e.g., Hannah and McGrew 1987; McGrew 1991). From the long-term perspective of an archaeologist, viewing the by-products of thousands of technological acts conducted by many generations of individuals compressed into a single assemblage, technology would still appear to track a changing environment. Yet this is hardly the kind of sophisticated strategizing many would consider a marker of fully modern humans.

I do not wish to argue that Mousterian tool makers learned only by imitation, or that they did not have mental models about appropriate technological responses to different needs. I suspect, in fact, that Middle Paleolithic artisans did engage in some degree of strategizing, though not necessarily strictly along the lines suggested by the model of technological provisioning. The point is that the findings of this research, and indeed of many other studies of technological "strategies" in the Paleolithic, are not up to the task of determining how these phenomena were actually organized on a microscopic scale, on the level of day-to-day behavior. The results tell us only that hominids operated within a domain in which time and energy, raw material, and transport costs placed some long-term constraints on technological actions. Actually sorting out the mechanisms that governed prehistoric behavior is a daunting, not to say impossible, task. Information travels through different "conduits" (genetic, experiential, symbolic) at different rates. In principle one could investigate the actual mechanism for selecting between alternative behavioral strategies by examining the scale and rapidity of responses to changing conditions, using an archaeological record with very high temporal resolution. However, such an undertaking is largely beyond the scope and ambition of the present study, and certainly exceeds the potential chronological resolution of the data base.

It should come as no surprise that an archaeological record consisting of a series of assemblages, each of which was deposited over tens or hundreds of years, does not provide us with a clear and immediate picture of the short-term strategies of Middle Paleolithic hominids. The kinds of technological variation described in this study represent change on an evolutionary time scale, a completely different frame of reference from the perspectives of the individual hominids who created the artifacts themselves (Binford 1986). Abstract discussions of aggregate responses to environmental changes taking place over thousands of years are certainly less appealing representations of the past than are reconstructions of the "daily lives" of human ancestors. On the other hand, biological evolution also unfolded over very long periods of time, and this is exactly the kind of long-term behavioral variation that must be gauged and understood if archaeological data are to be fully integrated into the investigation of human biocultural evolution.

Somewhat ironically, the very gradual nature of change over time in the subject assemblages does have some indirect implications for the nature of Middle Paleolithic technological strategies. Many researchers have remarked on the slow rates at which Mousterian and earlier manifestations of human culture changed (Binford 1973; Klein 1989b:311; Gamble 1986:248–50; White 1982), and the Pontinian is a case in point. Temporal trends in west-central Italy entail comparatively minor, incremental shifts in several dimensions of flake production technology and raw material economy that took place over 60,000 to 70,000 years. It seems to me that this remarkable stability must reflect a relatively high degree of *short-term* technological flexibility—as is also suggested, for example, by the wide variety of tactics for making flakes from small local pebbles. After all, a living system that was not flexible could not persist for long periods. At the same time, it is difficult to escape the impression that the short-term technological strategies of Middle Paleolithic hominids must have been influenced by a fairly restricted range of factors. In other words, Mousterian technologies could only have persisted if they were capable of responding to changing conditions, but could only have been so stable if the range of conditions to which they responded was limited.

To a certain extent, the stability of Middle Paleolithic technologies must be due to the fact that so many of the implements were used in manufacture and maintenance activities. The variety of functional demands placed on "tools to make tools" would seem to be comparatively restricted: the requirements of working wood or hide are determined largely by the physical properties of the materials themselves, and they vary little from place to place. However, this only begs the question, first posed in the opening chapter, of why other common components of modern human technologies, including weapons and other tools used for extracting and processing food, are so conspicuously absent from or poorly developed in Middle Paleolithic toolkits of stone and other durable materials. The decisions of modern human "technologists" are contingent on a vast range of factors belonging to both the energetic and social realms. Technological knowledge, skills, and the fruits of their application figure in a vast array of human interactions, competitive and otherwise. All of this is reflected in a great propensity for change and comparative instability over long periods of time. Toolmaking was unquestionably an integral and vital part of the Middle Paleolithic hominid adaptive repertoire. However, it is worth considering whether the remarkable constancy of the Middle Paleolithic (and earlier) records might not indicate that these technologies were somehow less extensively and thoroughly articulated with other domains of life than is the case with the technologies of modern humans.

Planning

Findings about the relationship between foraging, land use, and the manufacture and treatment of stone tools in the Pontinian Mousterian are also relevant to the general phenomenon of *technological planning*. A number of researchers have critically examined evidence that pre-modern hominids might have planned the manufacture and use of tools around future contingencies (e.g., Binford 1989; Geneste 1986; Roebroeks et al. 1988; Wilson 1988) and have come to quite disparate conclusions. Questions about whether Neandertals, or even early anatomically modern humans, habitually anticipated and coped with their future needs have even entered into discussions of modern human origins (Binford 1989; Clark and Lindly 1989a, 1989b; Mellars 1991; Trinkaus 1989b).

Discussions of planning among living humans and other animals focus on the mental processes and constructs involved in conceptualizing and responding to events yet to occur (see review in Parker and Milbrath 1993). Obviously, archaeologists cannot study mental phenomena directly. We can only study the material by-products of hominid behavior, and, more importantly, we can reconstruct only what ancient hominids did, not what they were capable of. What we commonly refer to as planning is not a unidimensional phenomenon, and the manufacture of tools in anticipation of needs depends on more than the native mental capacities of hominids. As argued in Chapter 3, anticipatory organization in the manufacture and use of tools is highly dependent on context. At the most general level, the need to plan around future contingencies varies as a direct function of an organism's dependence on technology: the more integral tool use is to daily survival, the less prepared an organism is to do without technological support, and the greater is the need to plan. Moreover, the benefits of looking ahead and the effectiveness of different strategies for coping with anticipated exigencies are contingent on factors both internal to and external to human cultural systems (Kuhn 1992b). If raw material is ubiquitous, for example, then there is little payoff to preparing for potential shortfalls in workable stone. More important for the discussion at hand, the effectiveness of alternative gambits for dealing with future needs depends on patterns of land use and mobility, on where and when people can anticipate encountering needs for tools.

A number of researchers have demonstrated quite convincingly that Mousterian hominids regularly moved artifacts and raw materials over distances in excess of 40 or 50 km, suggesting a level of anticipatory organization extending well beyond the duration of a single day's foraging (Kuhn 1992b:193; see synthesis in Roebroeks et al. 1988). If the data on artifact size are any indication, Mousterian tool makers in western Latium also sometimes moved tools around over appreciable distances, perhaps even exploiting sources of larger raw materials situated more than 50 km from the four cave sites. The real issue is not whether or not Mousterian tool makers planned, but why evidence for anticipatory organization varies over time. In the case of coastal Latium, the kinds of cortex present on artifacts (Table 5.11) make it abundantly clear that later Upper Paleolithic hominids transported artifacts from inland sources of nodular flints to the coast much more frequently than their Mousterian predecessors. The proportions of artifacts that are demonstrably exotic to the coastal pebble zone—specimens with nodular cortex—are an order of magnitude higher in the Upper Paleolithic assemblages. Even if we assume all "oversized" artifacts in the Mousterian assemblages to have come from outside the immediate environs of the cave sites—probably an exaggeration—their frequencies are not especially high. A comparison of Tables 5.11 and 5.14 shows that in the Mousterian assemblages the frequency of "oversized" pieces *likely* to have been imported varies from about 4% to around 20%, while in the Upper Paleolithic collections the percentage of cortical pieces that are *definitely* exotic varies from 17% to more than 40%.

From the lithic data at least, differences in the frequencies of exotic artifacts between Middle and Upper Paleolithic assemblages do not seem to reflect the use of qualitatively different provisioning strategies. It is likely that much or all of the "long-distance" transport of artifacts in Mousterian assemblages represents gear provisioned to individuals. The limited number of artifacts made on larger raw materials ("oversized" specimens) probably arrived at the caves as the inevitable "fallout" from transported toolkits, and not as a result of being carried to a specific location for use there. Provisioning of places, especially during the later Mousterian, took place on a much smaller scale, utilizing only pebbles found within a short radius of the individual caves. Available evidence from the Upper Paleolithic also suggests that most long-distance transport involved retouched tools, or artifacts that ended up being abandoned after such modification. The samples of Upper Paleolithic materials are not as complete as for the Mousterian, limiting analytical options. However, it is significant that frequencies of "exotic" non-pebble cortex are much higher for retouched tools than for cores in four Upper Paleolithic assemblages from the coastal zone or nearby foothills (Table 6.5). Had exotic material arrived as cores brought in to provision cave site occupations, we would expect a much closer agreement between retouched tools and cores, as in the case of the Nightfire Island site (see Chapter 2).

Within the sample of Mousterian cases, the frequencies of potentially exotic artifacts vary according to the mode of game procurement and, by implication, the duration of occupations. Assuming that most or all transported material arrived in sites as parts of toolkits carried by mobile individuals, the pronounced contrasts between raw material use in the Middle and Upper Paleolithic *could* be interpreted as an indication that Upper Paleolithic foragers were even more nomadic than Mousterian populations. If the Upper Paleolithic sites were all ephemerally used—transitory camps, for example—it would be logical for them to contain relatively high frequencies of transported gear. But this is a difficult argument to reconcile with the archaeological evidence.

It is worth pointing out that Upper Paleolithic archaeological deposits at Riparo Salvini and Grotta del Fossellone level 21 are quite substantial and dense. Each site yielded thousands of retouched tools from only a few cubic meters of sediment. If these deposits indeed resulted from ephemeral occupations, they contain the debris from a very large number of them. There is little information on Grotta Jolanda, while Le Grottacce represents an entirely different kind of animal, namely an open-air locality. More rigorous evaluation of the hypothesis that Upper Paleolithic foragers were much more mobile than their Mousterian counterparts requires some kind of independent control over patterns of site use. The kinds of data required to evaluate this hypothesis are available only for the Epigravettian. Unfortunately, little is currently known about the Aurignacian of the study area, other than what can be discerned from

Table 6.5. Frequencies of nodular cortex on Upper Paleolithic artifacts

Site	*Percentage (and ratio) of cortical pieces from nodular raw materials*: *Retouched tools*	*Cores*
Riparo Salvini (Epigravettian)	31.6 (66/209)	5.0 (8/159)
Grotta Jolanda (Epigravettian)	24.3 (9/37)	0.0 (0/14)
Grotta del Fossellone, stratum 21 (Aurignacian)	17.3 (83/481)	1.7 (1/58)
Le Grottacce (mixed)	25.0 (4/16)	0.0 (0/11)

a very few artifact assemblages from cave sites and open-air occurrences.

The animal bones from one Epigravettian site, Riparo Salvini, have been studied in much the same way as the Mousterian faunas. Salvini is thought to represent a temporary hunting camp, seasonally occupied (Avellino et al. 1989; Bietti and Stiner 1992). The fauna is characterized by a pattern of anatomical representation which is relatively complete and is, if anything, biased somewhat towards meat-bearing limb elements. Virtually all bones are highly fragmented and many are burned, suggesting that they were extensively processed for marrow (Bietti and Stiner 1992). Because the bones and teeth are so extensively broken up, it was impossible to procure a sufficiently large sample of teeth to estimate the age structure of the prey population at Riparo Salvini. However, prime-dominated age structures appear to be nearly universal in the record of modern human hunters in temperate and cold environments (Stiner 1990a), and a large sample of mandibles and maxillae from the pleni-contemporaneous Epigravettian site of Grotta Polesini, located in the Aniene valley east of Rome (Radmilli 1974), furnishes just this sort of pattern (Stiner 1994).

The faunal data from Riparo Salvini indicate a strong general similarity with Mousterian data from Grotta Breuil and Grotta di Sant'Agostino, in that all provide evidence for hunting and the processing and consumption of substantial amounts of meat at the shelters. Thus, within the limits of what we can learn from the fauna, it appears that Epigravettian foragers were pursuing many of the same subsistence options as the late Mousterian groups during what are considered to be relatively substantial occupations. Yet the Epigravettian groups deposited vastly larger numbers of tools of "exotic" raw materials at Riparo Salvini than were found in Breuil and Sant'Agostino. The Upper Paleolithic site even contains more exotics than those Mousterian assemblages (from Guattari and Moscerini) that are thought to represent especially brief occupations and situations of spatially extensive foraging—just the contexts where such "curated" gear are expected to be most common. Thus, the faunal evidence from Riparo Salvini runs contrary to the notion that especially high mobility in the Epipaleolithic explains the relative super-abundance of transported artifacts in comparison with the Middle Paleolithic.

Other indications of land use are also difficult to reconcile with the idea of very high mobility among late Upper Paleolithic foragers. Archaeologists often use variation in the size of open-air locations as a rough indicator of land use patterns: extensive, dense concentrations of material are generally considered to represent either large, prolonged occupations or frequent reoccupation of the same spot, whereas smaller, more diffuse accumulations are interpreted as the result of shorter stays (e.g., Munday 1976). In the study area, open-air localities yielding Mousterian artifacts tend to be small, but they are quite numerous and spatially ubiquitous. This would be most consistent with a highly dispersed land use pattern and few permanent or semi-permanent occupations. In contrast, Epigravettian artifacts are found in a comparatively few, comparatively large and dense surface concentrations (Bietti 1985; Bietti et al. 1989:258–59; Mussi and Zampetti 1984–87; Zampetti and Mussi 1984), suggestive of more substantial, prolonged occupational episodes. Similarly, Mousterian artifacts are also found in a great many caves and rockshelters, while Epigravettian assemblages are less numerous and restricted to a handful of natural shelters (Zampetti and Mussi 1984), perhaps in locations that afford particular strategic advantages. Riparo Salvini, the only substantial Epigravettian site on the entire modern coast of Latium, is situated at a point where the coastal plain narrows dramatically, a situation that could well have been ideal for ambushing game at some times of year (Bietti and Stiner 1992).

Within the study area there appears to be a real difference between the Middle and Upper Paleolithic in the nature and frequency of "anticipatory organization." The Mousterian data are *internally* consistent with respect to the inferred relationship between land use, planning, and technological provisioning, but observations about the Epipaleolithic, and perhaps about the Aurignacian as well, simply do not fit with the Middle Paleolithic pattern. There is good evidence that both Middle and

Upper Paleolithic hominids planned ahead, making, transporting, and maintaining artifacts in anticipation of future needs. Moreover, most of the long-term planning in all periods appears to have involved provisioning individuals with tools and tool blanks. (This may well be a partial function of raw material distributions, in that sources of nodular flints were simply too far away for it to be practical to transport significant quantities of unprocessed raw material all the way to the coastal caves.) However, even though basically the same provisioning strategy was used, both Aurignacian and Epigravettian foragers deposited a much larger proportion of stone from distant sources into the coastal caves than did Mousterian groups. There is no evidence to suggest that this was because later populations were moving around a great deal more; if anything, Epigravettian settlements were larger, more prolonged, suggestive of less (rather than more) frequent residential moves. At Riparo Salvini, late Epigravettian hunters were doing much the same thing as late Mousterian hunters, at least as far as subsistence activities are concerned, yet they deposited vastly greater numbers of exotic artifacts.

One could contend that the apparent discontinuities in artifact transport over time simply reflect changes in the sizes and shapes of human territories. If the tool makers of the Pontinian Mousterian had limited their activities largely to the coastal plains, they might not have had access to sources of larger pebbles or nodular flints. On the other hand, the use of nodular raw materials by Upper Paleolithic groups could simply reflect a reorganization of territorial boundaries, with more movement perpendicular to the coast, possibly in association with a pattern of seasonal transhumance (e.g., Barker 1975, 1981). However, addressing questions of prehistoric territorial boundaries is a tricky proposition. The most convincing attempts have been carried out in areas where there are many sources of high-quality raw material located at varying distances from the sites in question (e.g., Demars 1982; Geneste 1985, 1988; Larick 1986; Montet-White 1991). In such contexts the origins of artifacts are more likely to reflect where people normally ranged and where they had come from before using a particular location, rather than preferential uses of raw materials of differing quality. The study area by no means fits this description.

Aside from the raw material data, which can be explained in other ways, there is no strong evidence that Middle Paleolithic foragers confined their activities to the coastal zone. The only evidence commonly offered in support of the idea that the Mousterian populations of Monte Circeo and the Agro Pontino remained entirely within the coastal zone is the distinctive character of the Pontinian industries themselves (i.e., Piperno 1984). This argument is somewhat circular, however, as most or all of the typical attributes of the Pontinian assemblages, including their small size, heavy reduction, and even the forms of flake blanks, are directly or indirectly linked to the use of the small-pebble raw materials found only near the coast (Kuhn 1990a: 127–28).

Other aspects of the record are actually quite inconsistent with the notion that Mousterian populations lived and foraged exclusively along the coast and coastal plains for more than 70,000 years. For one thing, prolonged territorial stability would seem to presuppose a strong littoral or marine focus to subsistence, but, as discussed in Chapter 3, this is clearly not the case. None of the four cave sites was ever more than 10 km from the sea during the period of interest (Table 3.2), and during the earlier part of the Mousterian their situations would have resembled conditions found today. Shellfish *were* exploited during the Mousterian, but evidence for their use is confined to a single site, Grotta dei Moscerini. Moreover, even in the levels with the most abundant mollusk remains, the absolute quantities are small (Stiner 1993a), especially compared to Mesolithic shelters like nearby Riparo Blanc (Taschini 1964). Fish bones are completely absent from Mousterian deposits, and evidence of marine mammals is limited to a handful of fragmentary bones of monk seal, a species that enters caves on its own to give birth (Wirtz 1968).

There are also hints of a seasonal component to Mousterian use of the four cave sites. Based on a wide variety of evidence, including the ages and sexes of both carnivores and ungulates, as well as the presence of shed and unshed antler, Stiner argues that different modes of game procurement are associated with broadly differing seasons of hominid occupation. The hunted faunas from Grotta Breuil and Grotta di Sant'Agostino were probably amassed during the fall and winter, possibly extending into the early spring. At Grotta dei Moscerini at least, head-dominated, scavenged faunas appear to represent debris from spring or summer occupations. Although there is seasonal complementarity between the two classes of fauna, it is virtually impossible that the sample represents both halves of a single seasonal round, since the two basic patterns of ungulate procurement, and the groups of occupations attributed to different seasons, also separate in time. After all, it would be difficult to claim that the groups that used Grotta dei Moscerini in the summer spent the fall at Grotta Breuil or Sant'Agostino, since deposits at the latter site were formed some tens of thousands of years later. And if Mousterian groups were not using the coastal caves year round at any particular part of the sequence, they must have been somewhere else for at least part of the year.

If Mousterian populations did in fact move in and out of the coastal zone, one might expect a mixture of coastal and inland artifacts and raw materials nearer the margins of the coastal plains, even if small-pebble tools probably would not have been carried very far. Unfortunately, there are few well-documented sites from the foothills and eastern mountains. It is interesting that the small assemblage from Grotta della Cava (Segre-Naldini 1984), located in the hills just beyond the margins of the Pontine Plain, does contain a mixture of typical Pontinian tools on pebble flakes along with artifacts and larger implements made by different techniques using fine-grained local limestone.

In sum, we cannot exclude the eventuality that both Middle and late Upper Paleolithic populations had regular access to at least some sources of larger raw materials. Alternatively, I would argue that the facts at hand are most economically explained in terms of broadly different "strategies" of regional land use. In the opening chapter it was suggested that aspects of the Middle Paleolithic record, ranging from infrequent investment in features and facilities to the scarcity of evidence for the use of resources other than large game, could well reflect a tendency towards very frequent residential moves and repetitive but unpredictable reuse of "target" locations like cave sites. This same kind of land use pattern could also account for the comparative scarcity of exotic artifacts in assemblages in the coastal caves. More specifically, hypothetical variation in the duration of Mousterian cave site occupations, and technological responses to it, would represent localized variation within a larger theme of frequent, opportunistically organized mobility. In contrast, Epipaleolithic foragers may have engaged in more sharply differentiated, "logistical" kinds of movement (*sensu* Binford 1980).

If residential movement during the Mousterian was extremely frequent and opportunistic, it would have been difficult for hominids to know in advance just when they would enter the coastal zone, and less practical to "gear up" at sources of better raw materials. At the same time, we must recognize that the distances transported gear *appears* to have been moved are determined largely by the durability of tools and the length of time needed for people to move from one point to another (Kuhn 1992b). If Mousterian hominids took a long time to get to the coastal caves from places where larger flint was available, little of the transported toolkit would have survived long enough to reach the coast. In other words, regardless of the actual distance traversed, artifacts that are transported in the course of a series of short residential moves—determined perhaps by immediate foraging opportunities—are likely to break and wear out, and thus be discarded, before they travel very far. In contrast, Epipaleolithic occupations of the coastal zone may have been more markedly seasonal, part of a highly differentiated pattern of seasonal movement (e.g., Barker 1975, 1981; Bietti 1990). More seasonally and/or functionally restricted, *planned* use of the coastal zone could have better enabled later Upper Paleolithic populations to equip themselves in advance to cope with the poor coastal raw materials. Moreover, if the trip to the coast were made rapidly, rather than as a series of "short hops," a larger proportion of the transported toolkit would have survived to be used and discarded there.

Although only limited information relevant to general use of the landscape is available, the facts at hand at least warrant the hypothesis that the varying frequencies of transported, maintained artifacts in the Mousterian and the late Upper Paleolithic reflect contrasts in the extent and frequency of residential mobility. The nature and distribution of shelter and open-air sites suggest that Mousterian use of the coastal cave sites and surrounding areas was generalized and spatially diffused, as might be expected if residential mobility were very frequent and highly opportunistic. The small number of relatively dense Epigravettian sites could well have derived from a more functionally, seasonally, or strategically limited—and therefore predictable—use of the coastal zone.

It is important to emphasize that variation among Mousterian faunal assemblages with respect to seasonality and mode of procurement is completely consistent with a model of highly opportunistic land use. Depending on the possibilities that presented themselves and the success achieved in taking advantage of them, opportunistic foragers might spend very little time, or a very long period, in and around particular localities. With highly opportunistic subsistence, much variation in resource procurement would result directly from conditions specific to the various caves, or to the coastal zone itself, at different times. The very fact that such sharply contrasting patterns of ungulate exploitation persisted over the long periods of time represented by the thick geological layers which define the assemblages suggests that relatively stable environmental conditions, rather than more changeable human strategies, determined Mousterian game acquisition tactics. Depending on the location of the coast relative to the cave sites, different kinds of resource procurement opportunities would have been available in different seasons, for periods lasting as long as a particular high (or low) sea stand.

If the limited evidence for technological planning in the Middle Paleolithic simply reflects highly opportunistic patterns of land use that made it difficult to gear up for forays into raw-material-poor zones like the study area, it does beg the question of why foraging might have been organized so opportunistically. In attempting to account for differences between archaic *Homo* and modern humans, we too often structure the question in terms of what earlier hominids might have lacked. In an evolutionary sense, archaic *Homo sapiens* wasn't missing anything. Neandertals and Neandertal-grade hominids were a successful and long-lived (sub)species. The real issue concerns what (if anything) makes modern humans distinctive, and how the modern human patterns evolved.

I emphasize that portraying Mousterian land use as opportunistic is not the same as saying it was random. There is more than a rhetorical distinction between moving often and moving aimlessly. The contrasts are in the kind of information foragers use and act on. Where and how often modern foragers move depend on a variety of anticipated contingencies (Binford 1991:19) stemming to a large extent from temporal and spatial distribution of resources (e.g., Binford 1980; Dyson-Hudson and Smith 1978; Gamble 1978; Kelly 1983, 1992). Hunter-gatherers in the humid tropics may travel rapidly across a territory, moving from one resource "patch" to another as opportunities present themselves (Binford 1980; Kelly 1983). This is because animal and vegetable foods in tropical forests tend to be broadly and evenly dispersed, leaving little opportunity to organize subsistence around anticipated resource concentrations. In these modern cases, highly opportunistic land use patterns result from environmental circumstances that make it difficult and sometimes impractical to plan far into the future, and do not reflect in any way on the differential capacities of humans.

Because land use is closely attuned to resource distributions, it is possible that broad ecological change could account for some of the distinctions between Middle Paleolithic land use and the mobility of later foraging groups. Ecological factors would be especially relevant to accounting for differences between the Mousterian and the later Upper Paleolithic, especially around the time of the last glacial maximum (e.g., Gamble and Soffer 1990). However, it can't all be climate. The general characteristics of the European Mousterian which provide such ambiguous indications about land use patterns (scarcity of constructed features and facilities, lack of diet breadth, etc.) are extremely widespread. Yet Middle Paleolithic hominids survived, even prospered, in some very cold places, and they had to confront environments that were highly seasonal (Gamble 1986, 1987; Roebroeks et al. 1992). It remains to be seen whether they dealt with these kinds of rigorous conditions in the way later Upper Paleolithic foragers did. For example, it has been suggested that Middle Stone Age groups in southern Africa did not respond to climatic changes during the Middle and Upper Pleistocene in the same manner as modern human populations (Ambrose and Lorenz 1990). Also, while west-central Italy never experienced full glacial conditions, it was also never a tropical forest either. Conditions during the Dryas II, the approximate period during which the Epigravettian site of Riparo Salvini was occupied, might well have favored a differentiated, logistical type of land use; but conditions during the "Würm II," or the latter part of the Mousterian sequence (after 55,000 BP), were not so very mild either.

A wide variety of animals change their territories seasonally, but they do so largely out of instinct or as a response to the behavior of other species, not because they possess "up-to-date" information about opportunities in some other place. It would be safe to say that strategically organized, logistical foraging and mobility is unique to the genus *Homo*, if not to fully modern humans. What enables humans to make strategic decisions about where to go and what to look for is the communication and pooling of information. Sharing of knowledge within co-operating social groups permits individual decision makers to "experience" a much larger area than they could ever experience directly (Gamble 1982, 1986:57; Moore 1981; Smith 1981: 44). One person alone knows little more about the environment than any other large mammal. In fact, because of the weakness of many of our senses, we may actually know less about what surrounds us. However, a cooperating group of individuals, sharing experiential and received information, can draw on a base of knowledge unmatched by any other organism. Ultimately, then, the appearance of strategically organized systems of land use and resource exploitation could be due as much to fundamental, evolutionary shifts in culturally mediated modes of information storage and transmission (e.g., Gamble 1982, 1983; Whallon 1989) as to changes in either the ability of hominids to plan, or the seasonal and spatial structure of environments. Whether the evolutionary transition might reflect changes in the mode of information transmission itself (i.e., language) or the appearance of novel kinds of social formations and relationships is a topic of much current debate, and an issue that merits extensive research.

I have contrasted the Mousterian and the Epigravettian as a means of highlighting what I believe are distinctive and interesting aspects of Middle Paleolithic adaptations. With the information at hand I am presently in no position to explain changes in human land use systems or the evolution of hominid social organization from the late Mousterian to the later Upper Paleolithic. Quite obviously, there is a large gap between the Mousterian and the Epigravettian, and in Latium this range is filled by very few data. However, as was emphasized in the introductory chapters, the purpose of this study is not to explain why the Mousterian disappeared, but instead, by investigating the links between technology and foraging, to understand why it may have persisted so long. I have made a series of propositions about how we might understand pressures on Paleolithic technologies, and how Mousterian populations in one part of the world appear to have responded to those pressures within an evolutionary time frame. These findings tell us something about the evolutionary "raw material" in place in the Middle Paleolithic of Latium, and about adaptive change within the confines of what we call the Mousterian. Where things might have gone from there is quite another story.

Epilogue

Between roughly 110,000 and 35,000 years ago, groups of hominids, probably Neandertals, frequented a series of cave sites on the western coast of what is now the Italian peninsula. They lived by exploiting a variety of resources. In addition to the large terrestrial game animals that make up the bulk of the archaeological evidence, at various times they also ate shellfish, turtles, and perhaps the occasional seal. These populations used the small flint pebbles found along the coast to produce stone tools. The raw materials they found at hand forced tool makers to adopt unusual methods and techniques, resulting in artifacts that are distinctive and at the same time unequivocally Mousterian in character. Much later, another group of humans began to excavate in the Pleistocene deposits containing the bones and stones left by the earlier populations. These most recent visitors to the cave sites coined the term "Pontinian Mousterian" to refer to the lithic artifacts they found, in recognition of both the unique and the absolutely typical in them.

The Mousterian tool makers who used the Italian coastal caves made use of a variety of methods to turn out flakes and blanks for retouched implements. The diversity in core technology resulted in part from their use of small flint pebbles, as different procedures were needed to take advantage of the opportunities afforded by pebbles of diverse shapes. But only a fraction of the inter-assemblage variation in core reduction technology is attributable to differences in the forms of pebbles. The use of alternative approaches to flake production also appears to have been related to the fact that different methods produced varying numbers of flakes with unique shapes and functional characteristics.

Evidence shows that "decisions" about how to produce flakes and tool blanks, about whether and how extensively to modify implements, and about the transport of artifacts varied from site to site and from assemblage to assemblage. The extent to which cores were consumed is largely attributable to the ease with which raw material could be procured from a particular location. Other domains of technology, including methods of core reduction and tool resharpening, seem to be largely independent of raw material availability in these particular cases. Much of the variation in the manufacture and treatment of stone tools instead appears to be closely correlated with the procurement and exploitation of game. The connection between game use and stone technology is not functional, but "strategic" or economic. Two alternative tactics for obtaining medium-sized game observed in these Mousterian faunas—ambush hunting and scavenging—are symptomatic of different patterns of mobility and varying durations of cave site occupations. In turn, approaches to tool manufacture, modification, maintenance, and transport represent responses to the exigencies of keeping people supplied with tools and raw materials under different mobility regimes.

Scavenging targets dispersed, low-yield resources and results in relatively low returns. In the Mousterian of coastal Latium, it was part of a highly mobile, "extensive" foraging strategy, one that probably focused on a variety of small-package resources including mollusks and turtles. Under such conditions, Mousterian populations were forced to rely heavily on a strategy of provisioning individuals with toolkits they could carry with them. The manufacture of large, broad, "renewable" flake blanks using radial core technology and extensive modification and renewal of implements were concessions to the requirements of provisioning individuals with long-lasting tools. Hunting provided larger quantities of food per foraging event, enabling Mousterian foragers to provision more substantial or prolonged occupations with meat. Staying in one place longer made it more advantageous to stockpile local flint pebbles at cave sites, and lessened dependence on transported toolkits. Provisioning places with raw materials in turn relaxed the need to produce large flake blanks and resharpening them repeatedly. In these contexts, tool makers switched to more economical platform core techniques, producing fresh blanks from local pebbles when new edges were needed rather than resharpening old tools.

The linked suites of technological and faunal behaviors vary not only from site to site, but over time as well. Trends in both technology and subsistence appear to anticipate later, Upper Paleolithic developments, in that hunting and somewhat laminar (i.e., parallel) modes of flake production *partially* replaced scavenging and classic Mousterian centripetal core technology. However, it is also clear that the technological changes at least are highly localized. While shifts in core technology may extend to other sites in coastal Latium, Mousterian sites in the rest of Italy do not provide evidence for analogous tendencies. In the Pontinian cases, time trends seem to represent responses to changes in local environments occurring as sea levels varied over the course of the early Upper Pleistocene. As sea levels fell, the coastline moved farther away from the cave sites. At times of high sea levels, the caves would have opened either onto the beach or onto a system of coastal dunes and lagoons. When sea levels were lower, the caves would have been located in the midst of a coastal plain, several kilometers or more from the actual coast. These differing situations would have afforded widely varying resource procurement opportunities to Mousterian foragers, reflected in what they ate, how long they stayed in the cave sites, and how they managed their technology.

The observed shifts over time in Pontinian Mousterian technology are hardly revolutionary. Variation across time and space is best characterized as quantitative and incremental. Rather than containing radically different technological or faunal patterns, contrasts between assemblages mark greater or lesser reliance on alternative modes of ungulate procurement and different patterns of tool manufacture. However, it is interesting that the most recent Mousterian assemblage, from Grotta Breuil, dated to around 37,000 BP, may be something quite different. Data on core technology at least suggest a more fundamental qualitative change in how Mousterian tool makers did things. This single case might represent something like a local version of the Mousterian of Acheulean Tradition B or the Castelperronian of France—an "evolved," very late Mousterian. However, there remains for the present only one case, and we cannot generalize from a single example. Resolution of this issue must await full study of very recent Middle Paleolithic materials from other sites in the region.

This study demonstrates that the makers of Mousterian artifacts behaved over the long term in an economically rational manner, adjusting patterns of tool manufacture and use to fit the problems inherent in differing patterns of land use. Such findings reveal the constraints imposed on technology by subsistence adaptations. Showing how toolmaking and foraging are linked in this one distinctive context will eventually contribute to the explanation of long-term changes in human technology throughout the globe. However, saying that Middle Paleolithic tool makers acted in a manner consistent with economic models is a very different thing from asserting that Mousterian hominids were actively engaged in economic strategizing. How these responses were organized, how "choices" between technological alternatives were made, is an entirely separate issue, one lying far beyond the empirical potential of most archaeological data bases, this one included.

There is good evidence that while Mousterian foragers in coastal Latium did move artifacts around the landscape, they brought in significantly less raw material from distant sources of larger flints than did later Aurignacian and Epigravettian groups. I argue that changes in raw material utilization from Middle to Upper Paleolithic in the study area reflect differences in habitual planning behavior, but not necessarily differences in the planning abilities of hominids that produced the Mousterian and Upper Paleolithic records. Both the practicality of different strategies for coping with anticipated requirements for tools and the actual distances artifacts appear to have been transported depend in large part on how people move around the landscape, and how far in advance they can plan their next move. Mousterian groups may have moved often and somewhat unpredictably, while Epigravettian populations made fewer, more strategically planned moves over long distances. These findings do not exclude the possibility that Mousterian populations, as a group, tended to cope with future needs in a manner distinct from later populations. Within the data base at hand, however, evidence for differential planning seems best explained in terms of more universal limitations on the practicality and the archaeological visibility of alternative strategies for planning ahead.

In the previous three chapters, the discussion has progressed from a rather technical examination of Mousterian lithic technology to a consideration of the impacts that foraging and land use had on lithic raw material economy, ending with a brief series of speculations about social organization and the pooling of information within human groups. This is a long chain of inference, but not an unwarranted one. All of these realms—technology, subsistence and mobility, and social organization—are key elements in the evolution of the hominid line. Making connections between them, if at times even speculative ones, is part of the process of learning the right questions to ask. The study of human evolution is nothing if not open-ended. If the book ends with other questions, about the role of technology in the Middle Paleolithic and about why broad patterns of land use might have changed over time, then maybe there has been some progress.

REFERENCES CITED

Akerman, K. (1974). Spear making sites in the western desert, Western Australia. *Mankind* 9:310–13.

Albrecht, G. (1979). *Magdalénien-inventare von Petersfels: Siedlungsarchäologische Ergebnisse der Grabungen 1974–1976*. Archaeologica Venatoria (Tübinger Monographien zur Urgeschichte), 6.

Albrecht, G., and H. Berke (1988). The Bruder valley near Engen/Hegau: Varying land use in the Magdalenian. In *De la Loire à l'Oder. Les civilisations du paléolithique final dans le nord-ouest européen*, edited by M. Otte, pp. 465–73. BAR International Series, 444(ii). Oxford: British Archaeological Reports.

Allsworth-Jones, P. (1986). *The Szeletian and the Transition from Middle to Upper Paleolithic in Central Europe*. Oxford: Clarendon Press.

Ambrose, S., and K. Lorenz (1990). Social and ecological models for the Middle Stone Age in southern Africa. In *The Emergence of Modern Humans: An Archaeological Perspective*, edited by P. Mellars, pp. 3–33. Ithaca: Cornell University Press.

Anderson-Gerfaud, P. (1990). Aspects of behavior in the Middle Paleolithic: Functional analysis of stone tools from southwest France. In *The Emergence of Modern Humans: An Archaeological Perspective*, edited by P. Mellars, pp. 389–418. Ithaca: Cornell University Press.

Ansuini, P., M. La Rosa, and M. Zei (1990–91). Open air Mousterian sites in central-south coastal Latium. *Quaternaria Nova* 1:479–98.

Anzidei, A. P., A. Angelelli, A. Arnoldus-Huyzendveld, L. Caloi, M. R. Palombo, and A. Segre (1989). Le gisement pléistocène de la Polledrara di Cecanibbio (Rome, Italie). *L'Anthropologie* 93:749–82.

Ascenzi, A. (1990–91). A short account of the discovery of the Mount Circeo Neandertal cranium. *Quaternaria Nova* 1:69–80.

Audouze, F., D. Cahen, L. Keeley, and B. Schmider (1981). Le site magdalénien du Buisson Campin, Verberie (Oise). *Gallia Préhistoire* 24(1):99–143.

Avellino E., A. Bietti, L. Giacopini, A. Lo Pinto, and M. Vicari (1989). A new Dryas II site in southern Lazio—Riparo Salvini: Thoughts on the late Epigravettian in middle and southern Tyrrhenian Italy. In *The Mesolithic in Europe*, edited by C. Bonsall, pp. 516–37. Edinburgh: John Donald.

Ball, B. (1987). Eating rocks: Pebble cherts, fall-off curves and optimal diet theory. In *Archaeology in Alberta 1986*, edited by M. Magne, pp. 140–49. Occasional Papers, 31. Edmonton: Archaeological Survey of Alberta.

Bamforth, D. (1985). The technological organization of small group bison hunting on the Llano Estacado. *Plains Anthropologist* 30:243–58.

Bamforth, D. (1986). Technological efficiency and stone tool curation. *American Antiquity* 51:38–50.

Bamforth, D. (1991) Technological organization and hunter-gatherer land use: A California example. *American Antiquity* 56:216–34.

Barker, G. (1975). Prehistoric territories and economies in central Italy. In *Palaeoeconomy*, edited by E. S. Higgs, pp. 111–75. Cambridge: Cambridge University Press.

Barker, G. (1981). *Landscape and Society: Prehistoric Central Italy*. London: Academic Press.

Barra Incardona, A. (1969). Le nuove ricerche nelle cavernette e nei ripari dell'Agro Falisco. *Atti della Società Toscana di Scienze Naturali* 76:101–24.

Barton, C. M. (1990). Beyond style and function: A view from the Middle Paleolithic. *American Anthropologist* 92:57–72.

Bar-Yosef, O. (1988). The date of the south-west Asian Neandertals. In *L'homme de Néandertal*, Vol. 3: *L'anatomie*, edited by M. Otte, pp. 31–38. ERAUL 30. Liège: Université de Liège.

Bar-Yosef, O. (1989). Geochronology of the Levantine Middle Palaeolithic. In *The Human Revolution: Behavioural and Biological Perspectives on the Origins of Modern Humans*, edited by P. Mellars and C. B. Stringer, pp. 589–610. Edinburgh: Edinburgh University Press.

Bar-Yosef, O. (1991). The search for lithic variability among Levantine Epi-Paleolithic industries. In *25 ans d'études technologiques en préhistoire: Bilan et perspectives*, pp. 319–35. Actes des XIe Rencontres Internationales d'Archéologie et d'Histoire d'Antibes. Juan-les-Pins: Éditions APDCA.

Bar-Yosef, O., B. Vandermeersch, B. Arensburg, A. Belfer-Cohen, P. Goldberg, H. Laville, L. Meignen, Y. Rak, J. Speth, E. Tchernov, A. M. Tillier, and S. Weiner (1992). The excavations in Kebara cave, Mt. Carmel. *Current Anthropology* 33:497–550.

Bar-Yosef, O., B. Vandermeersch, P. Goldberg, H. Laville, L. Meignen, Y. Rak, E. Tchernov, and A. M. Tillier (1986). New data on the origin of modern man in the Levant. *Current Anthropology* 27:63–64.

Bateson, G. (1963). The role of somatic change in evolution. *Evolution* 17:529–39.

Baumler, M. F. (1988). Core reduction, flake production, and the Middle Paleolithic industry of Zobiste (Yugoslavia). In *Upper Pleistocene Prehistory of Western Eurasia*, edited by H. L. Dibble and A. Montet-White, pp. 255–74. University Museum Monograph 54. Philadelphia: The University Museum, University of Pennsylvania.

Beyries, S. (1987). *Variabilité de l'industrie lithique au moustérien: Approche fonctionelle sur quelques gisements français*. BAR International Series, 328. Oxford: British Archaeological Reports.

Beyries, S. (1988). Functional variability of lithic sets in the Middle Paleolithic. In *Upper Pleistocene Prehistory of Eurasia*, edited by H. Dibble and A. Montet-White, pp. 213–25. University Museum Monograph 54. Philadelphia: The University Museum, University of Pennsylvania.

Biddittu, I. (1971). Il paleolitico inferiore di Arce e Fontana Liri, Frosinone. *Archeologia, antropologia, etnologia* 101:251–52.

Biddittu, I. (1986). Il musteriano della Grotta dei Ladroni alla Ripagnola–Polignano a Mare (Bari). *Atti della XXV Riunione Scientifica dell'Istituto Italiano di Preistoria e Protostoria.*

Biddittu, I., P. Cassoli, and L. Malpieri (1967). Stazione musteriana in Valle Radice nel comune di Sora (Frosinone). *Quaternaria* 9:321–48.

Biddittu, I., P. Cassoli, F. Radicati di Brozolo, A. Segre, E. Segre Naldini, and I. Villa (1979). Anagni, a K-Ar dated Lower and Middle Pleistocene site, central Italy: A preliminary report. *Quaternaria* 21:53–71.

Biddittu, I., G. De Angelis, and A. Segre (1990–91). High altitude Mousterian sites in interior Latium: Monte Gennaro and Monte Pellecchia (Lucretili Regional Park, Rome). *Quaternaria Nova* 1:549–64.

Bietti, A. (1969). Due stazione di superficie del paleolitico superiore nella Pianura Pontina. *Bullettino di Paletnologia Italiana* 78:7–39.

Bietti, A. (1980). Un tentativo di classificazione quantitiva del "Pontiniano" laziale nel quadro delle industrie musteriane in Italia: Problemi di derivazione e di interpretazione culturale. *Rivista di Antropologia* 61: 161–602.

Bietti, A. (1982). The Mousterian complexes of Italy: A search for cultural understanding of the industrial assemblages. *Anthropos* 21:237–51.

Bietti, A. (1984). Primi risultati dello scavo nel giacimento epigravettiano finale di Riparo Salvini (Terracina, Latina). *Atti della XXIV Riunione Scientifica dell'Istituto Italiano di Preistoria e Protostoria*, pp. 195–205.

Bietti, A. (1985). Cultura e/o adattamento per il musteriano dell'Italia centro-tirrenica. *Atti della Società Italiana d'Ecologia* 5:929–33.

Bietti, A. (1990). The late Upper Paleolithic in Italy: An overview. *Journal of World Prehistory* 4:95–155.

Bietti, A. (1990–91). Is there a Pontinian culture? *Quaternaria Nova* 1:673–78.

Bietti, A., and S. Grimaldi (1990–91). Patterns of reduction sequences at Grotta Breuil: Statistical analyses and comparisons of archaeological vs. experimental data. *Quaternaria Nova* 1:379–406.

Bietti, A., and S. Kuhn (1990–91). Techno-typological studies on the Mousterian industry of Grotta Guattari. *Quaternaria Nova* 1:193–211.

Bietti, A., and M. Stiner (1992). Les modèles de subsistance et de habitat de l'epigravettien italien: L'exemple de Riparo Salvini (Terracina, Latium). In *Colloque international: Le peuplement magdalénien*, edited by H. Laville, J.-P. Rigaud, and B. Vandermeersch, pp. 137–52. Paris: Éditions CTHS.

Bietti, A., M. Brucchietti, and D. Mantero (1988b). Ricognizione sistematica di superficie nella piana di Fondi (Latina): Primi risultati. *Quaderni del Centro di Studio per l'Archeologia Etrusco-Italica* XVI: 389–96.

Bietti, A., S. Grimaldi, V. Mancini, P. Rossetti, and G.-L. Zanzi (1991). Chaînes opératoires et expérimentation: Quelques exemples du moustérien de l'Italie centrale. In *25 ans d'études technologiques en préhistoire: Bilan et perspectives*, pp. 109–24. Actes des XIe Rencontres Internationales d'Archéologie et d'Histoire d'Antibes. Juan-les-Pins: Éditions APDCA.

Bietti, A., S. Kuhn, A. Segre, and M. Stiner (1990–91). Grotta Breuil: Introduction and stratigraphy. *Quaternaria Nova* 1:305–24.

Bietti, A., G. Manzi, P. Passarello, A. Segre, and M. Stiner (1988a). The 1986 excavation campaign at Grotta Breuil (Monte Circeo, LT). *Quaderni del Centro di Studio per l'Archeologia Etrusco-Italica* XVI:372–88.

Bietti, A., F. Martini, and C. Tozzi (1983). L'epigravettien évolué et final de la zone moyenne et basse tyrrhénienne. *Rivista di Scienze Preistoriche* 38:319–49.

Bietti, A., P. Rossetti, and G.-L. Zanzi (1989). Cultural adaptations and environment: Two test-cases from southern central Tyrrhenian Italy. *Rivista di Antropologia* 67:239–64.

Binford, L. (1968). Post-Pleistocene adaptations. In *New Perspectives in Archaeology*, edited by S. Binford and L. Binford, pp. 313–42. Chicago: Aldine.

Binford, L. (1973). Interassemblage variability: The Mousterian and the "functional argument." In *The Explanation of Culture Change: Models in Prehistory*, edited by C. Renfrew, pp. 227–54. London: Duckworth.

Binford, L. (1977). Forty-seven trips: A case study in the character of archaeological formation processes. In *Stone Tools as Cultural Markers*, edited by R. V. S. Wright, pp. 24–36. Canberra: Australian Institute of Aboriginal Studies.

Binford, L. (1978a). *Nunamiut Ethnoarchaeology*. New York: Academic Press.

Binford, L. (1978b). Dimensional analysis of behavior and site structure: Learning from an Eskimo hunting stand. *American Antiquity* 43:330–61.

Binford, L. (1979). Organization and formation processes: Looking at curated technologies. *Journal of Anthropological Research* 35:255–73.

Binford, L. (1980). Willow smoke and dogs' tails: Hunter-gatherer settlement systems and archaeological site formation. *American Antiquity* 45:4–20.

Binford, L. (1981). *Bones: Ancient Men and Modern Myths*. New York: Academic Press.

Binford, L. (1984). *Faunal Remains from Klasies River Mouth*. New York: Academic Press.

Binford, L. (1986). In pursuit of the future. In *American Archaeology Past and Future*, edited by D. Meltzer et al., pp. 459–79. Washington, DC: Smithsonian Institution Press.

Binford, L. (1988). Étude taphonomique des restes fauniques de la Grotte Vaufrey, couche VII. In *La Grotte Vaufrey à Cenac et Saint-Julien (Dordogne): Paléoenvironnements, chronologie et activités humaines*, edited by J.-P. Rigaud, pp. 535–63. Mémoires de la Société Préhistorique Française, Tome 19.

Binford, L. (1990). Isolating the transition to cultural adaptations: an organizational approach. In *The Emergence of Modern Humans: Biocultural Adaptations in the Later Pleistocene*, edited by E. Trinkaus, pp. 18–41. Cambridge: Cambridge University Press.

Binford, L. (1991). When the going gets tough, the tough get going: Nunamiut local groups, camping patterns and economic organization. In *Ethnoarchaeological Approaches to Mobile Campsites: Hunter-Gatherer and Pastoralist Case Studies*, edited by C. Gamble and W. A. Boismer, pp. 25–137. Ann Arbor, MI: International Monographs in Prehistory.

Binford L., and S. Binford (1966). A preliminary analysis of functional variability in the Mousterian of Levallois facies. *American Anthropologist* 68:238–95.

Binford, L., and J. O'Connell (1984). An Alyawara day: The stone quarry. *Journal of Anthropological Research* 40:406–32.

Binford, L., and N. Stone (1985). "Righteous rocks" and Richard Gould: Some observations on a misguided "debate." *American Antiquity* 50:151–53.

Blanc, A. (1935a). Saccopastore II. *Rivista di Antropologia* 30:3–5.

Blanc, A. (1935b). Lo studio stratigrafico di pianure costiere. *Bollettino della Società Geologica Italiana* 54:277–88.

Blanc, A. (1935c). Sulla fauna quaternaria dell'Agro Pontino. *Atti della Società Toscana di Scienze Naturali* 44:108–10.

Blanc, A. (1935d). Sulle formazioni quaternarie di Nettuno e loro correlazione con la stratigrafia dell'Agro Pontino. *Bolletino della Società Geologica Italiana* 54:109–21.

Blanc, A. (1936). Sulla stratigrafia quaternaria dell'Agro Pontino e della bassa Versilia. *Bollettino della Società Geologica Italiana* 60:375–96.

Blanc, A. (1937a). Fauna a Ippopotamo ed industrie paleolitiche nel riempimento delle grotte litoranee del Monte Circeo, I: La Grotta delle Capre; II: La Grotta del Fossellone. *Rendiconti della Reale Accademia Nazionale dei Lincei* 25:88–93.

Blanc, A. (1937b). Nuovi giacimenti paleolitici del Lazio e della Toscana. *Studi Etruschi* 11:273–304.

Blanc, A. (1938). Una serie di nuovi giacimenti pleistocenici e paleolitici in grotte litoranee del Monte Circeo. *Rendiconti della R. Accademia Nazionale dei Lincei* 17:201–9.

Blanc, A. (1939a). L'uomo fossile del Monte Circeo, un cranio neandertaliano nella Grotta Guattari a San Felice Circeo. *Rivista di Antropologia* 23:1–18.

Blanc, A. (1939b). L'homme fossile au Monte Circé. *L'Anthropologie* 49:254–64.

Blanc, A. (1948). Notizie sui trovamenti e sul giacimento di Saccopastore e sulla sua posizione nel pleistocene laziale. *Paleontographia Italica* 42:3–23.

Blanc, A. (1954). Reperti fossili neandertaliani della Grotta del Fossellone al Monte Circeo: Circeo IV. *Quaternaria* 1:171–75.

Blanc, A. (1955a). Ricerche sul quaternario laziale. *Quaternaria* 2:187–200.

Blanc, A. (1955b). Scavi e ricerche al Monte Circeo e nella regione di Roma. *Quaternaria* 2:287–91.

Blanc, A. (1957). On the Pleistocene sequence of Rome: Paleoecologic and archeologic correlations. *Quaternaria* 4:1108–9.

Blanc, A. (1958–61). L'industria musteriana su calcare e su valva di *Meretrix chione* associata con fossili di elefante e rinosceronte, in nuovi giacimenti costieri del Capo di Leuca. *Quaternaria* 5:308.

Blanc, A., and A. Segre (1947). Nuovi giacimenti tirreniani e paleolitici sulla costiera tra Sperlonga e Gaeta. *Historia Naturalis* 2:1–2.

Blanc, A., and A. Segre (1953). *Excursion au Mont Circe: Livret-guide du IVe Congrès International INQUA, Roma–Pisa.* Rome.

Blanc, A., H. de Vries, and M. Follieri (1957). A first C14 date for the Würm I chronology on the Italian Coast. *Quaternaria* 5:83–93.

Bleed, P. (1986). The optimal design of hunting weapons. *American Antiquity* 51:737–47.

Bloom, A., W. Broecker, J. Chapell, R. Matthews, and K. Mesoella (1974). Quaternary sea level fluctuations on a tectonic coast: New ^{230}U/TH^{234} dates for the Huon peninsula, New Guinea. *Quaternary Research* 4:185–205.

Blumenschine, R. (1986). *Early Hominid Scavenging Opportunities: Implications of Carcass Availability in the Serengeti and Ngorongoro Ecosystems.* BAR International Series, 283. Oxford: British Archaeological Reports.

Boëda, É. (1986). *Approche technologique du concept Levallois et évolution de son champ d'application.* Thèse de Doctorat de l'Université de Paris X.

Boëda, É. (1991). Approche de la variabilité des systèmes de production lithique des industries du paléolithique inférieur et moyen: Chronique d'une variabilité attendue. *Technique et Culture* 17–18:37–79.

Boëda, É. (1993). Le débitage discoïde et le débitage Levallois récurrent centripète. *Bulletin de la Société Préhistorique Française* 90:392–404.

Boëda, É., J.-M. Geneste, and L. Meignen (1990). Identification de chaînes opératoires lithiques du paléolithique ancien et moyen. *Paléo* 2:43–79.

Boesch, C., and H. Boesch (1983). Optimization of nut-cracking with natural hammers by wild chimpanzees. *Behavior* 83: 265–286.

Bonner, J. (1980). *The Evolution of Culture in Animals.* Princeton: Princeton University Press.

Bordes, F. (1953). Essai de classification des industries "moustériennes." *Bulletin de la Société Préhistorique Française* 50:457–66.

Bordes, F. (1961a). *Typologie du paléolithique ancien et moyen.* Cahiers du Quaternaire No. 1, Institut du Quaternaire, Université de Bordeaux. Paris: CNRS.

Bordes, F. (1961b). Mousterian cultures in France. *Science* 134(3482):803–10.

Bordes, F. (1968). *The Old Stone Age.* New York: McGraw Hill.

Bordes, F. (1972). *A Tale of Two Caves.* New York: Harper and Row.

Bordes, F. (1984). *Leçons sur le paléolithique.* Cahiers du Quaternaire No. 7, Institut du Quaternaire, Université de Bordeaux. Paris: CNRS.

Bordes, F., and D. de Sonneville-Bordes (1970). The significance of variability in Paleolithic assemblages. *World Archaeology* 2:61–73.

Borzatti von Lowenstern, E. (1965). La grotta-riparo di Uluzzo C (campagna di scavi 1964). *Rivista di Scienze Preistoriche* 20(1):1–31.

Borzatti von Lowenstern, E. (1966). Alcuni aspetti del musteriano nel Salento. *Rivista di Scienze Preistoriche* 21(2):203–87.

Breuil, H., and A. Blanc (1936). Le nouveau crâne néandertalien de Saccopastore, Rome. *L'Anthropologie* 46:1–16.

Brink, J., M. Wright, B. Dawe, and D. Glaum (1986). *Final Report of the 1984 Season at Head-Smashed-In Buffalo Jump, Alberta.* Manuscript Series, No. 9. Edmonton: Archaeological Survey of Alberta.

Bronstein, N. (1977). Report on a replicative experiment in the manufacture and use of Western Desert microadzes. In *Puntutjarpa Rockshelter and the Australian Desert Culture*, by R. Gould, pp. 154–57. Anthropological Papers, Vol. 54. New York: American Museum of Natural History.

Burton, R. (1979). *The Carnivores of Europe.* London: B. T. Batsford.

Butzer, K. (1971). *Environment and Archaeology.* Chicago, New York: Aldine, Atherton.

Callahan, E. (1979). The basics of biface knapping in the eastern fluted point tradition: A manual for flintknappers and lithic analysts. *Archaeology of Eastern North America* 7:1–179.

Caloi, L., and M. Palombo (1988). Large Paleolithic mammals of Latium (central Italy): Palaeoecological and biostratigraphic implications. In *L'homme de Néandertal*, Vol. 2: *L'environnement*, edited by M. Otte, pp. 21–44. ERAUL 29. Liège: Université de Liège.

Caloi, L., and M. Palombo (1990–91). Les grands mammifères du pléistocène supérieur de Grotta Barbara (Monte Circeo, Latium méridional): Encadrement biostratigraphique et implications paléoécologiques. *Quaternaria Nova* 1:267–76.

Cann, R. (1988). DNA and human origins. *Annual Review of Anthropology* 17:127–143.

Cann, R., M. Stoneking, and A. Wilson (1987). Mitochondrial DNA and human evolution. *Nature* 325: 31–36.

Caspar, J. P. (1984). Matériaux lithiques de la préhistoire. In *Peuples chasseurs de la Belgique préhistorique dans leur cadre naturel*, edited by D. Cahen and P. Haesaerts, pp. 107–16. Brussels.

Chase, P. G. (1986a). *The Hunters of Combe Grenal: Approaches to Middle Paleolithic Subsistence in Europe*. BAR International Series, 286. Oxford: British Archaeological Reports.

Chase, P. G. (1986b). Relationships between Mousterian lithic and faunal assemblages at Combe Grenal. *Current Anthropology* 27:69–71.

Chase, P. G. (1989). How different was Middle Paleolithic subsistence? A zooarchaeological perspective on the Middle to Upper Palaeolithic transition. In *The Human Revolution: Behavioural and Biological Perspectives on the Origins of Modern Humans*, edited by P. Mellars and C. Stringer, pp. 321–37. Edinburgh: Edinburgh University Press.

Chase, P. G., and H. Dibble (1987). Middle Paleolithic symbolism: A review of current evidence and interpretations. *Journal of Anthropological Archaeology* 6:263–93.

Churchill, S. (1993). Weapon technology, prey size selection, and hunting methods in modern hunter-gatherers: Implications for hunting in the Paleolithic and Mesolithic. In *Hunting and Animal Exploitation in the Later Palaeolithic and Mesolithic of Eurasia*, edited by G. Larsen Peterkin, H. Bricker, and P. Mellars, pp. 11–24. Archaeological Papers of the American Anthropological Association, Vol. 4.

Churchill, S., and E. Trinkaus (1990). Neandertal scapular glenoid morphology. *American Journal of Physical Anthropology* 83:147–60.

Clark, G., and J. Lindly (1989a). Modern human origins in the Levant and Western Asia. *American Anthropologist* 91:962–85.

Clark, G., and J. Lindly (1989b). The case for continuity: Observations on the biocultural transition in Europe and Western Asia. In *The Human Revolution: Behavioural and Biological Perspectives on the Origins of Modern Humans*, edited by P. Mellars and C. Stringer, pp. 626–76. Edinburgh: Edinburgh University Press.

Clark, G., and L. Straus (1986). Synthesis and conclusions—Part 1: Upper Paleolithic and Mesolithic hunter-gatherer subsistence in northern Spain. In *La Rierra Cave: Stone Age Hunter-Gatherer Adaptations in Northern Spain*, edited by L. Straus and G. Clark, pp. 183–208. Anthropological Research Papers, No. 36. Tempe: Arizona State University Press.

Clutton-Brock, T., F. Guinness, and S. Albon (1982). *Red Deer: The Ecology of Two Sexes*. Chicago: University of Chicago Press.

Coles, J., and E. Higgs (1969). *The Archaeology of Early Man*. Harmondsworth: Penguin Books.

Collina-Girard, J., and A. Turq (1991). Le paléolithique moyen sur galets de la station de Planes, commune de Montayral (Lot-et-Garonne). *Paléo* 3:49–75.

Conato, V., and G. Dai Pra (1980). Livelli marini pleistocenici e neotettonica fra Civitavecchia e Tarquinia (Italia Centrale). Laboratoria di Geologia Ambientale, Internal Report No. MNTLT/5/80. Rome: CNEN.

Conard, N. (1990). Laminar lithic assemblages from the last Interglacial complex in northwestern Europe. *Journal of Anthropological Research* 46:243–62.

Cooper, H. (1954). Progressive modification of a stone artifact. *Records of the South Australia Museum* 11:91–103.

Copeland, L. (1975). The Middle and Upper Paleolithic of Lebanon and Syria in the light of recent research. In *Problems in Prehistory: North Africa and the Levant*, edited by F. Wendorf and A. Marks, pp. 317–50. Dallas: Department of Anthropology, Southern Methodist University.

Cremaschi, M. (1992). Ambiente e clima. In *Italia preistorica*, edited by A. Guidi and M. Piperno, pp. 3–39. Rome–Bari: Laterza.

Crew, H. (1975). *An Examination of the Variability of the Levallois Method: Its Implications for the Internal and External Relationships of the Levantine Mousterian*. Ph.D. dissertation, Dept. of Anthropology, Southern Methodist University. Ann Arbor, MI: UMI.

Dai Pra, G., A. Demoulin, and A. Ozer (1985). Mise en évidence de mouvements tectoniques récents par l'étude des encoches de corrosion et des épots tyrrhéniens dans le Latium méridional (Italie). 1st International Conference on Geomorphology, Manchester.

Delagnes, A. (1991). Mise en évidence de deux conceptions différentes de la production lithique au paléolithique moyen. In *25 ans d'études technologiques en préhistoire: Bilan et perspectives*, pp. 126–37. Actes des XIe Rencontres Internationales d'Archéologie et d'Histoire d'Antibes. Juan-les-Pins: Éditions APDCA.

Demars, P.-Y. (1982). *L'utilisation du silex au paléolithique supérieur: Choix, approvisionnement, circulation. L'exemple du Bassin de Brive*. Cahiers du

Quaternaire No. 5, Institut du Quaternaire, Université de Bordeaux. Paris: CNRS.

Dennel, R. (1992). Comment on Roebroeks et al. 1992. *Current Anthropology* 33:568.

Dibble, H. (1981). *Technological Strategies of Stone Tool Production at Tabun Cave, Israel*. Ph.D. dissertation, Dept. of Anthropology, University of Arizona. Ann Arbor, MI: UMI.

Dibble, H. (1987a). The interpretation of Middle Paleolithic scraper morphology. *American Antiquity* 52: 109–17.

Dibble, H. (1987b). Reduction sequences in the manufacture of Mousterian implements in France. In *The Pleistocene Old World: Regional Perspectives*, edited by O. Soffer, pp. 33–46. New York: Plenum Press.

Dibble, H. (1988). Typological aspects of reduction and intensity of utilization of lithic resources in the French Mousterian. In *The Upper Pleistocene Prehistory of Western Eurasia*, edited by H. Dibble and A. Montet-White, pp. 181–98. University Museum Monograph 54. Philadelphia: The University Museum, University of Pennsylvania.

Dibble, H. (1991). Local raw material exploitation and its effects on Lower and Middle Paleolithic assemblage variability. In *Raw Material Economies among Prehistoric Hunter-Gatherers*, edited by A. Montet-White and S. Holen, pp. 49–58. University of Kansas Publications in Anthropology, 19. Lawrence: University of Kansas.

Dibble, H., and P. Mellars, eds. (1992). *The Middle Paleolithic: Adaptation, Behavior, and Variability*. University Museum Monograph 72. Philadelphia: The University Museum, University of Pennsylvania.

Durante, S. (1974–75). Il terreniano e la malacofauna della Grotta del Fossellone (Circeo). *Quaternaria* 18:331–45.

Durante, S., and F. Settepassi (1974). Livelli marini e molluschi tirreniani alla Grotta delle Capre (Circeo). *Memorie dell'Istituto Italiano di Paleontologia Umana* 2:285–96.

Durante, S., and F. Settepassi (1976–77). Malacofauna e livelli marini tirreniani a Grotta Guattari, Monte Circeo (Latina). *Quaternaria* 19:35–69.

Dyson-Hudson, R., and E. Smith (1978). Human territoriality: An ecological reassessment. *American Anthropologist* 80:21–41.

Ebert, J. (1979). An ethnographical approach to reassessing the meaning of variability in stone tool assemblages. In *Ethnoarchaeology*, edited by C. Kramer, pp. 59–74. New York: Columbia University Press.

Ellis, C. (1989). The explanation of northeastern Paleoindian lithic procurement patterns. In *Eastern Paleoindian Lithic Resource Use*, edited by C. Ellis and J. Lothrop, pp. 139–64. Boulder: Westview Press.

Ewer, R. F. (1973). *The Carnivores*. London: Weidenfeld and Nicholson.

Farizy, C., ed. (1990). *Paléolithique moyen récent et paléolithique supérieur ancien en Europe*. Memoires du Musée de Préhistoire d'Ile de France, No. 3. Nemours: APRAIF.

Farizy, C., and F. David (1992). Subsistence and behavioral patterns of some Middle Paleolithic local groups. In *The Middle Paleolithic: Adaptation, Behavior, and Variability*, edited by H. Dibble and P. Mellars, pp. 87–96. University Museum Monograph 72. Philadelphia: The University Museum, University of Pennsylvania.

Federici, P. (1980). On the Riss glaciation of the Apennine. *Zeitschrift für Geomorphologie* 24:111–16.

Ferring, C. (1975). The Aterian in North African prehistory. In *Problems in Prehistory: North Africa and the Levant*, edited by F. Wendorf and A. Marks, pp. 113–26. Dallas: Department of Anthropology, Southern Methodist University.

Foley, R. (1987). *Another Unique Species: Patterns in Human Evolutionary Ecology*. New York: Longman Scientific and Technical.

Follieri, M., D. Magri, and L. Sadori (1988). 250,000-year pollen record from Valle di Castiglione (Roma). *Pollen et Spores* 30:329–56.

Fowler, C. (1992). *In the Shadow of Fox Peak: An Ethnography of the Cattail-Eater Northern Paiute People of Stillwater Marsh*. Cultural Resources Series, No. 5, Dept. of the Interior, Fish and Wildlife Service. Washington: US Government Printing Office.

Frank, A. (1969). Pollen stratigraphy of the Lake Vico (Central Italy). *Palaeogeography, Palaeoclimatology, Palaeoecology* 6:67–85.

Frayer, D., M. Wolpoff, A. Thorne, F. Smith, and G. Pope (1993). Theories of modern human origins: The paleontological test. *American Anthropologist* 95: 14–50.

Freeman, L. (1989). A Mousterian structural remnant from Cueva Morin. In *L'homme de Néandertal*, Vol. 6: *La subsistance*, edited by M. Otte, pp. 19–30. ERAUL 33. Liège: Université de Liège.

Frison, G. (1968). A functional analysis of certain chipped stone tools. *American Antiquity* 33:149–55.

Frison, G. (1991). The Clovis cultural complex: New data from caches of flaked stone and worked bone. In *Raw Material Economies among Prehistoric Hunter-Gatherers*, edited by A. Montet-White and S. Holen, pp. 321–34. University of Kansas Publications in Anthropology, 19. Lawrence: University of Kansas.

Gamble, C. (1978). Resource exploitation and the spatial patterning of hunter-gatherers: A case study. In *Social Organization and Settlement: Contributions for Anthropology, Archaeology and Geography*, Vol. 1, edited by D. Green, C. Haselgrove, and M. Spriggs, pp. 153–85. BAR International Series, S47. Oxford: British Archaeological Reports.

Gamble, C. (1982). Interaction and alliance in Paleolithic society. *Man* 17:92–107.

Gamble, C. (1983). Culture and society in the Upper Paleolithic of Europe. In *Hunter-Gatherer Economy in Prehistory: A European Perspective*, edited by G. Bailey, pp. 201–11. Cambridge: Cambridge University Press.

Gamble, C. (1984). Regional variation in hunter-gatherer strategy in the Upper Pleistocene of Europe. In *Hominid Evolution and Community Ecology*, edited by R. Foley, pp. 237–60. London: Academic Press.

Gamble, C. (1986). *The Palaeolithic Settlement of Europe*. Cambridge: Cambridge University Press.

Gamble, C. (1987). Man the shoveler: Alternative models for Middle Pleistocene colonization and occupation in northern latitudes. In *The Pleistocene Old World: Regional Perspectives*, edited by O. Soffer, pp. 81–98. New York: Plenum Press.

Gamble, C., and O. Soffer (1990). Pleistocene polyphony: The diversity of human adaptations at the Last Glacial Maximum. In *The World at 18,000 BP*, Vol. 2: *Low Latitudes*, edited by C. Gamble and O. Soffer, pp. 1–23. London: Unwin Hyman.

Geneste, J.-M. (1985). *Analyse lithique d'industries moustériennes du Périgord: Une approche technologique du comportement des groupes humains au paléolithique moyen*. Thèse de Doctorat de l'Université de Bordeaux I.

Geneste, J.-M. (1986). Systèmes d'approvisionnement en matières premières au paléolithique moyen et au paléolithique supérieur en Aquitaine. In *L'homme de Néandertal*, Vol. 8: *La mutation*, edited by M. Otte, pp. 61–70. ERAUL 35. Liège: Université de Liège.

Geneste, J.-M. (1988). Les industries de la Grotte Vaufrey: Technologie du débitage, économie et circulation de la matière première. In *La Grotte Vaufrey à Cenac et Saint-Julien (Dordogne): Paléoenvironnements, chronologie et activités humaines*, edited by J.-P. Rigaud, pp. 441–519. Mémoires de la Société Préhistorique Française, Tome 19.

Geneste, J.-M. (1989). Économie des ressources lithiques dans le moustérien du sud-ouest de la France. In *L'homme de Néandertal*, Vol. 6: *La subsistance*, edited by M. Otte, pp. 75–98. ERAUL 33. Liège: Université de Liège.

Geneste, J.-M., and J.-P. Rigaud (1989). Matières premières lithiques et occupation de l'espace. In *Variations des paléomilieux et peuplement préhistorique*, edited by H. Laville, pp. 205–18. Cahiers du Quaternaire No. 13, Institut du Quaternaire, Université de Bordeaux. Paris: CNRS.

Giacobini, G. (1990–91). Hyenas or cannibals: Fifty years of debate on Guattari Cave Neandertal cranium. *Quaternaria Nova* 1:593–604.

Gioia, P. (1988). Problems related to the origins of Italian Upper Paleolithic: Uluzzian and Aurignacian. In *L'homme de Néandertal*, Vol. 8: *La mutation*, edited by M. Otte, pp. 91–102. ERAUL 35. Liège: Université de Liège.

Gioia, P. (1990). An aspect of the transition between Middle and Upper Paleolithic in Italy: The Uluzzian. In *Paléolithique moyen recént et paléolithique supérieur ancien en Europe*, edited by C. Farizy, pp. 241–50. Mémoires du Musée de Préhistoire d'Ile de France, No. 3. Nemours: APRAIF.

Goodall, J. (1964). Tool using and aimed throwing in a community of wild chimpanzees. *Nature* 201:1264–66.

Goodyear, A. (1982). Toolkit entropy and bipolar reduction. Unpublished manuscript. Columbia, SC: Institute of Archaeology and Anthropology, University of South Carolina.

Goodyear, A. (1989). A hypothesis for the use of cryptocrystalline raw materials among Paleoindian groups of North America. In *Eastern Paleoindian Lithic Resource Use*, edited by C. Ellis and J. Lothrop, pp. 1–10. Boulder: Westview Press.

Gould, R. (1969). *Yiwara: Foragers of the Australian Desert*. New York: Charles Scribner's Sons.

Gould, R. (1977). *Puntutjarpa Rockshelter and the Australian Desert Culture*. Anthropological Papers, Vol. 54. New York: American Museum of Natural History.

Gould, R. (1980). *Living Archaeology*. Cambridge: Cambridge University Press.

Gould, R. (1985). The empiricist strikes back: Reply to Binford. *American Antiquity* 50:638–44.

Gould, R., and Saggers, S. (1985). Lithic procurement in central Australia: A closer look at Binford's idea of embeddedness in archaeology. *American Antiquity* 50:117–36.

Gould, S., and R. Lewontin (1979). The spandrels of San Marcos and the Panglossian paradigm: A critique of the adaptationist paradigm. *Proceedings of the Royal Society of London* 205:581–98.

Gramly, R. (1980). Raw material source areas and "curated" tool assemblages. *American Antiquity* 45:823–33.

Gramly, R. (1982). The Vail site: A Paleoindian encampment in Maine. *Buffalo Museum of Sciences Bulletin*, 30.

Graves, P. (1991). New models and metaphors for the Neandertal debate. *Current Anthropology* 32:513–542.

Grey, G. (1841). *Journals of Two Expeditions of Discovery in Northwest and Western Australia during the Years 1837, 38 and 39*. London: T. and W. Boon.

Grifoni, R. (1960). *Il micromusteriano della Grotta del Fossellone al Monte Circeo—Analisi dell'industria*. Tesi di Laurea, University of Rome.

Grün, R. (1988). The potential of ESR dating of tooth enamel. In *L'homme de Néandertal*, Vol. 1: *La chronologie*, edited by M. Otte, pp. 37–46. ERAUL 28. Liège: Université de Liège.

Grün, R., H. P. Schwarcz, and S. Zymella (1987). ESR dating of tooth enamel. *Canadian Journal of Earth Sciences* 24:1022–1037.

Guerreschi, A. (1992). La fine del pliestocene e gli inizi dell'olocene. In *Italia preistorica*, edited by A. Guidi and M. Piperno, pp. 198–237. Rome–Bari: Laterza.

Hahn, J. (1977). *Aurignacien: Das ältere Jungpaläolithikum in Mittel- und Osteuropa*. Fundamenta A9. Cologne: Herman Bohlau.

Hannah, A., and W. McGrew (1987). Chimpanzees using stone to crack open oil palm nuts in Liberia. *Primates* 28:31–46.

Harrold, F. (1983). The Chatelperronian and the Middle–Upper Paleolithic transition. In *The Mousterian Legacy*, edited by E. Trinkaus, pp. 123–40. BAR International Series, 164. Oxford: British Archaeological Reports.

Harrold, F. (1989). Mousterian, Chatelperronian and early Aurignacian in western Europe: Continuity or discontinuity? In *The Human Revolution: Behavioural and Biological Perspectives on the Origins of Modern Humans*, edited by P. Mellars and C. Stringer, pp. 677–713. Edinburgh: Edinburgh University Press.

Hayden, B. (1979). *Paleolithic Reflections: Lithic Technology and Ethnographic Excavations among Australian Aborigines*. Canberra: Australian Institute of Aboriginal Studies.

Hearty, P., and G. Dai Pra (1986). Aminostratigraphy of Quaternary marine deposits in the Lazio region of central Italy. *Zeitschrift für Geomorphologie*, Suppl. Bd. 62:131–40.

Henry, D. (1983). Adaptive evolution within the Epipaleolithic of the Near East. In *Advances in World Archaeology*, Vol. II, edited by F. Wendorf and A. Close, pp. 99–160. New York: Academic Press.

Henry, D. (1989). Correlations between reduction strategies and settlement patterns. In *Alternative Approaches to Lithic Analysis*, edited by D. Henry and G. Odell, pp. 139–58. Archaeological Papers of the American Anthropological Association, 1.

Henry, D. (1992). Transhumance during the late Levantine Mousterian. In *The Middle Paleolithic: Adaptation, Behavior, Variability*, edited by H. Dibble and P. Mellars, pp. 143–62. University Museum Monograph 72. Philadelphia: The University Museum, University of Pennsylvania.

Hitchcock, R. (1982). *The Ethnoarchaeology of Sedentism: Mobility Strategies and Site Structure among Foraging and Food Producing Populations in the Eastern Kalahari Desert, Botswana*. Ph.D. dissertation, Department of Anthropology, University of New Mexico. Ann Arbor, MI: UMI.

Hitchcock, R. (1987). Sedentism and site structure: Organizational change in Kalahari Baswara residential locations. In *Method and Theory for Activity Area Research*, edited by S. Kent, pp. 374–423. New York: Columbia University Press.

Hoffman, C. (1986). *The Punan: Hunter-Gatherers of Borneo*. Studies in Cultural Anthropology, 12. Ann Arbor, MI: UMI Research Press.

Hoffman, J. (1991). Folsom land use: Projectile point variation as a key to mobility. In *Raw Material Economies among Prehistoric Hunter-Gatherers*, edited by A. Montet-White and S. Holen, pp. 335–56. University of Kansas Publications in Anthropology, 19. Lawrence: University of Kansas.

Holdaway, S. (1989). Were there hafted projectile points in the Mousterian? *Journal of Field Archaeology* 16:79–85.

Holen, S. (1991). Bison hunting territories and lithic acquisition among the Pawnee: An ethnohistoric and archaeological study. In *Raw Material Economies among Prehistoric Hunter-Gatherers*, edited by A. Montet-White and S. Holen, pp. 399–411. University of Kansas Publications in Anthropology, 19. Lawrence: University of Kansas.

Hopkins, D. (1982). Aspects of the paleogeography of Beringia during the late Pleistocene. In *The Paleoecology of Beringia*, edited by D. Hopkins, J. Matthews, C. Schweger, and S. Young, pp. 3–21. New York: Academic Press.

Hughes, R. (1985). Obsidian Sourcing. In *Nightfire Island: Later Holocene Lakemarsh Adaptation on the Western Edge of the Great Basin*, by G. Sampson.

University of Oregon Anthropological Papers, 33. Eugene: University of Oregon.

Hunt, C., and W. Eisner (1991). Palynology of the Mezzaluna Core. In *The Agro Pontino Survey Project: Methods and Preliminary Results*, edited by A. Voorrips, S. Loving, and H. Kamermans, pp. 49–59. Studies in Prae- en Protohistorie 6, Instituut voor Prae- en Protohistorische Archaeologie Albert Egges van Giffen. Amsterdam: Universiteit van Amsterdam.

Ikeya, M. (1978). Electron spin resonance as a method of dating. *Archaeometry* 20:147–58.

Isaac, G. (1977). *Olorgesailie. Archaeological Studies of a Middle Pleistocene Lake Basin in Kenya*. Chicago: University of Chicago Press.

Jacob-Friesen, K. (1954). Eiszeitliche Elephantenjäger in der Lüneburger Heide. *Jahrbüch des Römisch-Germanischen Zentralmuseums Mainz* 3:1–22.

Jelinek, A. (1981). The Middle Paleolithic of the southern Levant from the perspective of Tabun Cave. In *Préhistoire du Levant*, edited by J. Cauvin and P. Sanlaville, pp. 265–80. Paris: CNRS.

Jelinek, A., A. Debénath, and H. Dibble (1989). A preliminary report on evidence related to the interpretation of economic and social activities of Neandertals at the site of La Quina (Charente), France. In *L'homme de Néandertal*, Vol. 6: *La subsistance*, edited by M. Otte, pp. 99–106. ERAUL 33. Liège: Université de Liège.

Johnson, J. (1986). Amorphous core technologies in the midsouth. *Midcontinental Journal of Archaeology* 2(2):136–51.

Kamermans, H. (1991). Faulted land: The geology of the Agro Pontino. In *The Agro Pontino Survey Project: Methods and Preliminary Results*, edited by A. Voorrips, S. Loving, and H. Kamermans, pp. 21–30. Studies in Prae- en Protohistorie 6, Instituut voor Prae- en Protohistorische Archaeologie Albert Egges van Giffen. Amsterdam: Universiteit van Amsterdam.

Kelly, R. (1983). Hunter gatherer mobility strategies. *Journal of Anthropological Research* 39:277–306.

Kelly, R. (1988). The three sides of a biface. *American Antiquity* 53:717–34.

Kelly, R. (1992). Mobility/sedentism: Concepts, archaeological measures, and effects. *Annual Reviews in Anthropology* 21:43–66.

Kelly, R., and L. Todd (1988). Coming into the country: Early Paleoindian hunting and mobility. *American Antiquity* 53:231–44.

Kent, S. (1991). The relationship between mobility strategies and site structure. In *The Interpretation of Archaeological Spatial Patterning*, edited by E. Kroll and T. Price, pp. 33–59. New York: Plenum Press.

Klein, R. (1977). The ecology of early man in southern Africa. *Science* 197: 115–127.

Klein, R. (1979). Stone age exploitation of animals in southern Africa. *American Scientist* 67:151–60.

Klein, R. (1989a). Biological and behavioral perspectives on modern human origins in southern Africa. In *The Human Revolution: Behavioural and Biological Perspectives on the Origins of Modern Humans*, edited by P. Mellars and C. Stringer, pp. 529–46. Edinburgh: Edinburgh University Press.

Klein, R. (1989b). *The Human Career: Human Biological and Cultural Origins*. Chicago: University of Chicago Press.

Knecht, H. (1993). Early Upper Paleolithic approaches to bone and antler projectile technology. In *Hunting and Animal Exploitation in the Later Palaeolithic and Mesolithic of Eurasia*, edited by G. Larsen Peterkin, H. Bricker, and P. Mellars, pp. 33–48. Archaeological Papers of the American Anthropological Association, Vol. 4.

Koldehoff, B. (1987). The Cahokia flake tool industry: Socio-economic implications for late prehistory in the central Mississippi valley. In *The Organization of Core Technology*, edited by J. Johnson and C. Morrow, pp. 151–86. Boulder, CO: Westview Press.

Koot, C. (1991). Marching through marshes: An historical survey of the Agro Pontino. In *The Agro Pontino Survey Project: Methods and Preliminary Results*, edited by A. Voorrips, S. Loving, and H. Kamermans, pp. 9–20. Studies in Prae- en Protohistorie 6, Instituut voor Prae- en Protohistorische Archaeologie Albert Egges van Giffen. Amsterdam: Universiteit van Amsterdam.

Kotsakis, T. (1990–91). Late Pleistocene fossil microvertebrates of Grotta Breuil (Monte Circeo, central Italy). *Quaternaria Nova* 1:325–32.

Kruuk, H. (1972). *The Spotted Hyaena*. Chicago: University of Chicago Press.

Kuhn, S. (1989). Hunter-gatherer foraging organization and strategies of artifact replacement and discard. In *Experiments in Lithic Technology*, edited by D. Amick and R. Mauldin, pp. 33–48. BAR International Series, 528. Oxford: British Archaeological Reports.

Kuhn, S. (1990a). *Diversity within Uniformity: Tool Manufacture and Use in the Pontinian Mousterian of Latium (Italy)*. Ph.D. dissertation, Dept. of Anthropology, University of New Mexico. Ann Arbor, MI: UMI.

Kuhn, S. (1990b). A geometric index of reduction for unifacial stone tools. *Journal of Archaeological Science* 17:583–93.

Kuhn, S. (1991a). New problems, old glasses: Methodological implications of an evolutionary paradigm for the study of Paleolithic technologies. In *Perspectives on the Past: Theoretical Biases in Mediterranean Hunter-Gatherer Research*, edited by G. A. Clark, pp. 243–57. Philadelphia: University of Pennsylvania Press.

Kuhn, S. (1991b). Unpacking reduction: Lithic raw-material economy in the Mousterian of West-Central Italy. *Journal of Anthropological Archaeology* 10: 76–106.

Kuhn, S. (1992a). Blank form and reduction as determinants of Mousterian scraper morphology. *American Antiquity* 57:115–28.

Kuhn, S. (1992b). On planning and curated technologies in the Middle Paleolithic. *Journal of Anthropological Research* 48:185–214.

Kuhn, S. (1993). Mousterian technology as adaptive response: A case study. In *Hunting and Animal Exploitation in the Later Palaeolithic and Mesolithic of Eurasia*, edited by G. Larsen Peterkin, H. Bricker, and P. Mellars, pp. 25–32. Archaeological Papers of the American Anthropological Association, Vol. 4.

Kuhn, S. (1994). A formal approach to the design and assembly of transported toolkits. *American Antiquity* 59:426–42.

Kuhn, S., and M. Stiner (1992). New research on Riparo Mochi, Balzi Rossi (Liguria): Preliminary results. *Quaternaria Nova* 2:77–90.

Kurten, B. (1976). *The Cave Bear Story: Life and Death of a Vanished Animal.* New York: Columbia University Press.

Laj-Pannocchia, F. (1950). L'industria pontiniana della Grotta di S. Agostino (Gaeta). *Rivista di Scienze Preistoriche* 5:67–86.

Laplace, G. (1964). Les subdivisions du leptolithique italien: Étude de typologie analytique. *Bullettino di Paletnologia Italiano* 73:25–63.

Laplace, G. (1966). *Recherches sur l'origine et l'evolution des complexes leptolithiques.* Mélanges d'Archéologie et d'Histoire, 4. Paris: École Française de Rome.

Laplace, G. (1977). Il Riparo Mochi ai Balzi Rossi di Grimaldi (fouilles 1938–1949): Les industries leptolithiques. *Rivista di Scienze Preistoriche* 32:3–131.

Larick, R. (1986). Perigord cherts: An analytical frame for investigating the movement of Paleolithic hunter-gatherers and their resources. In *The Scientific Study of Flint and Chert*, edited by G. de G. Sieveking and M. Hart, pp. 112–20. Cambridge: Cambridge University Press.

Larralde, S. (1990). *The Design of Hunting Weapons: Archaeological Evidence from Southwestern Wyoming.* Ph.D. dissertation, Department of Anthropology, University of New Mexico. Ann Arbor, MI: UMI.

Laville, H., J.-P. Rigaud, and J. Sackett (1980). *Rockshelters of the Perigord.* New York: Academic Press.

Leakey, M. (1971). *Olduvai Gorge: Excavations in Beds I and II, 1960–1963.* Cambridge: Cambridge University Press.

Lee, R. (1979). *The !Kung San: Men, Women and Work in a Foraging Society.* Cambridge: Cambridge University Press.

Le Gall, O. (1992). Poissons et pêches au paléolithique. *l'Anthropologie* 96:121–134.

Leonard, R., and G. Jones (1987). Elements of an inclusive evolutionary model for archaeology. *Journal of Anthropological Archaeology* 6:199–219.

Lewontin, R. (1961). Evolution and the theory of games. *Journal of Theoretical Biology* 1:382–403.

Lieberman, D. (1993). The rise and fall of seasonal mobility among hunter-gatherers. *Current Anthropology* 34:599–632.

Longo, E., F. Malegni, R. Mariani, A. Radmilli, and A. Segre (1981). Giacimento ad amigdale e resti umani de Castel di Guido a Roma. *Atti della XXIII Riunione Scientifica dell'Istituto Italiano di Preistoria e Protostoria.*

Longo, E., and A. Radmilli (1972). Nuovo giacimento con amigdale a Roma. *Rivista di Scienze Preistoriche* 27(2):403–9.

Lovejoy, O., and Trinkaus, E. (1980). Strength and robusticity in the Neandertal tibia. *American Journal of Physical Anthropology* 53:465–70.

Loving, S., H. Kamermans, C. W. Koot, and A. Voorrips (1990–91). The Pontinian on the plain: Some results from the Agro Pontino survey. *Quaternaria Nova* 1:453–77.

Lumley, H. de (1969). *Le paléolithique inférieur et moyen du Midi méditerranéen dans son cadre géologique.* 5th Supplement to *Gallia Préhistoire.* Paris: CNRS.

Lumley, H. de (1972). *La grotte moustérienne de l'Hortus.* Marseille: Études Quaternaires, 1.

Malatesta, A., ed. (1978). Torre in Pietra, Roma. *Quaternaria* 20:205–537.

Mallegni, F. (1990–91). Guattari 2 and 3: The stomatognatic apparatus. *Quaternaria Nova* 1:125–37.

Manzi, G., and P. Passarello (1988). From Casal de' Pazzi to Grotta Breuil: Fossil evidence from Latium before the appearance of modern humans. *Animal and Human Biology: Annual Report of the Dipartimento di Biologia Animale e dell'Uomo.* Rome: Università di Roma, pp. 111–43.

Manzi, G., and P. Passarello (1990–91). The human remains from Grotta Breuil (M. Circeo, Italy). *Quaternaria Nova* 1:429–39.

Marks, A. (1988a). The Middle to Upper Paleolithic transition in the southern Levant: Technological change as an adaptation to increasing mobility. In *L'homme de Néandertal*, Vol. 8: *La mutation*, edited by M. Otte, pp. 109–23. ERAUL 35. Liège: Université de Liège.

Marks, A. (1988b). The curation of stone tools during the Upper Pleistocene: A view from the central Negev, Israel. In *The Upper Pleistocene Prehistory of Eurasia*, edited by H. Dibble and A. Montet-White, pp. 275–86. University Museum Monograph 54. Philadelphia: The University Museum, University of Pennsylvania.

Marks, A., and D. Freidel (1977). Prehistoric settlement patterns in the Avdat/Aqev area. In *Prehistory and Paleoenvironments in the Central Negev, Israel*, Vol. 2, edited by A. Marks, pp. 131–58. Dallas: Department of Anthropology, Southern Methodist University.

Marshack, A. (1989). Evolution of the human capacity: the symbolic evidence. *Yearbook of Physical Anthropology*, Supplement 10, 32:1–34.

Marshack, A. (1990). Early hominid symbols and the evolution of the human capacity. In *The Emergence of Modern Humans: An Archaeological Perspective*, edited by P. Mellars, pp. 457–98. Ithaca: Cornell University Press.

Marshall, L. (1976). Sharing, talking and giving: Relief of social tensions among the !Kung. In *Kalahari Hunter-Gatherers: Studies of the !Kung San and Their Neighbors*, edited by R. Lee and I. DeVore, pp. 349–71. Cambridge, MA: Harvard University Press.

Martin, H. (1907–10). *Recherches sur l'évolution du moustérien dans le gisement de la Quina (Charente)*. Paris: Schleicher Frères.

Mathiassen, T. (1928). *Material Culture of the Iglulik Eskimos*. Report of the 5th Thule Expedition 6(1). Copenhagen: Gyldendal.

Maynard-Smith, J. (1982). *Evolution and the Theory of Games*. Cambridge: Cambridge University Press.

McCown, T., and A. Keith (1939). *The Stone Age of Mount Carmel*, Vol. 2: *The Fossil Human Remains from the Levalloiso-Mousterian*. Oxford: Clarendon Press.

McGill, R., J. Tukey, and W. Larson (1978). Variations of box plots. *American Statistician* 32:12–16.

McGrew, W. (1991). Chimpanzee material culture: What are its limits and why? In *The Origins of Human Behaviour*, edited by R. Foley, pp. 12–24. London: Unwin Hyman.

Meignen, L. (1988). Un exemple de comportement technologique différentiel selon des matières premières: Marillac, couches 9 et 10. In *L'homme de Néandertal*, Vol. 4: *La technique*, edited by M. Otte, pp. 71–80. ERAUL 31. Liège: Université de Liège.

Meignen, L., and O. Bar-Yosef (1988). Variabilité technologique au Proche-Orient: L'exemple de Kébara. In *L'homme de Néandertal*, Vol. 4: *La technique*, edited by M. Otte, pp. 81–95. ERAUL 31. Liège: Université de Liège.

Mellars, P. (1969). The chronology of Mousterian industries in the Perigord region of south-west France. *Proceedings of the Prehistoric Society* 35:134–71.

Mellars, P. (1973). The character of the Middle–Upper Paleolithic transition in Southwest France. In *The Explanation of Culture Change*, edited by C. Renfrew, pp. 255–76. London: Duckworth.

Mellars, P. (1988). The chronology of the south-west French Mousterian: A review of the current debate. In *L'homme de Néandertal*, Vol. 4: *La technique*, edited by M. Otte, pp. 97–120. ERAUL 31. Liège: Université de Liège.

Mellars, P. (1989). Major issues in the emergence of modern humans. *Current Anthropology* 30:349–85.

Mellars, P. ed. (1990). *The Emergence of Modern Humans: An Archaeological Perspective*. Ithaca: Cornell University Press.

Mellars, P. (1991). Cognitive changes and the emergence of modern humans. *Cambridge Archaeological Journal* 1:63–76.

Mellars, P., and C. Stringer, eds. (1989). *The Human Revolution: Behavioural and Biological Perspectives on the Origins of Modern Humans*. Edinburgh: Edinburgh University Press.

Metcalfe, D., and K. R. Barlow (1992). A model for exploring the optimal trade-off between field processing and transport. *American Anthropologist* 94: 340–56.

Miskovsky, J.-C. (1974). *Le quaternaire du Midi méditerranéen*. Études Quaternaires, Mémoire No. 3. Marseille: Université de Provence.

Montet-White, A. (1991). Lithic acquisition, settlements and territory in the Epigravettian of central Europe. In *Raw Material Economies among Prehistoric Hunter-Gatherers*, edited by A. Montet-White and S. Holen, pp. 205–20. University of Kansas Publications in Anthropology, 19. Lawrence: University of Kansas.

Moore, J. (1981). The effects of information networks in

hunter-gatherer societies. In *Hunter-Gatherer Foraging Strategies: Ethnographic and Archaeological Analyses*, edited by B. Winterhalder and E. Smith, pp. 194–217. Chicago: University of Chicago Press.

Mora, R., I. Muro, E. Carbonell, A. Cebria, and J. Martinez (1988). Chronostratigraphy of "Abric Romaní." In *L'homme de Néandertal*, Vol. 1: *La chronologie*, edited by M. Otte, pp. 53–59. ERAUL 28. Liège: Université de Liège.

Morrow, C. (1987). Blades and Cobden chert: A technological argument for their roles as markers of regional identification during the Hopewell period in Illinois. In *The Organization of Core Technology*, edited by J. Johnson and C. Morrow, pp. 119–50. Boulder, CO: Westview Press.

Morrow, C., and R. Jefferies (1989). Trade or embedded procurement? A test case from southern Illinois. In *Time, Energy and Stone Tools*, edited by R. Torrence, pp. 27–33. Cambridge: Cambridge University Press.

Movius, H. (1966). The hearths of the Upper Perigordian and Aurignacian horizons at the Abri Pataud, Les Eyzies (Dordogne), and their possible significance. *American Anthropologist* 68:296–325.

Müller-Beck, H. (1988). The ecosystem of the "Middle Paleolithic" (late Lower Paleolithic) in the upper Danube region. In *The Upper Pleistocene Prehistory of Western Eurasia*, edited by H. Dibble and A. Montet-White, pp. 232–54. University Museum Monograph 54. Philadelphia: The University Museum, University of Pennsylvania.

Munday, F. (1976). Intersite variability in the Mousterian occupation of the Avdat/Aqev Area. In *Prehistory and Paleoenvironments in the Central Negev, Israel*, Vol. 1, edited by A. Marks, pp. 113–40. Dallas: Department of Anthropology, Southern Methodist University.

Mussi, M. (1977–82). Musteriano a denticolati du ciottolo in località S. Andrea di Sabaudia (Prov. di Latina). *Origini* 11:45–70.

Mussi, M. (1990). Le peuplement de l'Italie à la fin du paléolithique moyen et au début du paléolithique supérieur. In *Paléolithique moyen récent et paléolithique supérieur ancien en Europe*, edited by C. Farizy, pp. 251–63. Mémoires du Musée de Préhistoire d'Ile de France, No. 3. Nemours: APRAIF.

Mussi, M., and D. Zampetti (1984–87). La presenza umana nella pianura Pontina durante il paleolitico medio e superiore. *Origini* 13:726.

Mussi, M., and D. Zampetti (1985) Il paleolitico delle Cavernette Falische: Una messa a punto. In *Studi di paletnologia in onore di S. M. Puglisi*, edited by M. Liverani, A. Palmieri, and R. Peroni, pp. 627–45. Rome: University of Rome, "La Sapienza."

Mussi, M., and D. Zampetti (1990–91). Le site moustérien de Grotta Barbara. *Quaternaria Nova* 1:277–87.

Myers, A. (1989). Reliable and maintainable technological strategies in the Mesolithic of mainland Britain. In *Time, Energy and Stone Tools*, edited by R. Torrence, pp. 78–91. Cambridge: Cambridge University Press.

Nelson, M. (1991). The study of technological organization. In *Archaeological Method and Theory*, Vol. 3, edited by M. Schiffer, pp. 57–100. Tucson: University of Arizona Press.

Oakley, K., P. Andrews, L. Keeley, and J. Clark (1977). A reappraisal of the Clacton spearpoint. *Proceedings of the Prehistoric Society* 43:13–30.

Orquera, L. A. (1984). Specialization and the Middle/Upper Paleolithic transition. *Current Anthropology* 25:73–98.

Osgood, C. (1940). *Ingalik Material Culture*. Yale University Publications in Anthropology, 22. New Haven.

Osgood, C. (1958). *Ingalik Social Culture*. Yale University Publications in Anthropology, 53. New Haven.

Otte, M., ed. (1986–89). *L'homme de Néandertal*, Vols. 1–8. ERAUL 28–35. Liège: Université de Liège.

Otte, M. (1991). Evolution in the relationship between raw materials and cultural tradition in the European Paleolithic. In *Raw Material Economies among Prehistoric Hunter-Gatherers*, edited by A. Montet-White and S. Holen, pp. 161–68. University of Kansas Publications in Anthropology, 19. Lawrence: University of Kansas.

Otte, M., J.-M. Evrard, and A. Mathis (1988). Interprétation d'un habitat au paléolithique moyen. In *The Upper Pleistocene Prehistory of Western Eurasia*, edited by H. Dibble and A. Montet-White, pp. 95–124. University Museum Monograph 54. Philadelphia: The University Museum, University of Pennsylvania.

Palma di Cesnola, A. (1965). Notizie preliminari sulla terza campagna di scavi nella Grotta del Cavallo (Lecce). *Rivista di Scienze Preistoriche* 25:3–87.

Palma di Cesnola, A. (1986). Panorama del Musteriano italiano. In *I Neandertaliani*, edited by D. Cochi Genick, pp. 139–74. Viareggio: Museo Preistorico e Archeologico "Alberto Carlo Blanc."

Parker, S., and C. Milbrath (1993). Higher intelligence, propositional language, and culture as adaptations for planning. In *Tools, Language and Cognition in Human Evolution*, edited by K. Gibson and T. Ingold, pp. 314–33. Cambridge: Cambridge University Press.

Parotto, M., and A. Praturlon (1975). Geological summary of the central Apennines. In *Structural Model of Italy*, edited by L. Ongiben, M. Parotto, and A. Praturlon, pp. 257–67. Quaderni de *La Ricerca Scientifica*, 90. Rome: CNR.

Parry, W., and R. Kelly (1987). Expedient core technology and sedentism. In *The Organization of Core Technology*, edited by J. Johnson and C. Morrow, pp. 285–309. Boulder, CO: Westview Press.

Perlès, C. (1976). Le feu. In *La préhistoire française*, Tome 1: *Les civilisations paléolithiques et mésolithiques de la France*, edited by H. de Lumley, pp. 679–83. Paris: CNRS.

Perlès, C. (1991). Économie des matières premières et économie du débitage: Deux conceptions opposées? In *25 ans d'études technologiques en préhistoire: Bilan et perspectives*, pp. 35–46. Actes des XIe Rencontres Internationales d'Archéologie et d'Histoire d'Antibes. Juan-les-Pins: Éditions APDCA.

Perlès, C. (1992). In search of lithic strategies: A cognitive approach to prehistoric chipped stone assemblages. In *Representations in Archaeology*, edited by J.-C. Gardin and C. Peebles, pp. 223–47. Bloomington: Indiana University Press.

Peterkin, G. L. (1993). Lithic and organic hunting technology in the French Upper Paleolithic. In *Hunting and Animal Exploitation in the Later Palaeolithic and Mesolithic of Eurasia*, edited by G. Larsen Peterkin, H. Bricker, and P. Mellars, pp. 49–68. Archaeological Papers of the American Anthropological Association, Vol. 4.

Peterson, R., J. Woolington, and T. Bailey (1984). Wolves of the Kenai peninsula, Alaska. *Wildlife Monographs*, 88.

Phillips, J. (1991). Refitting, edge-wear and *chaînes opératoires*: A case study from Sinai. In *25 ans d'études technologiques en préhistoire: Bilan et perspectives*, pp. 305–17. Actes des XIe Rencontres Internationales d'Archéologie et d'Histoire d'Antibes. Juan-les-Pins: Éditions APDCA.

Pigeot, N., M. Philippe, G. Le Licon, and M. Morgenstern (1991). Systèmes techniques et essai de technologie culturelle à Étiolles: Nouvelles perspectives. In *25 ans d'études technologiques en préhistoire: Bilan et perspectives*, pp. 168–85. Actes des XIe Rencontres Internationales d'Archéologie et d'Histoire d'Antibes. Juan-les-Pins: Éditions APDCA.

Piperno, M. (1976–77). Analyse du sol moustérien de la Grotte Guattari au Mont Circé. *Quaternaria* 19:71–92.

Piperno, M. (1984). L'acheuleano e il musteriano del Lazio. *Atti della XXIV Riunione Scientifica dell'Istituto Italiano di Preistoria e Protostoria*, pp. 39–53.

Piperno, M., and I. Biddittu (1978). Studio tipologico ed interpretazione dell'industria acheuleana e pre-musteriana dei livelli *m* e *d* di Torre in Pietra (Roma). *Quaternaria* 20:441–535.

Piperno, M., and G. Giacobini (1990–91). A taphonomic study of the paleosurface of Grotta Guattari (Monte Circeo, Latina, Italy). *Quaternaria Nova* 1:143–62.

Piperno, M., and A. Segre (1982). The transition from Lower to Middle Palaeolithic in central Italy: An example from Latium. In *The Transition from Lower to Middle Palaeolithic and the Origin of Modern Man*, edited by A. Ronen, pp. 203–21. BAR International Series, 151. Oxford: British Archaeological Reports.

Pitti, C., and C. Tozzi (1971). La Grotta del Capriolo e la Buca della Iena presso Mommio (Camaiore, Lucca). *Rivista di Scienze Preistoriche* 26:213–58.

Ploux, S. (1991). Technologie, technicité, techniciens: Méthode de détermination d'auteurs et comportements techniques individuels. In *25 ans d'études technologiques en préhistoire: Bilan et perspectives*, pp. 201–14. Actes des XIe Rencontres Internationales d'Archéologie et d'Histoire d'Antibes. Juan-les-Pins: Éditions APDCA.

Potts, R. (1988). *Hominid Activities at Olduvai*. New York: Aldine de Gruyter.

Ponzi, G. (1866). Sui manufatti di focaia rinvenuti all'Inviolatella nella campagna romana e sull'uomo all'epoca della pietra. *Atti della Reale Accademia dei Nuovi Lincei* 20.

Price, T. D. (1993). Issues in Paleolithic and Mesolithic research. In *Hunting and Animal Exploitation in the Later Palaeolithic and Mesolithic of Eurasia*, edited by G. Larsen Peterkin, H. Bricker, and P. Mellars, pp. 241–44. Archaeological Papers of the American Anthropological Association, Vol. 4.

Radmilli, A. M. (1974). *Gli scavi nella Grotta Polesini a Ponte Lucano di Tivoli e la più antica arte nel Lazio*. Florence: Sansoni Editore.

Rak, Y. (1986). The Neanderthal: A new look at an old face. *Journal of Human Evolution* 15:151–64.

Rasmussen, K. (1931). *The Netsilik Eskimos*. Report of the 5th Thule Expedition 8(1–2). Copenhagen: Gyldendal.

Rellini, U. (1920). Cavernette e ripari preistorici nell'Agro Falisco. *Monumenti Antichi, Accademia Nazionale dei Lincei* 26.

Rensink, E. (1987). The Magdalenian site Mesch-Steenberg (province of Limburg, The Netherlands): Manufacture of blades and maintenance of tools at an observation stand? In *The Big Picture: International Symposium of Refitting Stone Artifacts*, edited by E. Cziesla, E. Eikhoff, N. Arts, and D. Winter, pp. 165–76. Studies in Modern Archaeology, Vol. 1. Bonn: HOLOS.

Rensink, E., J. Kolen, and A. Spieksma (1991). Patterns of raw material distribution in the Upper Pleistocene of northwestern and central Europe. In *Raw Material Economies among Prehistoric Hunter-Gatherers*, edited by A. Montet-White and S. Holen, pp. 127–40. University of Kansas Publications in Anthropology, 19. Lawrence: University of Kansas.

Révillion, S. (1993). Question typologique a propos des industries laminaires du paléolithique moyen de Seclin (nord) et de Saint-Germain-des-Vaux/Port-Racine (Manche): Lames Levallois ou lames non Levallois? *Bulletin de la Société Préhistorique Française* 90:269–73.

Rigaud, J.-P. (1989). From the Middle to the Upper Paleolithic: transition or convergence? In *The Emergence of Modern Humans: Biocultural Adaptations in the Later Pleistocene*, edited by E. Trinkaus, pp. 142–53. Cambridge: Cambridge University Press.

Rigaud, J.-P., and J.-M. Geneste (1988). Utilisation de l'espace dans la grotte Vaufrey. In *La Grotte Vaufrey à Cenac et Saint-Julien (Dordogne): Paléoenvironnements, chronologie et activités humaines*, edited by J.-P. Rigaud, pp. 593–612. Mémoires de la Société Préhistorique Française, Tome 19.

Rigaud, J.-P., and J. Simek (1987). "Arms too short to box with God": Problems and prospects for Paleolithic prehistory in Dordogne, France. In *The Pleistocene Old World: Regional Perspectives*, edited by O. Soffer, pp. 47–61. New York: Plenum Press.

Robinson, W. S. (1951). A method for chronologically ordering archaeological deposits. *American Antiquity* 16:293–301.

Roebroeks, W. (1988). *From Find Scatters to Early Hominid Behavior: A Study of Middle Paleolithic Riverside Settlements at Maastricht-Belvédère (The Netherlands)*. Analecta Praehistorica Leidensia, 21. Leiden: Publications of the University of Leiden.

Roebroeks, W., J. Kolen, and E. Rensink (1988). Planning depth, anticipation and the organization of Middle Paleolithic technology: The "archaic natives" meet Eve's descendants. *Helinium* 28:17–34.

Roebroeks, W., N. J. Conard, and T. van Kolfschoten (1992). Dense forests, cold steppes, and the Palaeolithic settlement of northern Europe. *Current Anthropology* 33:551–86.

Rogers, L. L. (1981). A bear in its lair. *Natural History* 90:64–70.

Rolland, N. (1981). The interpretation of Middle Paleolithic variability. *Man* 16:15–42.

Rolland, N. (1986). Recent findings from La Micoque and other sites in south-western and Mediterranean France: Their bearing on the "Tayacian" problem and Middle Paleolithic emergence. In *Stone Age Prehistory: Studies in Honour of Charles McBurney*, edited by G. Bailey and P. Callow, pp. 121–51. Cambridge: Cambridge University Press.

Rolland, N. (1988). Observations on some Middle Paleolithic time series in southern France. In *The Upper Pleistocene Prehistory of Western Eurasia*, edited by H. Dibble and A. Montet-White, pp. 161–80. University Museum Monograph 54. Philadelphia: The University Museum, University of Pennsylvania.

Rolland, N., and H. Dibble (1990). A new synthesis of Mousterian variability. *American Antiquity* 55:480–99.

Ruff, C., E. Trinkaus, A. Walker, and C. Larsen (1993). Postcranial robusticity in *Homo*, I: Temporal trends and mechanical interpretation. *American Journal of Physical Anthropology* 91:21–53.

Sala, B. (1990). Loess fauna in deposits of shelters and caves in the Veneto region and examples in other regions of Italy. In *Loess Aeolian Deposits and Related Paleosols in the Mediterranean Region*, edited by M. Cremaschi, pp. 139–50. *Quaternary International*, 5.

Sampson, C. G. (1985). *Nightfire Island: Later Holocene Lakemarsh Adaptation on the Western Edge of the Great Basin*. Anthropological Papers 33. Eugene: University of Oregon.

Schick, K. (1987). Modeling the formation of Early Stone Age artifact concentrations. *Journal of Human Evolution* 16:789–807.

Schild, R. (1987). The exploitation of chocolate flint in Poland. In *The Human Uses of Flint and Chert*, edited by G. de G. Sieveking and M. Newcomer, pp. 137–49. Cambridge: Cambridge University Press.

Schwarcz, H., and R. Grün (1993). Electron spin resonance (ESR) dating and the origin of modern man. In *The Origin of Modern Humans and the Impact of Chronometric Dating*, edited by M. Aitken, C. Stringer, and P. Mellars, pp. 40–49. Princeton: Princeton University Press.

Schwarcz, H. P., W. Buhay, R. Grün, M. Stiner, S. Kuhn, and G. H. Miller (1990–91). Absolute dating of sites in coastal Lazio. *Quaternaria Nova* 1:51–67.

Schwarcz, H. P., A. Bietti, W. M. Buhay, M. Stiner, R. Grün, and A. Segre (1991). On the reexamination of Grotta Guattari: Uranium-Series and Electron-Spin-Resonance dates. *Current Anthropology* 32: 313–16.

Segre, A. (1953). Risultati preliminari dell'esplorazione econometrica del Basso Tirreno. *La Ricerca Scientifica* 29(9).

Segre, A. (1959). Giacimenti pleistocenici con fauna e industria litica a Monte Argentario (Grossetto). *Rivista di Scienze Preistoriche* 14:1–18.

Segre, A. (1983). Geologia quaternaria e paleolitico nella bassa valle dell'Aniene, Roma. *Rivista di Antropologia* 62 (supplement):87–98.

Segre, A., and A. Ascenzi (1984). Fontana Ranuccio: Italy's earliest Middle Pleistocene hominid site. *Current Anthropology* 25:230–33.

Segre, A., I. Biddittu, and P. Cassoli (1984). Il bacino di Sora (Frosinone) e i suoi giacimenti musteriani. *Atti della XXIV Riunione Scientifica dell'Istituto Italiano di Preistoria e Protostoria*, pp. 149–54.

Segre-Naldini, E. (1984). Il musteriano di Grotta della Cava, Sezze Romano (Latina). *Atti della XXIV Riunione Scientifica dell'Istituto Italiano di Preistoria e Protostoria*, pp. 142–47.

Sept, J. (1992). Was there no place like home? A new perspective on early hominid archaeological sites from the mapping of chimpanzee nests. *Current Anthropology* 33:187–207.

Sergi, S. (1929). La scoperta di un cranio del tipo Neanderthal presso Roma. *Rivista di Antropologia* 28: 457–62.

Sergi, S. (1931). Le crâne Néanderthalien de Saccopastore, Rome. *L'Anthropologie* 41:241–47.

Sergi, S. (1939). Il cranio neandertaliano del Monte Circeo. *Rendiconti delle Sedute della Reale Accademia Nazionale dei Lincei*, Ser. 6. 29:672–85.

Sergi, S. (1954). La mandibola neandertaliana Circeo II. *Rivista di Antropologia* 41:305–44.

Sergi, S. (1955). La mandibola neandertaliana Circeo III (mandibola B). *Rivista di Antropologia* 42:337–403.

Sergi, S., and A. Ascenzi (1974). *Il cranio neandertaliano di Monte Circeo (Circeo I)*. Rome.

Sevink, J., P. Vos, W. E. Westehoff, A. Stierman, and H. Kamermans (1982). Sequence of marine terraces near Latina (Agro Pontino, Central Italy). *Catena* 9: 361–78.

Sevink, J., A. Remmelzwaal, and O. Spaargaren (1984). *The Soils of Alzio and Adjacent Campania*. Publicatie van het Fysisch Geografisch en Bodemkundig Laboratorium, No. 38. Amsterdam: Universiteit van Amsterdam.

Sevink, J., J. Duivenvoorden, and H. Kamermans (1991). The soils of the Agro Pontino. In *The Agro Pontino Survey Project: Methods and Preliminary Results*, edited by A. Voorrips, S. Loving, and H. Kamermans, pp. 31–47. Studies in Prae- en Protohistorie 6, Instituut voor Prae- en Protohistorische Archaeologie Albert Egges van Giffen. Amsterdam: Universiteit van Amsterdam.

Shackleton, N., and N. Opdyke (1973). Oxygen isotope and paleomagnetic stratigraphy of equatorial Pacific core V28–238: Oxygen isotope temperatures and ice volumes on a 10^5 and 10^6 year scale. *Quaternary Research* 3:39–55.

Shea, J. (1989). A functional study of the lithic industries associated with hominid fossils in the Kebara and Qafzeh caves, Israel. In *The Human Revolution: Behavioural and Biological Perspectives on the Origins of Modern Humans*, edited by P. Mellars and C. Stringer, pp. 611–25. Edinburgh: Edinburgh University Press.

Shott, M. (1986). Settlement mobility and technological organization: An ethnographic examination. *Journal of Anthropological Research* 42:15–51.

Shott, M. (1989a). Technological organization in Great Lakes Paleoindian assemblages. In *Eastern Paleoindian Lithic Resource Use*, edited by C. Ellis and J. Lothrop, pp. 221–38. Boulder: Westview Press.

Shott, M. (1989b). On tool-class use lives and the formation of archaeological assemblages. *American Antiquity* 54:9–30.

Silberbauer, G. (1981). *Hunter and Habitat in the Central Kalahari*. Cambridge: Cambridge University Press.

Simán, K. (1991). Patterns of raw material use in the Middle Paleolithic of Hungary. In *Raw Material Economies among Prehistoric Hunter-Gatherers*, edited by A. Montet-White and S. Holen, pp. 49–58. University of Kansas Publications in Anthropology, 19. Lawrence: University of Kansas.

Simek, J. (1987). Spatial order and behavioral change in the French Paleolithic. *Man* 61:25–40.

Simek, J. (1988). Analyse spatiale de la distribution des objets de la couche VIII de la Grotte Vaufrey. In *La Grotte Vaufrey à Cenac et Saint-Julien (Dordogne): Paléoenvironnements, chronologie et activités humaines*, edited by J.-P. Rigaud, pp. 569–92. Mémoires de la Société Préhistorique Française, Tome 19.

Smart, P., and D. Richards (1992). Age estimates for the late Quaternary high sea-stands. *Quaternary Science Reviews* 11:687–96.

Smith, E. (1981). The application of optimal foraging

theory to the anaysis of hunter-gatherer group size. In *Hunter-Gatherer Foraging Strategies: Ethnographic and Archaeological Analyses*, edited by B. Winterhalder and E. Smith, pp. 36–65. Chicago: University of Chicago Press.

Smith, F. (1983). A behavioral interpretation of changes in craniofacial morphology across the archaic/modern *Homo sapiens* transition. In *The Mousterian Legacy*, edited by E. Trinkaus, pp. 141–64. BAR International Series, 164. Oxford: British Archaeological Reports.

Smith, F., A. Falsetti, and S. Donnelly (1989). Modern human origins. *Yearbook of Physical Anthropology* 32:35–68.

Smith, P. (1976). Dental pathology in fossil hominids: What did Neanderthals do with their teeth? *Current Anthropology* 17:149–51.

Smyth, R. B. (1878). *Aborigines of Victoria: With Notes Relating to the Habits of the Natives of Other Parts of Australia and Tasmania*. 2 vols. Melbourne: Government Printing Office.

Snedecor, G., and W. Cochran (1980). *Statistical Methods*. Seventh edition. Ames: Iowa State University Press.

Spencer, M., and D. Demes (1993). Biomechanical analysis of masticatory system configuration in Neandertals and Inuits. *American Journal of Physical Anthropology* 91:1–20.

Spuhler, J. (1988). Evolution of mitochondrial DNA in monkeys, apes and humans. *Yearbook of Physical Anthropology* 31:15–48.

Stefansson, V. (1914). *The Stefansson–Anderson Arctic Expedition of the American Museum: Preliminary Ethnographic Report*. Anthropological Papers, Vol. 16(1). New York: American Museum of Natural History.

Stephens, D., and J. Krebs (1986). *Foraging Theory*. Princeton: Princeton University Press.

Stevenson, M. (1985). The formation of artifact assemblages at workshop/habitation sites: Models from Peace Point in northern Alberta. *American Antiquity* 50:63–81.

Stiner, M. (1990a). The use of mortality patterns in archaeological studies of hominid predatory adaptations. *Journal of Anthropological Archaeology* 9: 305–51.

Stiner, M. (1990b). *The Ecology of Choice: Procurement and Transport of Animal Resources by Upper Pleistocene Hominids in West-central Italy*. Ph. D. dissertation, Dept. of Anthropology, University of New Mexico. Ann Arbor, MI: UMI.

Stiner, M. (1990–91). Ungulate exploitation during the terminal Mousterian of west-central Italy: The case of Grotta Breuil. *Quaternaria Nova* 1:333–50.

Stiner, M. (1991a). A taphonomic perspective on the origins of the faunal remains of Grotta Guattari (Latium, Italy). *Current Anthropology* 32:103–17.

Stiner, M. (1991b). Food procurement and transport by human and non-human predators. *Journal of Archaeological Science* 18:455–82.

Stiner, M. (1991c). The community ecology perspective and the redemption of "contaminated" faunal records. In *Perspectives on the Past: Theoretical Biases in Mediterranean Hunter-Gatherer Research*, edited by G. Clark, pp. 229–42. Philadelphia: University of Pennsylvania Press.

Stiner, M. (1992). Overlapping species "choice" by Italian Upper Pleistocene predators. *Current Anthropology* 33:433–51.

Stiner, M. (1993a). Small animal exploitation and its relation to hunting, scavenging, and gathering in the Italian Mousterian. In *Hunting and Animal Exploitation in the Later Palaeolithic and Mesolithic of Eurasia*, edited by G. Larsen Peterkin, H. Bricker, and P. Mellars, pp. 107–26. Archaeological Papers of the American Anthropological Association, Vol. 4.

Stiner, M. (1993b). Modern human origins—faunal perspectives. *Annual Review of Anthropology* 22:55–82.

Stiner, M. (1994). *Honor among Thieves: A Zooarchaeological Perspective on Neandertal Ecology*. Princeton: Princeton University Press.

Stiner, M., and S. Kuhn (1992). Subsistence, technology, and adaptive variation in Middle Paleolithic Italy. *American Anthropologist* 94:12–46.

Stoneking, M., and R. Cann (1989). African origin of human mitochondrial DNA. In *The Human Revolution: Behavioural and Biological Perspectives on the Origins of Modern Humans*, edited by P. Mellars and C. Stringer, pp. 17–30. Edinburgh: Edinburgh University Press.

Straus, L. (1982a). Comment on R. White 1982. *Current Anthropology* 23:185–86.

Straus, L. (1982b). Carnivores and cave sites in Cantabrian Spain. *Journal of Anthropological Research* 38:75–96.

Straus, L. (1983). From Mousterian to Magdalenian: Cultural evolution viewed from Vasco-Cantabrian Spain and Pyrenean France. In *The Mousterian Legacy*, edited by E. Trinkaus, pp. 73–111. BAR International Series, 164. Oxford: British Archaeological Reports.

Straus, L. (1990a). The early Upper Palaeolithic of southwest Europe: Cro-Magnon adaptations in the Iberian peripheries, 40,000–20,000 BP. In *The Emergence of Modern Humans: An Archaeological Perspective*, edited by P. Mellars, pp. 276–302. Ithaca: Cornell University Press.

Straus, L. (1990b). The original arms race: Iberian perspectives on the Solutrean phenomenon. In *Feuilles du pierre: Les industries à pointes foliacées du paléolithique supérieur européen*, edited by J. Kozlowski, pp. 425–47. ERAUL 42. Liège: Université de Liège.

Straus, L. (1993). Upper Paleolithic hunting tactics and weapons in western Europe. In *Hunting and Animal Exploitation in the Later Palaeolithic and Mesolithic of Eurasia*, edited by G. Larsen Peterkin, H. Bricker, and P. Mellars, pp. 83–94. Archaeological Papers of the American Anthropological Association, Vol. 4.

Straus, L., and B. Vierra (1989). Preliminary investigation of the concheiro of Vidigal. *Mesolithic Miscellany* 10:2–11.

Stringer, C. (1988). Palaeoanthropology: The dates of Eden. *Nature* 331:565–66.

Stringer, C., and P. Andrews (1988). Genetic and fossil evidence for the origins of modern humans. *Science* 239:1263–68.

Stringer, C., and C. Gamble (1993). *In Search of the Neanderthals*. New York: Thames and Hudson.

Tagliacozzo, A. (1992). I mammiferi dei giacimenti pre- e protostorici italiani. Un inquadrimento paleontologico e archeozoologico. In *Italia preistorica*, edited by A. Guidi and M. Piperno, pp. 68–102. Rome–Bari: Laterza.

Tanaka, J. (1980). *The San: Hunter-Gatherers of the Kalahari*. Tokyo: University of Tokyo Press.

Tankersley, K. (1991). A geoarchaeological investigation of distribution and exchange in the raw material economies of Clovis groups in eastern North America. In *Raw Material Economies among Prehistoric Hunter-Gatherers*, edited by A. Montet-White and S. Holen, pp. 285–304. University of Kansas Publications in Anthropology, 19. Lawrence: University of Kansas.

Taschini, M. (1964). Il livello mesolitico del Riparo Blanc al Monte Circeo. *Bullettino di Paletnologia Italiana* 17:65–88.

Taschini, M. (1967). Il protopontiniano rissiano di Sedia del Diavolo e di Monte delle Gioie (Roma). *Quaternaria* 9:301–19.

Taschini, M. (1970). La Grotta Breuil al Monte Circeo, per una impostazione dello studio del pontiniano. *Origini* 4:45–78.

Taschini, M. (1972). Sur le paléolithique de la Plaine Pontine (Latium). *Quaternaria* 16:203–23.

Taschini, M. (1979). L'industrie lithique de Grotta Guattari au Mont Circé (Latium): Définition culturelle, typologique et chronologique du pontinien. *Quaternaria* 12:179–247.

Tavoso, A. (1984). Reflexions sur l'économie des matières premières au moustérien. *Bulletin de la Société Préhistorique Française* 81:79–82.

Tavoso, A. (1988). L'outillage du gisement de S. Francesco à Sanremo (Ligurie, Italie): Nouvel examen. In *L'homme de Néandertal*, Vol. 8: *La mutation*, edited by M. Otte, pp. 193–210. ERAUL 35. Liège: Université de Liège.

Tchernov, E. (1992). Evolution of complexities, exploitation of the biosphere and zooarchaeology. *Archaeozoologia* 5:9–42.

Templeton, A. (1993). The "Eve" hypothesis: A genetic critique and reanalysis. *American Anthropologist* 95:51–72.

Tixier, J., M. L. Inizian, and H. Roche (1980). *Préhistoire de la pierre taillée*, Vol. 1: *Terminologie et technologie*. Valbonne: CREP.

Tooby, J., and I. DeVore (1987). The reconstruction of hominid behavioral evolution through strategic modeling. In *The Evolution of Human Behavior: Primate Models*, edited by W. G. Kinsey, pp. 183–237. Albany: State University of New York Press.

Torrence, R. (1983). Time budgeting and hunter-gatherer technology. In *Hunter-Gatherer Economy in Prehistory: A European Perspective*, edited by G. Bailey, pp. 11–22. Cambridge: Cambridge University Press.

Torrence, R. (1989). Re-tooling: Towards a behavioural theory of stone tools. In *Time, Energy and Stone Tools*, edited by R. Torrence, pp. 57–66. Cambridge: Cambridge University Press.

Toth, N. (1985). The Oldowan reassessed: A close look at early stone artifacts. *Journal of Archaeological Science* 12:101–20.

Toth, N., and T. White (1990–91). Assessing the ritual cannibalism hypothesis at Grotta Guattari. *Quaternaria Nova* 1: 213–222.

Tozzi, C. (1970). La Grotta di S. Agostino (Gaeta). *Rivista di Scienze Preistoriche* 25:3–87.

Trinkaus, E. (1981). Neandertal limb proportions and cold adaptation. In *Aspects of Human Evolution*, edited by C. Stringer, pp. 187–224. London: Taylor and Francis.

Trinkaus, E. (1983a). Neandertal postcrania and the adaptive shift to modern humans. In *The Mousterian Legacy*, edited by E. Trinkaus, pp. 165–200. BAR International Series, 164. Oxford: British Archaeological Reports.

Trinkaus, E. (1983b). *The Shanidar Neandertals*. New York: Academic Press.

Trinkaus, E. (1986). The Neandertals and modern human origins. *Annual Review of Anthropology* 15:193–218.

Trinkaus, E. (1989a). Issues concerning human emergence in the later Pleistocene. In *The Emergence of Modern Humans: Biocultural Adaptations in the Later Pleistocene*, edited by E. Trinkaus, pp. 1–17. Cambridge: Cambridge University Press.

Trinkaus, E., ed. (1989b). *The Emergence of Modern Humans: Biocultural Adaptations in the Later Pleistocene*. Cambridge: Cambridge University Press.

Trinkaus, E. (1992). Paleontological perspectives on Neandertal behavior. In *Cinq millions d'années: L'aventure humaine*, edited by M. Toussaint, pp. 151–76. ERAUL 56. Liège: Université de Liège.

Trinkaus, E., and P. Shipman (1993). *The Neandertals*. New York: Knopf.

Turq, A. (1989). Approche technologique et économique du faciès moustérien de type Quina: Étude préliminaire. *Bulletin de la Société Préhistorique Française* 86:244–56.

Turq, A. (1992). Raw material and technological studies of the Quina Mousterian in Perigord. In *The Middle Paleolithic: Adaptation, Behavior, and Variability*, edited by H. Dibble and P. Mellars, pp. 75–85. University Museum Monograph 72. Philadelphia: The University Museum, University of Pennsylvania.

Valladas, H., J. L. Horon, G. Valladas, B. Arensburg, O. Bar-Yosef, A. Belfer-Cohen, P. Goldberg, H. Laville, L. Meignen, Y. Rak, E. Tchernov, A. M. Tillier, and B. Vandermeersch (1987). Thermoluminescence dates for the Neanderthal burial sites at Kebara (Mount Carmel), Israel. *Nature* 330:159–60.

Valladas, H., J. L. Reyss, J. L. Horon, G. Valladas, O. Bar-Yosef, and B. Vandermeersch (1988). Thermoluminescence dating of Mousterian "Proto-Cro-Magnon" remains from Israel and the origin of modern man. *Nature* 331:614–16.

Vandermeersch, B. (1981). *Les hommes fossiles de Qafzeh (Israël)*. Paris: CNRS.

Vander Wall, S. (1990). *Food Hoarding in Animals*. Chicago: University of Chicago Press.

Van Peer, P. (1991). Interassemblage variability and Levallois styles: The case of the northern African Middle Paleolithic. *Journal of Anthropological Archaeology* 10:107–51.

Vicino, G. (1972). Gli scavi preistorici nell'area dell'ex casinò dei Balzi Rossi (nota preliminare). *Rivista Ingauna e Intemelia*, N.S. 27:77–97.

Vincent, A. (1988a). Remarques préliminaires concernant l'outillage osseux de la Grotte Vaufrey. In *La Grotte Vaufrey à Cenac et Saint-Julien (Dordogne): Paléoenvironnements, chronologie et activités humaines*, edited by J.-P. Rigaud, pp. 529–33. Mémoires de la Société Préhistorique Française, Tome 19.

Vincent, A. (1988b). L'os comme artifact au paléolithique moyen: Principes d'étude et premiers résultats. In *L'homme de Néandertal*, Vol. 4: *La technique*, edited by M. Otte, pp. 185–96. ERAUL 31. Liège: Université de Liège.

Vitagliano, S. (1984). Nota sul pontiniano della Grotta dei Moscerini, Gaeta (Latina). *Atti della XXIV Riunione Scientifica dell'Istituto Italiano di Preistoria e Protostoria*, 8–11 Ottobre 1982, pp. 155–64.

Vitagliano, S., and M. Piperno (1990–91). Lithic industry of level 27 *beta* of the Fossellone Cave (S. Felice Circeo, Latina). *Quaternaria Nova* 1:289–304.

Volkman, P. (1983). Boker Tachtit: Core reconstructions. In *Prehistory and Paleoenvironments in the Central Negev, Israel*, Vol. 3, edited by A. Marks, pp. 123–90. Dallas: Department of Anthropology, Southern Methodist University.

Volkman, P., and T. Kaufman (1983). A reassessment of the Emireh point as a possible type fossil for the technological shift from Middle to Upper Paleolithic in the Levant. In *The Mousterian Legacy*, edited by E. Trinkaus, pp. 35–52. BAR International Series, 164. Oxford: British Archaeological Reports.

Volman, T. (1984). Early prehistory of southern Africa. In *Southern African Prehistory and Paleoenvironments*, edited by R. Klein, pp. 169–220. Rotterdam: A. A. Balkema.

Voorrips, A., S. Loving, and H. Kamermans, eds. (1991). *The Agro Pontino Survey Project: Methods and Preliminary Results*. Studies in Prae- en Protohistorie 6, Instituut voor Prae- en Protohistorische Archaeologie Albert Egges van Giffen. Amsterdam: Universiteit van Amsterdam.

Wengler, L. (1990). Économie des matières premières et territoire dans le moustérien et l'atérien maghrébin: Exemples du Maroc oriental. *L'Anthropologie* 94(2): 321–34.

Wengler, L. (1991). Choix des matières premières lithiques et comportement des hommes au paléolithique moyen. In *25 ans d'études technologiques en préhistoire: Bilan et perspectives*, pp. 139–58. Actes des XIe Rencontres Internationales d'Archéologie et d'Histoire d'Antibes. Juan-les-Pins: Éditions APDCA.

Whallon, R. (1989). Elements of cultural change in the later Paleolithic. In *The Human Revolution: Behavioural and Biological Perspectives on the Origins of Modern Humans*, edited by P. Mellars and C. Stringer, pp. 432–54. Edinburgh: Edinburgh University Press.

White, J., N. Modjeska, and I. Hipuya (1977). Group definitions and mental templates: An ethnographic experiment. In *Stone Tools as Cultural Markers*, edited by R. Wright, pp. 380–90. Canberra: Australian Institute of Aboriginal Studies.

White, R. (1982). Rethinking the Middle/Upper Paleolithic transition. *Current Anthropology* 23:169–92.

White, T., and N. Toth (1991). The question of ritual cannibalism at Grotta Guattari. *Current Anthropology* 32:118–24.

Wiessner, P. (1982). Risk, reciprocity and social influences on !Kung San economics. In *Politics and History in Band Societies*, edited by E. Leacock and R. Lee, pp. 61–84. Cambridge: Cambridge University Press.

Wilmsen, E., and J. Denbow (1990). Paradigmatic history of San-speaking peoples and current attempts at revision. *Current Anthropology* 31:489–524.

Wirtz, W. O. (1968). Reproduction, growth and development, and juvenile mortality in the Hawaiian monk seal. *Journal of Mammalogy* 49:229–38.

Willoughby, P. (1987). *Spheroids and Battered Stones in the African Early and Middle Stone Age*. BAR International Series, 321. Oxford: British Archaeological Reports.

Wilson, L. (1988). Petrography of the Lower Paleolithic tool assemblage of the Caune de l'Arago. *World Archaeology* 19:376–87.

Wobst, H. M. (1978). The archaeo-ethnology of hunter-gatherers or the tyranny of the ethnographic record in archaeology. *American Antiquity* 43:303–9.

Wolpoff, M. (1989). Multiregional evolution: The fossil alternative to Eden. In *The Human Revolution: Behavioural and Biological Perspectives on the Origins of Modern Humans*, edited by P. Mellars and C. Stringer, pp. 62–108. Edinburgh: Edinburgh University Press.

Wynn, T., and W. McGrew (1989). An ape's view of the Oldowan. *Man* 24:383–98.

Yellen, J. (1977). *Archaeological Approaches to the Present*. New York: Academic Press.

Zampetti, D., and M. Mussi (1984). Structures d'habitat et utilisation du territoire au paléolithique supérieur dans le Latium (Italie centrale): État de la question. In *Upper Paleolithic Settlement Patterns in Europe*, edited by H. Berke, J. Hahn, and C.-J. Kind, pp. 69–78. Verlag Archaeologica Venatoria. Tübingen: Institut für Urgeschichte der Universität Tübingen.

Zei, M. (1953). Esplorazione di grotta nei pressi di Sezze-Romano. *Bullettino di Paletnologia Italiana* 8:102–7.

Zei, M. (1970). Ricerche nel Lazio meridionale. *Rivista di Scienze Preistoriche* 25:323–25.

INDEX